Innovative Disposal Technology and Management Practice for Medical Waste

Yang Chen · Qinzhong Feng · Zhongkui Zhang

Innovative Disposal Technology and Management Practice for Medical Waste

SSTP

Springer

Yang Chen
College of Resources and Environment
University of Chinese Academy of Sciences
Beijing, China

Qinzhong Feng
College of Resources and Environment
University of Chinese Academy of Sciences
Beijing, China

Zhongkui Zhang
Henan Liying Environmental Technology
Co., Ltd.
Beijing, China

ISBN 978-981-99-6785-8 ISBN 978-981-99-6786-5 (eBook)
https://doi.org/10.1007/978-981-99-6786-5

Jointly published with Shanghai Scientific and Technical Publishers
The print edition is not for sale in China (Mainland). Customers from China (Mainland) please order the print book from: Shanghai Scientific and Technical Publishers.

The translation was done with the help of artificial intelligence (machine translation by the service DeepL.com). A subsequent human revision was done primarily in terms of content.

This Springer imprint is published by the registered company Springer Nature Singapore Pte Ltd.
The registered company address is: 152 Beach Road, #21-01/04 Gateway East, Singapore 189721, Singapore

Paper in this product is recyclable.

Preface I

Medical waste refers to the waste with direct or indirect infection, toxicity, and other hazards generated by medical and health institutions in medical treatment, prevention, health care, and other related activities, and are listed in the National Hazardous Waste List. With the social progress and the continuous development of science and technology, the management and disposal of medical waste has attracted more and more attention from the international community. The most important hazardous waste characteristic of medical waste is infectiousness. In order to eliminate the infectious nature of medical waste, and protect the environment and human health, incineration has been commonly used for many years internationally, but the incineration process of medical waste is prone to dioxins, heavy metals and acid gases, and other pollutants. Among them, dioxins are known as "the most toxic toxins on earth", which is a key pollutant controlled by the Stockholm Convention on Persistent Organic Pollutants (POPs). With the advancement and development of technology, non-incineration technologies for medical waste such as high-temperature steam, chemical, microwave, and high-temperature dry heat are gradually being applied worldwide as new non-dioxin-producing medical waste treatment technologies. However, these technologies can produce pollutants such as odor and volatile organic compounds (VOCs) during their application. In addition, improper application of these technologies can result in incomplete disinfection, making it impossible to completely eliminate the infectiousness of medical waste.

Therefore, in-depth implementation of POPs Convention in China, especially since the beginning of 2020, in response to the onslaught of the COVID-19 epidemic and the rapid consumption of masks, gowns, and other medical materials, resulting in a surge of infectious medical waste, how to further implement the national implementation plan of the POPs Convention to promote the adoption of best available technologies and best environmental practices (BAT/BEP), promote the process of sustainable environmental management of medical waste, and establish a medical waste disposal technology and management model suitable for China's national conditions, is a major issue that must be faced.

This book comprehensively analyzes the new development trend of international medical waste disposal technology and combines the current situation of medical waste management and disposal in China under the new situation of epidemic prevention and control as well as the requirements of international conventions. Furthermore, this book explores the development trend of medical waste disposal technology in China as well as the corresponding optimal medical waste disposal technology and management model. The book covers the screening and evaluation of medical waste treatment and disposal technologies, source separation and reduction of medical waste, optimization and management of medical waste incineration, optimization and management of non-incineration medical waste treatment technologies, innovation of medical waste treatment technologies, specific case studies and future development trends, and so on, and comprehensively implementing the concept of life cycle and the whole process management.

The authors of this book propose the following innovative thoughts and suggestions for the implementation of the relevant results and the future research space:

(1) China should fully integrate the characteristics of medical waste as well as local characteristics, adapt to local conditions, and continue to improve the city-centered medical waste management model, promote the management and disposal of medical waste in a separate manner, and promote regional medical waste management and disposal problems.
(2) The optimization of medical waste disposal technology and management model should take life cycle management as the basic factor of medical waste management, and incorporate the concept of the whole process management into the application process of medical waste disposal technology to effectively solve the problem of infectious control and pollution control of medical waste disposal.
(3) Medical waste disposal technology and management model optimization requires the establishment of practical technology optimization evaluation methods. The application of performance evaluation methods should be further implemented to strengthen the supervision of the operation of facilities and promote the safe and harmless management and disposal of medical waste.
(4) Medical waste management and disposal is a system project; the core issue is to eliminate infectious nature and control the secondary pollution generated in the process of medical waste disposal, in terms of its disposal process, to promote the organic unity of reduction; harmlessness and resource utilization is the ultimate goal of promoting the system project.
(5) Medical waste disposal technology is in a continuous process of development; new disposal technologies such as mobile treatment technology, microwave disinfection technology, plasma technology, etc. are less polluting, efficient treatment, but there is also the applicability of technology and management model which needs to explore the space for upgrade. Accelerate the research and development of new technologies to promote the development of harmless, efficient, low-cost treatment and disposal technologies for medical waste.

(6) Intelligent supervision during the medical waste disposal process reduces the close contact between medical waste and humans through intelligent operation of medical waste disposal as far as possible and promotes the efficient and safe management and disposal of medical waste.

The research results of this book can provide support for the research and development, application and promotion in the field of medical waste in China. They also play an important role in the promotion of better implementation of the POPs Convention in China. It can provide support for the construction of engineering, operation of facilities, supervision and management, and environmental monitoring/testing-related standards by the exploration and adoption of BAT/BEP in the field of medical waste.

We acknowledge support during the writing of our book from Department of Science and Technology Standards of the Ministry of Ecology and Environment, the Center for Foreign Cooperation and Exchange of the Ministry of Ecology and Environment, the Technical Center for Solid Waste and Chemical Management of the Ministry of Ecology and Environment, the Shenyang Institute of Environmental Science, the Institute of Environmental Planning of the Ministry of Ecology and Environment, the Institute of Hospital Management of the Health Commission, the Chinese Center for Disease Control and Prevention, and the Journal of Environmental Engineering for their guidance and assistance, as well as to Jiangsu Ltd., Chongqing Zhide Thermal Industry Co., Henan Province Liying Environmental Protection Technology Co. Ltd, Youyi International Environmental Protection Technology (Beijing) Co., Ltd., Hangzhou Dadiwei Kang Medical Environmental Protection Co, and National Science Library, Chinese Academy of Sciences.

This book is edited by Yang Chen, Qinzhong Feng, and Zhongkui Zhang. The specific division of labor is as follows: Chap. 1 was written by Shuyan Wang, Xiaoli Zhou, Weiwei Shang, Wei Jin, Shuguang Jin, and Xiaohong Xu; Chap. 2 was written by Yuanxun Zhang, Yang Chen, Zhiyuan Ren, Liyuan Liu, Yuanhao Zhang, Wei Jin, and Shuguang Jin; Chap. 3 was written by Qinzhong Feng, Shujuan Shan, Yang Zheng, Ning Sun, Xi Liu, Hui Liu, Jingwei Li, Jun Bao, Yue Gong, and Jianbo Guo; Chap. 4 was written by Xiaodong Yu, Gang Chen, Ding Qian, Tianming Zhang, Pingping Fang, Kai Wang, Kaiyue Wang, and Shitong Yang; Chap. 5 was written by Rongzhi Chen, Zhongkui Zhang, Xuehui Zou, Shan Yang, Ermin Yan, Ming Zhu, Qian Weng, Zhaojun Wang, Huilai Chen, Kai Wang, Huanli Yang, Hongfei Long, and Tongzhe Wang; Chap. 6 was written by Zheng Zhang, Shuyan Wang, Liang Cheng, Yaohong Chen, Qingwei Meng, Xiangming Shi, Xiujin Zhang, and Jun Chen; Chap. 7 was written by Sushuang Hou, Daoguang Xu, Fenghua Mo, Xi Chen, Shaohong Hu, Yining Chen, and Yitao Zhang; Chap. 8 was written by Chunxia Guo, Yang Chen, Shuyan Wang, Rongzhi Chen, Xiaoli Zhou, Dandou Xu, Hao Cui, and Camillus Uchenna Okonkwo. The whole book was drafted by Qinzhong Feng, Zhongkui Zhang, and Wei Jin, with the participation of Litian Zhang and Shujuan Shan, and finalized by Yang Chen.

Due to the author's level, there are inevitably shortcomings in the book, and the above innovative ideas and thinking may have some limitations; we welcome readers to discuss and communicate with us when reading this book, and provide valuable comments so that we can improve in the next step.

Beijing, China
September 2020

Yang Chen

Preface II

Medical waste is a kind of hazardous waste with great environmental risks, and its generation and disposal have drawn great concerns around the world. The development of Chinese medical waste technology research and management started in the 1980s, and since then China has been paying more and more attention to medical waste management, and medical waste has been listed as an important hazardous waste in the National Hazardous Waste List, and a series of laws, regulations, and standards have been issued to regulate the management and disposal of medical waste. At present, Chinese technical level and management system have made significant progress.

In terms of the development of medical waste treatment and disposal technology in China, the incineration disposal technology centered on rotary kiln and fixed bed, and non-incineration treatment technology centered on high-temperature steam, chemical, microwave, and high-temperature dry heat, have been widely applied, presenting a new situation where multiple technologies coexist with incineration and non-incineration technologies. However, there are still many problems in the specific implementation process of these technologies, how to effectively control the production of toxic and harmful secondary pollutants such as dioxins and mercury, how to suppress the harm of odorous, volatile organic compounds and other atmospheric pollutants, and how to avoid the risk of incomplete effects, etc. Therefore, practical and feasible whole-process management measures must be taken to achieve the organic combination of source control, process control, and end control of medical waste.

In the construction of the medical waste management system, the national environmental protection and health departments have issued dozens of documents related to medical waste management (including laws, regulations, standards, norms, guidelines, etc.). In particular, since the outbreak of the COVID-19 in early 2020, for medical waste management and disposal, the Ministry of Ecology and Environment and other relevant departments first issued the "new coronavirus infection of COVID-19 medical waste emergency disposal management and technical guidelines (for trial implementation)" and other technical management documents, providing a basis for technical guidance for grassroots medical personnel, medical waste disposal personnel, and relevant departments; local levels of ecological environment. The competent departments also operate under the unified leadership of the local government at this level. They work in coordination with health and other departments to enhance the coordination mechanism for emergency disposal.

The occurrence of the COVID-19 epidemic has demonstrated that the safe disposal of medical waste is an important component of the national security system and occupies an important position in the ecological environment safety system, urban emergency management, and crisis management. In early 2020, the central government clearly put forward "to fight the battle of pollution prevention and control, to promote sustained improvement in the quality of the ecological environment, to accelerate the completion of medical waste, hazardous waste collection and treatment facilities short board". This requires the relevant parties to fully absorb and learn from the experience, lessons, and inspiration of medical waste disposal in the process of the COVID-19 epidemic, and take effective measures to make up for the shortcomings of China's medical waste technology and management system as soon as possible.

With the rapid development of the Chinese medical waste technology and management system, we have shifted from "learning from foreign experience" to independent innovation and develop a medical technology management system with Chinese characteristics.

The research team led by Prof. Chen Yang of the University of Chinese Academy of Sciences has been devoted to the research and development of innovative medical waste disposal technologies and management practices since the SARS outbreak in 2003, and has made a lot of academic achievements in the fields of medical waste technology innovation and management system construction, and has guided the construction of several pilot demonstration projects. He has led and participated in the preparation of a series of national, industry, group, and local standards and specifications.

The book "Innovative Medical Waste Disposal Technologies and Management Practices" comprehensively analyzes the new domestic development trends and in-depth analyzes the scope and characteristics of various medical disposal technologies, combining with China's current situation and international convention compliance requirements. The book ultimately proposes the concept of life cycle and overall process management. The book is not only a reference for professionals in relevant industries, but also provides support for the research and development of new medical waste treatment technologies and the summary of new experiences in medical waste management. It is also expected to promote the sustainable environmental management process of medical waste disposal in China in the future.

Beijing, China
December 2020

Yang Chen
Qinzhong Feng
Zhongkui Zhang

Contents

Chapter 1
The Development of Medical Waste Disposal Technology and Management Needs

Based on the analysis of the composition and characteristics of medical waste, this chapter will provide a comprehensive introduction to the international demand for medical waste management and disposal, and explore the development trend of medical waste disposal technology and the corresponding optimal medical waste disposal technology and management model in China, taking into account the current situation of medical waste management and disposal in China and the requirements of international conventions.

1.1 Composition and Characteristics of Medical Waste

1.1.1 Composition of Medical Waste

The amount of medical waste generated by medical and health institutions in China is large. According to the survey, in 2018, 200 large and medium-sized cities in China produced 81.7 $\times$ 10^4 t of medical waste, with an average daily generation of 0.2238 $\times$ 10^4 t. At the end of 2018, the number of beds in medical and health institutions was 8,404,100, and the number of beds in medical and health institutions per 1,000 population increased from 2.5 to 6.0 in 2002. In recent years, the amount of medical waste generated in China has grown rapidly, from 33.6 $\times$ 10^4 t in 2010 to 98.0 $\times$ 10^4 t in 2018, with an average annual growth rate of 14.3%. After systematic analysis and study of relevant domestic literature, basic information on the basic composition, basic chemical composition and physicochemical properties of medical waste were obtained, which are shown in Tables 1.1, 1.2 and 1.3.

From Tables 1.1, 1.2 and 1.3, it can be seen that the composition and type of medical waste are complex, so the disposal of medical waste needs to be thoroughly disintegrated or safe the macromolecules in it, and to promote environmentally sound and safe management and disposal.

Y. Chen et al., *Innovative Disposal Technology and Management Practice for Medical Waste*, https://doi.org/10.1007/978-981-99-6786-5_1

Table 1.1 Basic components of medical waste

Organic matter content/%	Organs	0.05
	Plastic	17.91
	Cotton swabs	9.36
	Paper	22.08
	Fabric	11.53
	Total	60.93
Inorganic content/%	Glass	26.66
	Metal	3.70
	Total	30.36
Other/%		8.71

Table 1.2 Basic chemical composition of medical waste

Composition	Carbon	Oxygen	Hydrogen	Large amount of other elements
Percentage/%	50	20	6	24

Table 1.3 Physicochemical properties of medical waste

Generation volume/ (kg/bed/d)	Density/ (kg/m^3)	Humidity/ %	Combustible sex/%	Calorific value of waste/(kcal/ kg)		Aluminum content volume/%	Mercury content volume/ (mg/kg)	Calcium content volume/ (mg/kg)	Lead content volume/ (mg/kg)
				Low	High				
0.5	110	35	90	3500	3900	0.4	2.5	1.5	28

In November 2021, China released the "medical waste classification catalog (2021 version)", the medical waste is divided into five categories, and increased the principles of medical waste classification, recommended collection methods and exemptions from management requirements, see Table 1.4. In the "medical waste exemption management list", for meeting the corresponding conditions, medical waste can be listed in the link in accordance with the exemptions from the provisions of the implementation of exemptions from management, see Table 1.5.

From Table 1.4, it is clear that the management of medical waste by category is of great significance to strengthen the control of infectiousness of medical waste. The determination of the composition or characteristics of medical waste will play an important role in the determination of waste classification and collection programs and the subsequent selection of disposal processes and technologies. Therefore, it is an inevitable choice to classify and manage medical waste.

Table 1.4 Medical waste classification catalog

Category	Characteristics	Common component or waste name	Collection method
Infectious waste	Waste carrying pathogenic microorganisms, with the risk of spreading infectious diseases	1. Medical waste other than sharps contaminated by patient blood, bodily fluids, excreta, etc.; 2. Disposable medical devices discarded after use, such as syringes, infusion sets, dialyzers, etc.; 3. Abandoned pathogen culture media, specimens, preservation solutions and containers for bacterial and viral strains in pathogenic microbiology laboratories; Discarded blood, serum, secretions, and other specimens and containers from other laboratories and departments; 4. Waste generated by quarantining infectious disease patients or suspected infectious disease patients	1. Collect in medical waste packaging bags that comply with the "Standard for Medical Waste Packaging Bags, Containers, and Warning Signs" (HJ421); 2. Abandoned pathogen culture media, specimens, preservation solutions for bacterial and viral strains, and their containers in the pathogenic microbiology laboratory should be sterilized by pressure steam or other methods at the production site, and then collected and treated as infectious waste; 3. Medical waste generated from the isolation of infectious disease patients or suspected infectious disease patients should be packaged in double-layer medical waste packaging bags
Pathologic waste	Treatment process generated by the human body waste and medical laboratory animal carcasses, etc.	1. Surgery and other medical services generated in the process of waste human tissue, organs 2. Discarded human tissue and pathological wax blocks after pathological sectioning 3. Discarded tissues and carcasses of medical laboratory animals 4. Embryonic tissues under 16 weeks gestational age or weighing less than 500 g, etc. 5. Placenta of pregnant women diagnosed, suspected of infectious diseases, or carrying infectious disease pathogens	1. Collect in medical waste packaging bags that comply with the "Standard for Medical Waste Packaging Bags, Containers, and Warning Signs" (HJ421); 2. The placenta of confirmed or suspected infectious disease pregnant women or pregnant women carrying infectious disease pathogens should be packed in double-layer medical waste packaging bags; 3. It can be used for anti-corrosion or low-temperature storage

(continued)

Table 1.4 (continued)

Category	Characteristics	Common component or waste name	Collection method
Injury waste	Discarded medical sharp tools that can stab or cut the human body	1. Discarded metal sharps, such as needles, suture needles, acupuncture and moxibustion needles, probes, puncture needles, scalpels, scalpel, surgical saws, skin knives, steel nails and guide wires; 2. Discarded glass sharps, such as cover glass slides, glass slides, glass ampoules, etc.; 3. Discarded sharp objects of other materials	1. Collect in sharp tool boxes that comply with the "Standard for Medical Waste Special Packaging Bags, Containers, and Warning Signs" (HJ421); 2. When the sharp tool box reaches 3/4 full, it should be tightly sealed and transported and stored according to the process
Pharmaceutical waste	Expired, obsolete, deteriorated or contaminated waste drugs	1. Discarded generic drugs; 2. Discarded cytotoxic and genotoxic drugs; 3. Discarded vaccines and blood products	1. A small amount of pharmaceutical waste can be incorporated into infectious waste, but it should be indicated on the label; 2. Batch discarded pharmaceutical waste shall be collected and handed over to qualified medical waste disposal units or hazardous waste disposal units for disposal
Chemical waste	Waste chemicals with toxicity, corrosiveness, flammability, and reactivity	Waste hazardous chemicals included in the National Hazardous Waste List, such as formaldehyde, xylene, etc.; Hazardous wastes from non specific industry sources, such as mercury containing sphygmomanometer, mercury containing thermometers, abandoned dental amalgam materials and their residues	1. Collect in a container, label and indicate the main ingredients; 2. Collect and hand over to qualified medical waste disposal units or hazardous waste disposal units for disposal

Table 1.5 Medical waste exemption management list

Name	Exemption segment	Conditions of exemption	Content of exemption
Medical waste generated by medical institutions with a total number of beds below 19 (inclusive)	Transportation (including delivery and trans-shipment)	Separate and collect as required	Transportation or transfer process not in accordance with medical waste management
Medicine bottles (penicillin bottles, ampoules, etc.)	Complete chain	1. Not contaminated by the patient's blood, body fluids and excretions	The whole process is not in accordance with medical waste management
		2. Vials containing non-cytotoxic, non-genotoxic drugs	
		3. Packaging containers to meet the requirements of anti-penetration and anti-puncture	
Syringes for dispensing	Complete chain	1. Non-cytotoxic, non-genotoxic drug dispensing use	The whole process is not in accordance with medical waste management
		2. Recycling meets closed-loop safety management	
Cotton swabs, cotton balls	Complete chain	Individual patients not collected in accordance with medical waste segregation due to disinfection and hemostasis by compression	The whole process is not in accordance with medical waste management
Post-use disinfectants such as discarded glutaraldehyde, o-phthalaldehyde, etc.	Complete chain	Into the healthcare organization's internal sewage treatment system to treat and meet the requirements of the discharge standard	The whole process is not in accordance with medical waste management

(continued)

Table 1.5 (continued)

Name	Exemption segment	Conditions of exemption	Content of exemption
Containers for disinfectants, dialysis solutions, medical test reagents, etc.	Complete chain	Recycling meets closed-loop safety management	The whole process is not in accordance with medical waste management
Discarded medical fabrics in non-communicable disease areas	Complete chain	After harmless treatment	The whole process is not in accordance with medical waste management

Note ① The exemption list included in this table is for waste that meets the definition of medical waste, but is non-risky or low-risk, and can be exempted from management as medical waste when the relevant conditions are met

② Other waste, such as infusion bottles (bags), disposable medical overpack items, discarded herbal medicines and herbal decoction residue, medicine cups containing drugs, paper towels, wet wipes, diapers and other disposable sanitary products, the use of urinals and other waste, because it is not medical waste, it is not included in this table

1.1.2 *Characteristics of Medical Waste*

Medical waste has hazardous waste characteristics such as infectious, toxic, corrosive and combustible, but the most important hazardous waste characteristic is infectious (Diaz et al. 2008). Infectious waste is mainly due to the fact that such waste has a greater risk to the environment and personnel. There is a higher risk of transmission of pathogenic microorganisms to people in contact with them, causing a large area of secondary cross-infection, so it should be strictly controlled, especially the infectious disease hospital ward waste which is also defined as medical waste. Pathologic waste mainly includes discarded human tissues, organs, etc. and medical laboratory animal carcasses produced during the treatment process, including pathological specimens, based on ethical and moral concepts and relevant national policies, large human limbs are sent to crematoriums for incineration. Injury waste mainly includes various sharp instruments, needles, surgical instruments, glass ampoules and other glass items produced during the treatment process. After leaving the hospital, such sharp instruments, if not effectively managed, are also very likely to cause bodily harm to waste disposal personnel and the general public, and then cause the occurrence of related diseases. Pharmaceutical waste mainly refers to expired, eliminated, deteriorated or contaminated general drugs, carcinogenic, teratogenic, cytotoxic, genotoxic and blood products in bulk expired drugs, residual drugs after use by patients. This type of waste has a greater potential harm to the population and the environment, if improperly disposed of, directly to the human body causing serious harm, into the environment may contaminate the soil, water or air, or worse, some unscrupulous people directly acquired drugs, after packaging, and then into society, causing great harm to patients. Chemical waste mainly refers to medical toxic, corrosive, flammable and explosive chemicals, medical laboratories and related medical laboratory waste and reagents after use, mercury-containing waste is classified in this type of waste.

In addition, during a major epidemic, the type and quantity of infectious waste can increase dramatically and is highly infectious. Take medical waste during the Corona Virus Disease 2019 epidemic (hereinafter referred to as "COVID-19 epidemic"), as an example, it includes: medical waste from designated medical institutions, fever clinics and isolation observation points, sludge from wastewater treatment; patient excrement from medical institutions without wastewater treatment systems after disinfection; isolated or suspected patients admitted to medical institutions Produced by the domestic waste; by confirmed or suspected patients' blood, body fluids, excreta contaminated items, such as discarded cotton swabs, gauze, disposable sanitary products and disposable medical devices; new coronavirus-related teaching, research and other medical activities generated in the specimen strain, culture media, etc. In addition, medical institutions in the treatment of new coronavirus infection of pneumonia patients and suspected patients in the fever clinic and wards (rooms) generated by garbage and waste are to be collected in accordance with the classification of medical waste. Hospitals set up isolation areas, isolation areas generated by the garbage and waste should also be included in the management of medical waste.

In order to eliminate the infectious nature of medical waste to protect the environment and human health, the disposal method of incineration has been commonly used internationally for many years. However, the incineration process of medical waste is prone to produce dioxins and heavy metals and other pollutants. Among them, dioxins are known as “the most toxic poison on earth”. It is a chlorine-containing, highly toxic organic chemical substances, almost non-existent in nature, only through chemical synthesis to produce, and is currently one of the most terrible chemicals manufactured by humans.

The World Health Organization (WHO) has been committed to the management of hospital waste as special waste, actively advocating countries to establish a comprehensive medical waste disposal and whole process management system to ensure human health and environmental safety, with the aim of continuously exploring sustainable medical waste disposal technologies and effective and efficient medical waste management measures, promoting the development and application of the best feasible technologies, and promoting the environmentally sound medical waste. The aim is to continuously explore sustainable medical waste disposal technologies and effective and efficient medical waste management measures, promote the research and development and application of the best available technologies, and promote the process of sound medical waste management and disposal (WHO 2005).

In fact, the definition of medical waste is similar around the world, for example, the U.S. Environmental Protection Agency (EPA) considers medical waste as medical services such as diagnosis, treatment and immunization for humans or animals, as well as medical research, biological experiments and biological products production processes, which are hazardous or may generate multiple health risks (infectious or potentially infectious) of solid waste. As in other countries around the world, in order to promote the management and disposal of medical waste, China has implemented legislation for the management and disposal of medical waste. 2003 promulgated the “Regulations on Medical Waste Management” (State Council Decree No. 380) states, “Medical waste refers to medical and health institutions in the medical, preventive, health care and other related activities generated in the direct medical waste, is the direct or indirect infectious, toxic and other hazardous waste generated by medical and health institutions in medical, preventive, health care and other related activities”.

After more than 10 years of development, China’s urban medical waste management and disposal has gradually established and improved the city-based medical waste management system, medical waste harmless and safe disposal capacity has been strengthened, but there are still many problems for the management and disposal of medical waste in remote areas. Remote areas refer to the nearest centralized medical waste disposal facilities are relatively far away from the transport distance (such as more than 200 km), inconvenient traffic, and the region’s medical waste generation is less (such as less than 1 t/d) of the region. Remote areas are basically more backward areas, medical institutions are mainly health centers, clinics, etc., rarely carry out large-scale surgery. Therefore, in the type of waste generated pathologic waste is less, mainly infectious waste, injury waste, pharmaceutical waste and chemical waste, including identifiable human tissue, blood products, patient or animal excrement, cotton swabs and other similar waste medical needles from treatment centers, broken glass and other medical sharps that have been in contact with infectious patients. Based on the hazardous characteristics of medical waste in remote areas, it is best to focus on the control of infectious waste to reduce the risk of disease transmission from medical waste; in terms of treatment technology, priority is given to non-incineration treatment technology.

Other hazardous characteristics of medical waste such as toxicity, corrosiveness and flammability may have direct negative impacts on the natural environment such as soil, water bodies and the atmosphere. At the same time, improper disposal of medical waste can also pose a threat to public health, including the health of medical professionals. For example, in terms of occupational health risks, about100 sharps injuries occur each year among the 8.8 million health care workers in the United States, and many are infected with diseases such as AIDS and hepatitis B as a result. Rising potential risks from the transfer of medical waste pose a growing threat to human health.

Therefore, in the case of medical waste, eliminating its infectious nature is the primary goal, but the disposal process of medical waste can also produce serious secondary pollution if the technical level of equipment is low and the management measures are not perfect. Such as the incineration process will produce dioxins, acid gas and other toxic and harmful substances, medical waste non-incineration process will also produce volatile organic compounds (VOCs), odor and other environmental pollution problems.

1.2 International Requirements for Medical Waste Management and Disposal

1.2.1 POPs Convention Requirements for Medical Waste Management and Disposal

The POPs Convention states that the incineration of medical waste is prone to produce PCDD (polychlorinated dibenzo-phenylene), PCDF (polychlorinated dibenzofuran), HCB (hexachlorobenzene), PCBs (polychlorinated biphenyls) and other trace heavy metal pollutants (POPs). Among them, dioxins are listed in Annex C of the POPs Convention as POPs substances unintentionally produced and emitted by human beings. In terms of dioxin reduction time, according to the Convention, each Party shall develop an implementation plan within two years of the Convention's entry into force, and propose phased implementation of best available techniques for new sources of priority sources, including medical waste incineration, no later than four years after the entry into force of the Convention for that Party.

The POPs Convention sets out five main objectives to be achieved: ① elimination of the 12 POPs; ② support the transition to safer alternatives to POPs; ③ take action to control more POPs; ④ remove stockpiles of POPs and clean up equipment containing POPs; ⑤ mobilize all efforts to work towards the elimination of POPs.

The above objectives ① and ② are directly related to medical waste. In objective ①, PCDD, PCDF, PCB and HCB are the main pollutants emitted during the incineration and disposal of medical waste. Studies have shown that the main cause of these problems is the presence of hydrocarbons and aromatic polymers in medical waste, which occur after incineration under non-ideal conditions in the presence of elemental chlorine and metals such as copper, iron and aluminum. Therefore, in the management and disposal of medical waste, the POPs Convention requires the use of BAT/BEP to effectively reduce the emission of POPs. The general preventive measures of BAT/ BEP are given in Annex C of Part V of the Convention, and useful measures related to medical waste management include: ① the use of technologies that can easily achieve reduction; ② the use of less hazardous substances to implement medical waste disposal; ③ improve waste management, stop incinerating waste in the open air and in other uncontrolled ways; ④ when considering the construction of new waste disposal measures, priority should be given to incineration alternatives; ⑤ reduce the emissions of pollutants in medical waste disposal.

The Convention's Code of Action gives details on the concrete implementation and promotion of BAT/BEP. The main contents related to medical waste management and disposal are summarized as follows: ① promote government legislation. Governments should formulate relevant policies and supporting management methods to achieve the management and disposal of medical waste in accordance with the law and in line with the requirements of the Convention; ② establish a medical waste classification catalog to determine the types of medical waste related to the generation of POPs; ③ promote the reduction of medical waste at source, strictly distinguish between domestic waste and medical waste, and reduce the use of disposable medical supplies in the medical process; ④ reduce the use of chlorine-containing medical supplies; ⑤ establish an effective medical waste management system, in the process of classification, collection, packaging, transfer, temporary storage, minimize packaging, reuse as much as possible under the premise of safety packaging can be used to reduce plastic packaging; ⑥ scientific classification of medical waste, should be selected and disposal methods to adapt to the classification of medical waste; ⑦ caution and reduce construction of medical waste incinerators, and reduce the amount of medical waste disposed of by incineration; ⑧ in the premise of meeting public health and safety, as far as possible to promote resource recycling; ⑨ attention to scientific and technological progress, promote the use of new technologies to replace the outdated and unreasonable medical waste disposal technology; ⑩ Convention parties and national governments should support the development and promotion of waste minimization technology research, promote the development and application of environmentally friendly disposal technologies.

In order to promote the implementation of the Convention, based on Annex C of Part V of the POPs Convention, the POPs Convention Secretariat of the United Nations Environment Programme has developed the Best Available Techniques and Best Environmental Practices Guidelines (hereinafter referred to as the "BAT/BEP Guidelines"), which require Parties to adoptBATx and BEPx as soon as possible.

According to the BAT/BEP guidelines, incineration is the most mature technology for medical waste disposal, but the Convention also treats incinerators as disposal devices with greater environmental risks, and imposes strict requirements on the process design and operating parameters of mainstream incineration technologies, such as rotary kilns, pyrolysis, fluidized beds and other incinerators, as well as on the generation, collection, classification, storage, transportation, treatment and final disposal process of hazardous wastes. The detailed requirements are put forward. The guidelines also discuss alternative technologies for medical waste disposal incineration, such as high-temperature steam, microwave and chemical treatment, etc., and put forward alternative technologies that should be given priority when building new disposal facilities. The core concept of the BAT/BEP guidelines is to balance two aspects, namely, technical issues and management issues, that is, how to promote the effective combination of technology and management in the context of reducing emissions of dioxins and other pollutants. Since 2000, driven by external pressure to comply with international conventions such as the POPs Convention and internal pressure to improve environmental quality, countries and regions have implemented more stringent air pollutant emission standards for incineration facilities, and non-incineration facilities have largely replaced the original small-scale incineration facilities, and new facilities are also primarily using non-incineration technologies that do not emit toxic and hazardous substances.

The POPs Convention entered into force for China on November 11, 2004, and on April 14, 2007, the State Council approved the implementation of the "National Implementation Plan for the Stockholm Convention on Persistent Organic Pollutants", with corresponding provisions for new and existing sources. For new sources, the implementation plan states that "by 2010, the establishment and improvement of new source emission standards for key industries, and the inclusion of dioxins in the pollution control indicators. The implementation plan also points out that "to reduce and eliminate unintentional POPs emissions" actions, with the specific goal: "By 2008, the basic establishment of unintentional POPs key industries to effectively implement the BAT/BEP management system, to achieve the application of BAT to new sources of key industries, and promote BEP". For existing sources, the implementation plan points out that "by 2010 to initially establish and improve the emission standards for existing sources in key industries, the inclusion of dioxins in the pollution control indicators". Medical waste incineration is an unintentional source of POPs emissions. In January 2012, China has released the "Medical Waste Treatment and Disposal Pollution Prevention and Control Best Available Technology Guide (Trial)" according to the implementation plan, which introduced the BET for medical waste incineration and disposal, and indicated the direction (Chen et al. 2008).

1.2.2 Basel Convention Requirements for Medical Waste Management and Disposal

The Basel Convention states that medical waste is infectious and is an important component of hazardous waste. It is listed in Annex 1 of the Basel Convention on Transboundary Movements of Hazardous Wastes (Basel Convention), and its core objective is to achieve environmentally sound management (ESM), protect human health, and reduce its damage to the environment. In order to promote the implementation of the Convention, the Secretariat of the United Nations Convention has prepared the Technical Guidelines for the Environmentally Sound Management of Medical Waste. The guidelines put forward requirements on how to avoid and prevent waste generation, how to implement separation, collection, labeling and transfer, how to transport and temporary storage, and how to recycle. In terms of specific disposal methods, the guidelines also put forward corresponding recommendations for different types of waste disposal methods recommended, such as thermal treatment, chemical disinfection treatment, irradiation treatment, and incineration treatment. Waste disposal must have reliable supervision and safeguards to ensure that the treatment process to meet the corresponding standard requirements. In addition, the guide is also on the emergency plan and emergency management and other corresponding discussions.

1.2.3 World Health Organization Requirements for Medical Waste Management

In 2014, the World Health Organization (WHO) released the "Manual for the Sound Management *of Wastes from Health-Care Activities* (revised)" (*Safe Management of Wastes from Health-Care Activities*), which divides the wastes generated from medical activities into two categories, one is general waste that is not hazardous, and the other is hazardous medical waste, specifically including infectious waste, pathological waste, injury waste, chemical waste, cytotoxic waste (including genotoxic waste), radioactive waste and pharmaceutical waste.

In response to the above waste management issues, WHO has put forward corresponding requirements in terms of national legislation and management action plans, medical waste management within medical institutions, medical waste reduction measures, medical waste collection, sorting, packaging, transportation, treatment, disposal, priority activities to be supported, and development strategies.

The report states that national legislation is the basis for ensuring the safe disposal of medical waste, and that legal provisions should include waste classification, collection, packaging, transportation and disposal, as well as the responsibilities and training of various personnel; this must be complemented by government documents and relevant technical guidance. Medical waste management is a continuous process that requires periodic monitoring and evaluation. State-recommended medical disposition methods should be updated in due course according to the latest advances in scientific research. The basic formulation of the WHO action plan on national legislation and management has become an important legislative management guide for the advancement of medical waste management in countries around the world.

The results of this study suggest that good management of medical waste within health care facilities depends on clear organizational structures, adequate legislation and necessary economic conditions, and the involvement of trained personnel. In terms of policies and measures to promote waste minimization. It is proposed that: ① reduce the use of raw materials, choose less dangerous and environmentally friendly products, and use physical cleaning methods as much as possible; ② strengthen management control measures in medical institutions; ③ strengthen the management of chemical and pharmaceutical storage; ④ in terms of classification, collection, packaging, storage and transportation of medical waste, it is proposed that source classification and labeling should be done, and the implementation of regular collection of waste, fixed-point temporary storage of waste, etc.

The report also suggests that safe and reliable methods of medical waste disposal are of paramount importance; effective medical waste management and disposal objectively requires the cooperation and coordination of the whole society; the establishment of a set of legal and regulatory systems, personnel training, and raising public awareness are important conditions for the successful implementation of medical waste management and disposal.

In terms of guiding principles for activities related to medical waste management supported by WHO, the following directions are proposed: ① prevention of potential risks to medical personnel and public health from medical waste by promoting the adoption of efficient environmental management policies; ② support for global reduction of harmful emissions, reduction of disease and slowing down of global warming; ③ support for activities related to compliance with the Stockholm Convention; ④ support for activities related to compliance with the Basel Convention. Support activities related to the implementation of the Basel Convention on hazardous and other wastes.

In terms of strategies related to medical waste management, it is proposed that in order to better promote the medical waste management approach that each country decides to adopt, WHO has proposed immediate, medium and long term medical waste management strategies.

Immediate objectives include: ① to encourage the reuse of plastic medical injection products; ② choose medical devices that do not contain polyvinyl chloride (PVC); ③ promote the development and application of reuse methods (such as plastic, glass products, etc.); ④ research and promote new technology upgrades, the use of alternative technologies for small-scale incinerators; ⑤ countries with economies in transition and developing countries in medical waste management should adopt safe management and Disposal mode; ⑥ incineration is an acceptable disposal technology, but to grasp the key issues of the application of incinerators, including effective waste reduction and classification, incinerators away from densely populated areas, satisfactory engineering design, reasonable spatial layout, appropriate operational management, regular maintenance and training of personnel.

Medium-term goals include: ① further reduce the number of unnecessary syringes and reduce the amount of medical waste disposal; ② study the impact of PCDD/Fs exposure on human health; ③ carry out medical waste incineration, and in due course, medical waste disposal risk assessment work.

Long-term goals include: ① Promote the application of non-incineration technologies in medical waste treatment to reduce health risks associated with unsafe disposal of medical waste and exposure to PCDD/Fs; ② Support countries in developing effective medical waste management guidelines; ③ Support countries in developing and implementing national plans, policies, and regulatory systems; ④ promote the application of the principles of medical waste management set forth in the Basel Convention; ⑤ support the promotion of resource allocation among countries in terms of human resources and funding for the promotion of medical waste management.

1.2.4 International Requirements for Medical Waste Management and Emergency Disposal During Major Infectious Disease Outbreaks

During major infectious disease outbreaks, the relevant sources of solid waste living may be contaminated with pathogens. Interrupting the spread of pathogens during solid waste disposal is the key to effective containment of an epidemic. In the case of a major infectious disease outbreak, the surge in the number of people carrying pathogens will lead to a corresponding increase in medical and quasi-medical behavior for epidemic prevention, resulting in a rapid increase in the amount of medical waste generated; and temporary medical institutions and centralized isolation and observation sites will also produce a large amount of new care medical waste, making the amount of medical waste increase to several times the norm. For example, during the outbreak of the new crown epidemic in Wuhan, the daily amount of medical waste generated during the most serious epidemic was 247.3 t, which was 5–6 times the normal amount. In addition, during the epidemic, a large number of personal protective equipment, such waste has a high risk of pathogen contamination; and patients, people with a history of contact with the epidemic area must be observed in medical isolation, and the waste they produce is also susceptible to pathogen contamination, and must be properly handled and disposed off. In addition to routine attention to the medical institution's feces, medical sewage and sewage treatment sludge, during the epidemic, special attention should be paid to the medical institution's non-patient areas and general residential areas generated by the feces, sewage and sewage treatment sludge, because they may also be endowed with pathogens. During major epidemics, some countries and international organizations have also had the practice of adjusting the coverage of medical waste management in order to cut off the pathway of pathogen transmission. For example, France stipulates that medical waste during an epidemic should include sources of: ① waste generated by medical institutions; ② medical waste generated by decentralized epidemic prevention and medical departments; ③ medical waste generated by patients who self-medicate at home. In the United Kingdom, medical waste management during an epidemic should include: ① waste generated from patients with confirmed or suspected disease (regardless of whether the causative agent is known), and waste that may contain pathogens; and ② waste generated from patients with undiagnosed but suspected infection and potential risk of infection. During the Ebola epidemic, the EU regulations for medical waste management are: ① sharps, dressings and other supplies discarded after use in the care of suspected or confirmed Ebola patients; ② discarded supplies for clinical laboratory testing of samples from suspected or confirmed Ebola patients; ③ contaminated space for suspected or confirmed Ebola virus (e.g. wards, aircraft, ambulances and other vehicles, airports and other transportation facilities, residences, etc.) cleaning generated waste; ④disposable personal protective equipment (PPE) removed and discarded after working in a suspected or confirmed Ebola-contaminated environment.

In August 2020, the United Nations Environment Programme released the report *Waste Management during the COVID-19 Pandemic* (Response to Recovery—Medical Waste Management, noting that safe international rules and guidelines for medical waste are available and have been widely cited and followed by most countries. This provides a good basis for the management of medical waste during a COVID-19 pandemic (Tsukiji et al. 2020). However, the significant increase in medical waste generation during the epidemic placed an additional burden on the current medical waste management and disposal system. During the emergency response phase of the epidemic, regional and national governments should conduct a rapid assessment of the current medical waste management and disposal system to identify available capacity and gaps in their respective countries or cities, and develop emergency management plans based on the current situation to try to increase the number of feasible treatment technologies to achieve maximum medical waste disposal capacity. In the recovery phase of an outbreak, existing medical waste policies and regulatory frameworks can be reviewed and reflected upon, and recovery period and preparedness plans can be developed, leading to a resilient or sustainable medical waste and municipal solid waste management system.

The document also proposes that in order to better cope with the increase in infectious waste during the epidemic, the current medical waste management system can be adjusted in the following ways: ① Proper segregation, packaging and storage of solid waste that may be contaminated by pathogens, double bags can be used; ② adjust the collection frequency according to the priority (organic waste, infectious waste, etc.) and minimize the collection of recyclable items; ③ proper handling of medical waste use personal protective equipment, pay attention to hand hygiene and take other precautions to ensure the health and safety of waste disposal workers; ④ encourage all waste disposal workers, including first-line managers, to use personal protective equipment; ⑤ from the perspective of sustainability, pay special attention to the informal sector (which plays an important role in solid waste management under normal circumstances), for example, by reducing the risk of infection transmission, enhancing social security, occupational safety and health, insurance, etc., to achieve continuity in medical waste management. In April 2020, WHO issued a guidance document on personal and public health protection and medical waste management related to the COVID-19 outbreak (*Water, Sanitation, Hygiene and Waste Management for the COVID-19 Virus*), recommending ways and means to dispose of medical waste involved in the outbreak on-site while ensuring safe collection and sorting.

1.3 Foreign Medical Waste Disposal Technology Development and Management Model

1.3.1 Foreign Medical Waste Disposal Technology and Development

The types of medical waste disposal technologies are quite broad, and some foreign institutions classify medical waste disposal technologies into thermal treatment technologies, chemical treatment technologies, radiation treatment technologies, and biological treatment technologies according to the mechanism of medical waste disposal technologies to achieve the harmlessness of medical waste. However, medical waste disposal technologies can be generally classified into two categories: incineration and non-incineration (Windfeld and Brooks 2015).

1.3.1.1 Medical Waste Incineration Disposal Technology Development

Incineration disposal technology is generally medical waste into the incineration chamber, under the action of high-temperature flame, medical waste after drying, gasification, incineration of three stages, so that it decomposes into gas and residue, and the final discharge of smoke and residue for the necessary harmless treatment (Yang et al. 2012). Due to the high temperature of incineration (higher than 850 °C), the harmful substances in medical waste and germs are completely destroyed or destroyed, so incineration is a thorough medical waste treatment method. Incineration disposal technology has the advantages of large volume reduction, disinfection and sterilization effect, stability and safety. In a variety of incineration technology, according to its different working principles and combustion methods can be divided into small single combustion chamber incinerator, mechanical grate incinerator, rotary kiln incinerator, controlled gas incinerator, two-stage pyrolysis gasification incineration (batch), vertical pyrolysis gasification incinerator, electric arc furnace method, or a combination of technologies. According to the incineration method to divide, there are peroxide combustion method, pyrolysis gasification method. Many domestic scholars believe that pyrolysis gasification incineration is more suitable for the current state of China's medical waste treatment of an incineration technology, the reason is:

(1) The medical waste is pyrolyzed and gasified at a temperature of 450–600°C to produce combustible gas and cracked coke, then the combustible gas and cracked coke enter the incineration chamber for full incineration, and the heat generated in the incineration chamber is used to pyrolyze the new waste in the pyrolysis furnace, which integrates pyrolysis and incineration and is conducive to the stable operation of the incineration;
(2) Pyrolysis is carried out under the conditions of oxygen deprivation and removal of acidic gases such as chlorine, which can greatly inhibit the generation of dioxins;
(3) The pyrolysis incineration method requires a smaller air coefficient, the amount of flue gas produced is greatly reduced, the required flue gas cleaning device is also smaller, and the overall cost is smaller than that of conventional incineration, which can be used in most areas in China.

Incineration disposal technology according to the furnace type are rotary kiln type, reciprocating grate furnace, chain furnace, vertical rotary furnace, etc. At present, the domestic application of more furnace type including fixed bed incinerator, mechanical grate incinerator, pyrolysis incinerator, rotary kiln incinerator, fluidized bed incinerator, segmental combustion incinerator, etc.; foreign applications are the most widespread controlled gas incinerator.

Medical waste incineration facilities usually consist of hazardous waste feed systems, incineration systems, flue gas cleaning systems, and residue treatment systems. Waste preparation and supply, waste incineration, and air pollution control facilities can vary from one medical waste disposal facility to another. As a result, medical waste incineration disposal technology presents a combination of different forms. Incineration technology is the most widespread, longest-established and most mature medical waste disposal technology in the world because of its applicability and thorough disposal. Incineration technology is the most commonly used disposal technology in the world, its application has a history of more than 100 years, the incineration method can completely destroy the bacteria and viruses in the waste, the degree of harmlessness is complete, the performance of the residue is stable, and the treatment of waste is unrecognizable, capacity reduction is relatively large, with a small area, treatment is complete and other advantages, so it was once a widespread technology (Alvim-Ferraz and Afonso 2005).

Disadvantages of medical waste incineration disposal technology is mainly in the high cost, serious air pollution, easy to produce dioxins, polycyclic aromatic compounds, PCBs and other highly toxic substances and hydrogen chloride, hydrogen fluoride and sulfur dioxide and other harmful gases, the need to configure a complete tail gas purification system, residue and fly ash is hazardous.

1.3.1.2 Medical Waste Non-incineration Treatment Technology Development

Medical waste non-incineration technology refers to low-temperature heat treatment technology, chemical disinfection treatment technology, radiation treatment technology and biological treatment technology. Low heat treatment technology generally refers to wet heat (steam) disposal and dry heat disposal technology, wet heat treatment of the thermal medium can be directly with the help of high-temperature saturated steam, but also indirectly through the microwave steam generation. However, medical waste non-incineration treatment technologies are customarily divided according to their processes, such as high-temperature steam, microwave, chemical disinfection, electronic irradiation, biological treatment and other different types of treatment and disposal technologies. From the current international common non-incineration treatment technology, high-temperature steam, microwave, chemical are the most application of non-incineration treatment technology.

(1) High temperature steam treatment technology for medical waste

This technology is a process in which medical waste is placed in a metal pressure vessel (autoclave, with sufficient pressure strength) and superheated steam is used in a certain way to kill pathogenic microorganisms in it. The steam needs to be in full direct contact with the medical waste at a certain temperature (130–190 °C) and pressure (100–500 kPa) for a certain period of time to ensure that the pathogenic microorganisms present in the medical waste are killed. The sterilization effect mainly depends on the temperature, steam contact time and the degree of steam penetration. These factors are related to the type of medical waste, packaging, density and loading. Its advantages are that it requires less space; the process equipment is simple; it is easy to operate and does not require special training for the operator; and the sterilization is rapid and complete. The disadvantages are that the sterilization effect is affected by the degree of contact between the surface of the waste and steam, the temperature and pressure of steam, the skill level of the operator, etc.; the requirements for packaging are high, and special packaging and special treatment are often required; toxic volatile organic compounds and toxic waste liquids are easily produced during the process, and there are environmental problems such as odor and drainage; it is not suitable for treating pathologic waste, liquid waste, surgical cuttings, and volatile chemical substances (Dominica 2001).

(2) High-temperature dry thermal treatment technology for medical waste

The technology is the medical waste grinding, exposed to negative pressure and high temperature environment and stay for a certain period of time, the general temperature of 160–200 °C, treatment time of 20–30 min, the use of conduction procedures to efficiently conduct heat to the medical waste tó be treated, so that the pathogenic microorganisms with its protein denaturation and coagulation, and then lead to the death of pathogenic microorganisms in medical waste, so that medical waste is harmless to achieve the purpose of safe disposal. The system configuration of high-temperature dry heat treatment facility should include feeding unit, crushing unit, high-temperature dry heat treatment unit, discharge unit, heating unit, automatic control unit, waste gas treatment unit and waste water/liquid treatment unit. In terms of pollutant emissions, the technology has good comparative advantages and produces relatively less VOCs and odors. In the scope of application, like other non-incineration technologies, it is not applicable to the treatment of pathological waste, pharmaceutical waste and chemical waste.

(3) Chemical treatment technology for medical waste

The technology has a long history of disinfection and sterilization and more widespread application. The process of chemical treatment technology generally involves mixing the crushed medical waste with chemical disinfectants (such as sodium hypochlorite, ethylene oxide, glutaraldehyde, lime powder, etc.) and leaving it for a sufficient period of time, during which organic substances are decomposed and infectious pathogens are killed or inactivated. The maximum contact between disinfectant and medical waste is the prerequisite to guarantee the treatment effect. Usually rotating crushing equipment is used to increase the degree of crushing to ensure that the disinfectant can penetrate it; a small amount of water is added to the crushing process to absorb the heat generated by the crushing on the one hand; on the other hand, water can also be used as a medium for chemical reactions. The chemical disinfection process is suitable for the treatment of liquid medical waste and pathological aspects of waste, and has recently been gradually used for the treatment of medical waste that cannot be sterilized by heating or wetting. In addition, some newly developed technologies combine chemical disinfection with heat sterilization to reduce treatment time and improve treatment effectiveness. Chemical disinfection methods are generally divided into dry chemical disinfection and wet chemical disinfection. For dry chemical disinfection, generally has a process equipment and operation is relatively simple; one-time investment is small, low operating costs; high rate of waste reduction. Site selection is convenient, can be mobile processing; operation is simple and convenient, the operating system can be shut down at any time, will not produce waste liquid or waste water and exhaust gas emissions, environmental pollution is very small advantages. However, the requirements for the crushing system are high; the pH monitoring (automation level) of the operation process is very demanding. For the wet chemical disinfection method, it generally has the advantages of low one-time investment, low operating costs, and simple process equipment and operation. However, the treatment process will have waste liquid and exhaust gas generation, most of the disinfection solution is harmful to humans, high requirements for operators, the operator's labor intensity and other disadvantages. In general, chemical disinfection is not suitable for treating chemotherapy waste, radioactive waste, volatile and semi-volatile organic compounds, etc.

(4) Medical waste microwave treatment technology

The technology refers to the principle of using a certain frequency and wavelength of microwave action to kill most microorganisms by microwave excitation of prebroken and wetted waste to generate heat and release steam. Microwaves and the right amount of moisture are the two basic conditions for generating heat for sterilization. Microwave technology for medical waste can be divided into the following five steps: ① water and waste stirring vibration; ② loading facilities to transfer the wetted waste to the shredding equipment, crushed into pieces; ③ steam injection, and the wetted waste is transferred to the irradiation chamber has been equipped with a microwave generator; ④ the waste in which the irradiation for about 20 min, the microwave will be the moisture in the waste heated to 95 °C, thus completing the sterilization of medical waste ⑤ The waste is compressed in a special container and sent for disposal (landfill or incineration). This technology can significantly reduce the volume of waste, and does not produce acidic gases and gaseous pollutants such as dioxins during the treatment process. However, the effect of sterilization is affected by the source strength of electromagnetic waves, the length of radiation duration, the degree of waste mixing, the amount of water content in the waste and other factors; the operator may be affected by bacteria and electromagnetic waves, resulting in occupational hazards; high construction and operating costs; not applicable to the treatment of pathologic waste, pharmaceutical waste and chemical waste.

Currently, there are more than 40 non-incineration technologies and more than 70 equipment providers worldwide, located in the United States, Europe, the Middle East, and Australia, with some still being explored, refined, and used in conjunction with other methods. Although the disposal capacity of non-incineration technologies for medical waste varies, the degree of automation varies, and the degree of reduction varies, all of these technologies use one or more of the following methods to achieve the disposal of medical waste, and the following conclusions can be drawn: (i) microwave, high-temperature steam and other auxiliary heating methods to heat the waste to more than 90–95 °C, (ii) disinfection of medical waste by exposing it to chemicals such as sodium hypochlorite, chlorine dioxide and calcium oxide; (iii) disinfection of medical waste by exposing it to radiation sources. In conclusion, compared with incineration disposal technology, medical waste non-incineration treatment technology has been applied and developed worldwide because of its embodied low construction and treatment costs, low difficulty in meeting treatment standards, high public acceptability, and no international convention requirements.

1.3.1.3 Trends in the Application of Medical Waste Disposal Technology

The potential drivers for updating and improving medical waste disposal technologies involve technical, environmental, economic, and social aspects. However, in terms of global development trends, medical waste disposal technologies show a continuous process of learning from experience and innovation, with countries around the world moving in a similar but distinctive development direction. Before people are aware of the special infectious and polluting nature of medical waste, landfills were the main disposal technology, and medical waste was often disposed of in landfills together with domestic waste during this period. With the awareness of the special infectious and polluting hazards of medical waste, incineration technology became the most mainstream disposal technology. When people realized that the incineration of medical waste produces dioxins and other highly toxic gases, non-incineration technology developed rapidly, especially the development of high-temperature steam, microwave and chemical disinfection as the representative of the application of non-incineration technology is faster.

In the United States, before 1997, the main method of disposal of medical waste incineration. In 1997, the U.S. Environmental Protection Agency issued a new incinerator standard to improve the dioxin emission standard limit requirements. According to the standard, the size of the incinerator, according to <91, 91–230 and ≥230 kg/h incinerator into small, medium and large three cases, and the implementation of different management requirements. For large and medium-sized incinerators should be implemented for trial burn testing, while for small incinerators and rural incinerators without annual testing and performance testing. In order to promote the implementation of new standards, but also to incur certain costs, such as for new or existing incinerators, the need to invest or additional investment of 150,000–250,000 U.S. dollars, equipped with the necessary pollution control facilities. And for medium and large incinerators, the need to invest 300,000–500,000 U.S. dollars to complete the necessary upgrades. On this basis, according to the U.S. Best Available Technology (MACT) requirements, to promote the construction and operation in order to meet the latest standards requirements. In addition, the tightening of incineration standards has also led to a significant increase in the cost of medical waste disposal, which has also resulted in a significant reduction in the number of incinerators. Medical waste disposal technology began to shift to disinfection treatment, starting in 1998, a large number of non-incineration treatment facilities began to build, that year, the establishment of 1516 non-incineration treatment facilities, including high-temperature steam 931, chemical disinfection 173, hot steam disinfection 92, microwave treatment 254, other new technologies 61. The number of medical waste incineration disposal facilities has been reduced from 6,200 in 1998 to more than 100. Currently, the United States is further closing medical waste incineration facilities, and further construction of non-incineration treatment facilities. Most U.S. states require pre-treatment of medical waste prior to disposal, either on-site or through a qualified facility. On-site treatment is usually considered non-infectious and can be mixed with regular waste in some cases. However, there are still some states that require medical waste to be handled separately. All states allow on-site disposal of medical waste using appropriate methods. Some states require disposal companies to have permits or disposal plans, while others only require incineration facilities to have air emission permits. In the United States, the primary methods of medical waste treatment and disposal are incineration, high-temperature steam, and mechanical/chemical, microwave, and irradiation technologies. According to the U.S. EPA, 90% of medical waste is incinerated. Regardless of the treatment/disposal method used, treated wastewater can generally be discharged underground or into the sewer. Treatment facilities can engage private businesses to collect, treat and dispose of medical waste.

In the EU, incineration was the main method of medical waste disposal until the 1990s. Germany used to use oxidative incineration processes to dispose off medical waste, and in 1984 there were at least 554 small incineration facilities in use in hospitals in Germany. In Germany now, using non-incineration technology is mainly high-temperature steam technology, there are now more than 500 internal hospital high-temperature steam treatment facilities, as well as four commercial centralized medical waste disposal centers. The United Kingdom before 1980, there are about 700 hospitals built medical waste incinerators, non-incineration technology has now treated more medical waste than incineration technology. In Ireland and the autonomous region of Northern Ireland, medical waste disposal methods using steam-based non-incineration technology, using disinfection technology to dispose of 10,000 tons of medical waste per year. Since 1995, Slovenia has been using mobile hot steam disinfection of medical waste disposal, the country currently has three mobile hot steam disinfection treatment facilities. Spain 96% of medical waste disposed of by sterilization (mainly high-temperature steam), only 4% of medical waste disposed of by incineration. Poland also began to restrict the incineration of medical waste containing PVC plastic, and the establishment of new regulations to promote the recycling of medical waste and promote the application of alternative technologies to incineration. In particular, after 2004, most EU countries signed the POPs Convention to promote the use of non-incineration treatment technology results in the most appropriate medical waste disposal technology options for EU countries. In order to promote relevant work, the Health Care Waste (HCWH) and WHO are collaborating to encourage the use of non-incineration technologies in Argentina, India, Latvia, Lebanon, the Philippines, Senegal, Vietnam, and seven other countries. The aim is to reduce dioxins and mercury pollutants through the implementation of the project. Additionally, this initiative serves as a reference for other countries in implementing the POPs Convention.

Medical waste is intermediate between municipal waste and hazardous waste, and there is no regulation of its risk level or potential risk. In Spain, the management of medical waste includes sorting, collection, packaging, indoor transfer, indoor storage, and outdoor collection and transfer, including final disposal. The most commonly used disposal method is incineration, although the incineration process may have an impact on the environment; alternative technologies to incineration have been proposed in some regions, such as autoclaving, steam sterilization, and chemical treatment methods. Although microwave sterilization is currently considered the most economically competitive alternative to incineration, and there is no legal alternative technology.

In Canada, the main methods of medical waste treatment and disposal are as follows:

Steam treatment. High-pressure steam treatment method mainly treats microbiological laboratory waste, human blood and body liquid waste, waste sharps and non-anatomical animal waste.

Chemical treatment. Chemical treatment methods can deal with microbiological laboratory waste, human blood and body liquid waste, waste sharps.

Incineration disposal. Incineration is the main method of disposal of anatomical and non-anatomical biomedical wastes. Human tissue, organs and body parts (but not teeth, hair and nails) must be disposed of by incineration.

Discharge to sewer. Untreated blood, excreta, and secretions may be discharged directly into sewers.

In Japan, according to the statistical results in 2000, in its licensed disposal of medical waste manufacturers, the use of incineration technology is 360, high-temperature incineration technology is 7, high-temperature sterilization technology is 3, dry sterilization technology is 6, the use of other technologies is 6. In Japan, the mainstream incineration technology continues to evolve, mostly using pyrolysis gasification melting furnace, and the extensive use of selective catalytic reduction technology to remove dioxins. After incineration, the weight of medical waste is reduced to 1/6, the volume is reduced to 1/10–1/20, the best working conditions to achieve basically harmless disposal.

In India, the main disposal technology is still incineration, and on the basis of incineration as the main disposal technology, there is a tendency to use a combination of incineration and non-incineration treatment equipment to deal with different types of medical waste, in order to solve the problem of applicability of different disposal technologies for medical waste (Patil and Shekdar 2001). For example, cytotoxic and pharmaceutical waste is incinerated, polyethylene material is used instead of PVC material as much as possible, chlorine-containing substances are reduced, and the source of dioxin production is reduced. India requires each integrated medical waste treatment plant equipped with the following facilities: high-temperature steam/microwave, incineration facilities, shredding facilities, secure landfill facilities, sharps storage/reuse facilities and wastewater treatment facilities, etc., on the basis of which, according to the applicability of different technologies for the waste to take different disposal methods; for disinfection and shredding of plastic waste to be recycled or disposed off using sanitary landfill methods; For sterilized sharps should be disposed off by sanitary landfill or recycling; for incineration fly ash should be disposed off by safe landfill; for oil and grease, sanitary landfill should be used for disposal; for treated wastewater, it can enter the sewerage network or be recycled. It can be said that India in strict medical institutions within the classification system of medical waste, based on different types of waste to take different treatment methods, with a strong focus.

From the evolution of global medical waste disposal technology and the development of history, basically have experienced from incineration to high-temperature steam, microwave and chemical disinfection and other non-incineration treatment technology gradually change the process. High-temperature steam treatment technology, microwave treatment technology and other non-incineration technology has been widespread applied. The main reasons for the following five aspects are as follows;

(1) After 1990, health and environmental research institutions in developed countries concluded after extensive research that the disposal of medical waste by incineration would lead to serious environmental pollution, and the lack of necessary training and financial supplements in the operation of incineration facilities made the operation of incineration facilities more risky. Therefore, it also promotes further public attention to the safety of medical waste collection,

storage, transportation and disposal processes, and also promotes the research and application of new medical waste disposal methods.

(2) Incineration technology has been used in developed countries for decades, and the relevant medical waste incineration disposal equipment has reached the renewal period, with the increasing maturity of non-incineration treatment technology, its construction costs and operating costs compared with incineration technology reflects a greater advantage.

(3) Europe and the United States and other countries have promulgated more stringent incinerator standards, new and existing medical waste incinerators must meet the corresponding pollutant emission limit requirements.

(4) As the POPs Convention entered the implementation phase in May 2004, the Convention requires all signatory Countries to reduce the generation of by-products such as dioxins, and medical waste incineration is an important source of these by-products, and is undoubtedly one of the main elements restricted by the Convention.

(5) There is a growing tendency for the public to oppose tail gases such as dioxins from the incineration process, and although there are environmental problems with non-incineration treatment technologies, they are much more acceptable from the public's point of view than incineration.

Of course, whether it is incineration technology or non-incineration treatment technology, any one disposal technology is not a panacea, have their advantages and disadvantages and specific application range. Therefore, how to better use the scientific mode of technology application, and effectively promote the safety and timeliness of technology applications, a scientific management model to match the application of non-incineration technology is an inevitable choice to promote optimal management of medical waste. However, from both economic and technical perspectives, especially from the perspective of dioxin emission reduction, non-incineration technologies should be preferred for medical waste treatment, and it is worthwhile to reflect on the past thinking of paying too much attention to incineration technologies.

1.3.2 Foreign Medical Waste Disposal Management Model

The level of medical waste management and disposal in a country or region is closely related to its level of economic development, the degree of governmental attention, sound laws, regulations and implementation of monitoring systems, adequate financial and human support, strong scientific research capabilities, the community's environmental awareness and systematic training of medical waste management and disposal personnel (Almuneef and Memish 2003).

In the United States, medical waste management is mainly managed at the state level, supplemented by federal regulation (Lee and Huffman 1991). Since 1991, the U.S. EPA has delegated the direct management of medical waste to the state governments, and the federal government is mainly responsible for the overall framework of medical waste and the development of relevant standards and norms, while the specific implementation is completed by the state government organizations (Qinghua 2008). At present, only the "Model Guidelines for State Medical Waste Management" issued in 1992 remains as the guiding document for medical waste management in each state, and the other medical waste-related regulations issued before it all expired after 1991. In response to the disposal of medical waste with infectious risk, the National Safety Council organized the Centers for Disease Control and Prevention, the National Environmental Protection Agency, the U.S. Department of Transportation and other departments to jointly prepare the Planning Guide for the Disposal of Solid Waste Contaminated with Class A Infectious Waste Materials and the Guide for the Management of Class A Infectious Solid Waste in 2017 and 2019, respectively, to guide the departments in developing specific plans for medical waste collection, transportation, disposal, etc. However, the state government regulations still play an important role in the actual implementation. In terms of disposal technology, it is usually done in the form of recommendations, which do not have the force of law, but if the guidelines become statutory provisions, they have the force of law (Zhipeng 2003). In order to promote the standardization of medical waste incineration facilities, the EPA Office of Air Quality Programs and Standards (OAQPS) developed emission standards specifically for new medical waste incinerators, making them an integral part of the U.S. Clean Air Act. In order to meet this requirement, the need to configure the appropriate tailpipe pollution control equipment, to be configured on the basis of the original incineration of the second stage furnace, to carry out regular monitoring of the incineration tailpipe, the incinerator temperature and other operating parameters to be continuously monitored. The regulations also provide for the training and qualification of operators, to develop appropriate waste management plans, the establishment of reporting and record-keeping system. According to this standard, all existing medical waste incinerators must comply with the hospital/medical waste incinerator standards by 2002, which involves a much stricter and more complex management and technical processes. The U.S. EPA combined with the hazardous waste and medical waste incineration and disposal facilities license management system, the facility's commissioning, quality control, performance testing and other specific out of the detailed requirements. For the incineration facility commissioning management, the U.S. EPA has developed technical guidelines for commissioning, and put forward specific requirements for performance testing during the commissioning period, after the commissioning phase of the implementation of performance testing and evaluation of qualified before the issuance of hazardous waste business license, and officially put into operation. Performance testing process requires examination of the incineration facility's limit operating conditions and normal operating conditions of the relevant parameters of the test problem, on the basis of which to determine the type of waste incineration and disposal facilities can dispose of and their corresponding working parameters and the main operating parameters of the facility. The above management system of the

United States both for the U.S. medical waste management and disposal industry technology, equipment and disposal facilities construction of the corresponding requirements, but also for the development of medical waste disposal technology to provide legal, policy support and guidance.

From 1990 to 2005, the EU countries have made a lot of changes to the existing medical waste management and disposal regulations, from the generation, classification to the final disposal of medical waste have made more scientific regulations. Medical waste is collected according to its material, place of generation, and degree of pollution, and each type is disposed off in a way that meets its own characteristics, in order to achieve the best treatment effect and the best environmental protection. In medical waste incineration disposal, according to the EU Waste Incineration Directive (2000/76/EC), medical waste incinerator dioxin emissions must meet the emission limit of 0.1 ngTEQ/Nm^3, to the phase-out of small incineration disposal equipment and the development and application of non-incineration treatment technology provides a necessary and wide development space (Coutinho et al. 2000). In addition, Europe also for medical waste disposal facilities of various types of pollution control, operating permits, supervision and management, technical assessment, operation of facilities have made more detailed provisions, reflecting the transition from end-to-end control to the whole process management, aimed at promoting the whole process management to provide the basis and conditions (Muhlich et al. 2003).

Japan's medical waste management is the responsibility of the Environmental Recycle Bureau under the Ministry of the Environment, which guides and regulates medical waste treatment and disposal nationwide through the issuance of policies and regulations and the preparation of related standards and technical guidelines (Li et al. 2003). At the local government level, there are environmental bureaus that are responsible for the implementation of relevant policies in their jurisdictions. In the 1970s, the Japanese government first enacted the core legislation in the field of solid waste, the Law Concerning Waste Disposal and Cleaning, which sets out the requirements for the disposal of medical waste. For medical waste with infectious risk, Japan promulgated the "Handbook of Infectious Waste Disposal Based on the Waste Disposal Law" in 1922, and revised the handbook in 2018 to clarify the responsible parties for infectious waste treatment and disposal, and the revised handbook has become the benchmark document leading the treatment of infectious waste in Japan (Miyazaki and Une 2005). Under the systematic policy and clear administrative responsibilities, Japan has established a set of management system for the whole process of medical waste from generation, collection, transportation, storage, treatment to final disposal. In order to minimize the risk of pathogen transmission, Japan requires that infectious waste be disposed off at the facility where it is generated as much as possible, and therefore, it is clearly stated in its regulations that "the site where medical waste is generated is fully responsible for the treatment and disposal of infectious waste". In contrast, non-infectious wastes within hospitals, such as those generated by logistics and administrative departments, should be treated differently from infectious wastes and treated as general household wastes, which can be reused. Japan also has strict management and registration of medical waste disposal facilities, requiring the use of special vehicles for medical waste delivery, and the use of special incineration facilities for medical waste disposal, the disposal process should be regularly monitored regularly or online for tail gas emissions and working conditions, and the residue after incineration is disposed off in the final disposal site for landfill. In terms of supervision and management of medical waste disposal, the supervision of the disposal process is carried out jointly by the prefectures, municipalities, cities, towns and villages, the National Federation of Industrial Waste, the Association for the Promotion of Proper Medical Waste Disposal Plan and civil society, and a monitoring system is set up for easy-to-offend locations and a telephone and fax reporting system for local residents, and appropriate penalties are imposed on violators.

In 1961, 1963, and 1977, Korea enacted the Dirt Cleaning Act, the Public Hazard Prevention Act, and the Environmental Conservation Act, which regulated waste pollution prevention and control, but did not separately classify medical waste, which was almost always mixed with domestic waste (Jang et al. 2006). Until 1981, the Korean Ministry of Health and Welfare issued some regulations on medical waste disposal, defining the scope of the concept of medical waste, specifying the treatment process, guiding the classification, and encouraging the construction of private incinerators (Chartier et al. 2014). Although these management regulations are related to medical waste management, they are still not institutionalized and systematized. In 1986, Korea introduced the comprehensive Waste Management Act as the basic law for waste pollution prevention. In the face of the problem of medical waste management, the Korean National Assembly revised the Waste Management Act in 1999. The Act defines medical waste as "designated waste", i.e., hazardous substances that may cause pollution to the surrounding environment, such as waste oil and acid, or substances that cause harm to humans in workplace waste. Accordingly, medical waste is subject to the hazardous waste regulations under the Waste Management Act and is required to be collected, transported or stored in special containers for medical waste to prevent risks such as various infections caused by medical waste. In addition, the Korean Ministry of Environment has issued a series of regulations on the definition, classification, packaging, tracking and disposal of medical waste. In December 2019, Korea revised and released the Guidelines for Medical Waste Classification and Management, which systematically describes the definition, classification and management of medical waste, the approval process of treatment plans and medical waste disposal. According to the Waste Management Act and the Guidelines for the Classification and Management of Medical Waste, Korea has established a whole process management system for medical waste, which transmits information on its discharge, collection, transportation, and disposal to the national integrated waste management system in real time through the electronic tag identification of special containers for medical waste for unified statistics, supervision, and management. The Waste Management Law clearly stipulates that all enterprises with the discharge, transportation and disposal of waste must be included in the national waste management system for management. National and local regulatory agencies can use the system to track the whole process of real-time monitoring of medical waste generation, transportation, treatment and final disposal, which can prevent the occurrence of illegal treatment and disposal of medical waste. Korea requires all medical waste to be processed in incineration facilities or sterilization shredders, and reuse is prohibited. However, according to the Korean Pharmaceutical Affairs Act, only companies that use placenta as a raw material for pharmaceutical processing are allowed to reuse it after receiving permission. Medical waste disposal manufacturers are only allowed to incinerate and dispose of medical waste, and those who dispose of it in hospitals themselves can choose to incinerate or sterilize and shred it. Liquid medical waste mixed with patients' blood or secretions, excreta, etc., is disposed off directly into water pollution control facilities in accordance with the Law Concerning the Protection of Water Quality and Water Ecology. At the same time, Korea is actively promoting the reduction of medical waste at the source, reducing the amount of medical waste discharged due to improper classification and operation at the source by exhaustive classification of medical waste and strengthening the training and education of medical institutions and other related staff to prevent the mixing of other wastes. Korea also guides and encourages producers to reduce waste emissions and enterprises to participate in medical waste collection, transportation and disposal by strengthening economic incentives and the role of market mechanisms, such as medical waste facility construction, in order to rapidly increase the construction of medical waste disposal facilities and improve disposal capacity.

In India, the Ministry of Environment and Forests requires medical waste facilities to meet technical standards from operation to discharge, including incinerator standards, high temperature steam standards, microwave standards, and landfill standards. Although incineration is still the main method of medical waste disposal in India, non-incineration technologies are gradually attracting attention and are being considered by the relevant agencies.

Along with the world's advancing wave of medical waste disposal technologies, developing countries are gradually improving the current technology application management system (Lee et al. 2004), but there are large gaps in the capacity building of disposal facilities, policies, regulations and standards guiding the application of technology, as well as in the promotion of relevant training systems and environmental awareness in developing countries, and continue to improve through increased investment, through the upgrading of technology, through enhanced training and environmental awareness. We will continue to improve by increasing investment, upgrading and transformation of technology, and enhancing training and environmental awareness (Diaz et al. 2005).

In developed countries, best practices for medical waste are generally promoted by strengthening legislation and preparing supporting standards and guidelines for feasible methods of medical waste collection, transportation, storage, and disposal (Insa et al. 2010). The proper disposal of medical waste is also promoted through the use of BAT, such as alternative technologies, in order to minimize the risks to human health and the environment. In developing countries, the issue of medical waste has not received sufficient attention, and medical waste is still treated and disposed off together with domestic waste, posing a greater threat to operators, the public, and the environment (Mbongwe et al. 2008). In addition, no comprehensive measures have been taken to promote the management of medical waste within medical institutions. It can be said that sustainable and stable scientific management and disposal of medical waste is a daunting task globally (Blenkharn 2006). International research on medical waste management and disposal models is still in the process of continuous development. Governments in all countries have invested a lot of human, material and financial resources to carry out relevant research work. A more scientific and sustainable medical waste disposal technology application and management model is the goal that all countries are striving for.

In the medical waste incineration and disposal technology application management model, various countries around the medical waste disposal pollution control has carried out a lot of work (Alagöz and Kocasoy 2008). In order to control the environmental problems caused by medical waste incineration process (Aristizbal et al. 2007), foreign countries on the one hand to promote the compliance process continue to use BAT/BEP, improve the incineration process, reduce the emission of toxic substances; on the other hand, also continue to tighten the corresponding pollution control standards in order to better manage the medical waste incineration process (Alvim Ferraz et al. 2000). From the current international situation in general, first of all, incineration is the primary source of dioxin production (Addink and Altwicker 2004), and is also the main source of mercury production, the incineration process can also produce other metal pollutants such as lead, cadmium, arsenic and chromium, as well as halogenated hydrocarbons, acidic gases that cause acid rain, greenhouse gases, etc. In addition, the incineration process is very strict on temperature, time and oxygen requirements, and once the problem or if the incineration of medical waste does not take BAT/BEP, there is a potential risk of generating and releasing the above-mentioned high concentrations of pollutants, posing a great threat to human health and the environment. Second, the United States and the European Union have set strict air emission standards for incinerators, making it necessary to upgrade equipment and install monitoring devices at most facilities, which greatly increases the investment costs and operating costs of incineration. However, even if the installation of air pollution control equipment and play a role, but only the transfer of pollutants from the atmosphere to the fly ash, and thus create a need for further treatment of hazardous waste streams, subject to further environmentally sound disposal, indicating that the problem of pollutant emissions is not fundamentally solved, but only the transfer of pollutants from one medium to another. The incineration of fly ash is highly hazardous (Zhao et al. 2008), but the current regulations are very weak. Therefore, the unfavorable factors of incineration disposal technology have led to a declining trend in its application worldwide.

In the medical waste non-incineration treatment technology application mode, non-incineration treatment technology cannot dispose off all medical waste, such as high-temperature steam, microwave and chemical disinfection, only infectious waste, sharps and part of the pathologic waste, while chemical waste and pharmaceutical waste is not applicable to non-incineration treatment technology. If the classification and disposal is not properly managed, the infectious problem of medical waste is not easy to completely eliminate. In addition, the non-incineration process of medical waste also produces secondary pollutants such as VOCs, odor and dust, and residual liquid and wastewater are generated during the disposal process, so the pollution control problems in the disposal process cannot be ignored. Therefore, foreign production of medical waste non-incineration treatment facilities are equipped with a more complete odor and VOCs pollution control system, and through automated control, to achieve the safe management of medical waste disposal process. In addition, improper application and management of non-incineration treatment technologies can pose potential risks to human health and the surrounding environment, and improper treatment can still lead to the spread of disease.

The application of medical waste disposal technology is not only a matter of end-of-pipe disposal, in fact, the management and disposal of medical waste is a systematic project, specifically reflected in the source classification of medical waste, the internal management process of medical institutions, the applicability of disposal technology to strengthen the internal management and disposal of medical waste in medical institutions.

Classification of medical waste is the starting point of medical waste management, but also the key link. It is relatively well accepted that the emissions of air pollutants during the incineration and disposal of medical waste are closely related to the classification measures. Medical waste includes many PVCs and is therefore a likely precursor for dioxin generation. In addition, the presence of metals such as copper and iron in medical waste is likely to be a catalyst for dioxin generation and will increase the probability of dioxin generation. Therefore, an important purpose of medical waste classification is to use different disposal technologies to dispose of medical wastes that are compatible with them, in order to achieve good disinfection, low environmental pollution and low disposal costs.

Both international conventions and countries around the world have given high priority to the separation of medical waste. In essence, strict segregation of medical waste management can effectively exclude domestic waste from being mixed with medical waste, reduce the amount of medical waste generation and final disposal, reduce the emission of dioxins and other pollutants, and, of course, effectively control the direct harm caused by infectivity. In terms of a standardized medical waste management, it should include the processes of collection, sorting, packaging, transportation, treatment and disposal. However, the relevant work in developing countries is generally at a relatively backward stage of development, and there is a gap compared with developed Countries. In terms of the amount of medical waste generated, only about 10–25% of hospital waste is medical waste, which is correspondingly infectious and dangerous. Countries around the world are also taking different approaches to the classification of medical waste. On the one hand, the classification aims to promote the infection control of medical waste, and the classification of medical waste is also linked to the disposal of medical waste, which requires special attention.

Due to the economic capacity of developing countries, their medical waste treatment facilities generally have a single, lack of and rudimentary phenomenon (Abdulla et al. 2008), resulting in the classification of medical waste and the final disposal of poor connectivity, the concept and technology of medical waste disposal is often used in developed countries more than a decade ago and now has been eliminated methods (Bdour et al. 2007). The classification of medical waste, due to the constraints of the final disposal technology, often only has the effect of reducing the mixing of domestic waste, and its main characteristics are as follows:

(1) Lack of operational guidelines to guide the matching of medical waste sorting and final disposal techniques.
(2) The concept of medical waste disposal lags behind, staying at the lower level of disposal of medical waste through incineration or landfill, ignoring the serious pollution that may be brought by unscientific and unclean terminal disposal methods.
(3) Lack of financial support, resulting in waste separation and final disposal of the hardware part of the technical specifications cannot be implemented, forming an incomplete policy level, the implementation and operation of the lack of embarrassing situation, such as India, the laws and regulations are more complete, supporting policies have been developed, but due to the high requirements, lack of funding, more than 60% of the country's hospitals cannot meet the standards.
(4) Small-scale incinerators basically do not meet the requirements and are the main source of emissions of dioxins and other pollutants.
(5) Since the disposal methods of medical waste in developing countries are mainly landfills and incineration, the correlation between good waste separation and end-of-pipe disposal is not high, which makes it difficult to push and to implement accurate separation.

It can be said that, from the overall development trend of medical waste disposal in the world, the first is to establish a perfect management model for the application of medical waste disposal technology to achieve material substitution of medical supplies, medical waste classification, medical waste collection, storage, transportation, disposal of the whole life cycle of the management.

1.4 China's Medical Waste Disposal Technology Development and Management Model

1.4.1 Development History of Medical Waste Disposal Technology Applications in China

In the application of medical waste disposal technology, it is of great historical significance that in 2004 the State Council approved the "National Hazardous Waste and Medical Waste Disposal Facilities Construction Plan". Through the implementation of the plan, 331 centralized medical waste disposal facilities were built, with an investment of 6.89 billion Yuan. The distribution of medical waste disposal facilities after the completion of the plan is shown in Fig. 1.1.

From Fig. 1.1, by the end of 2018, the number of incineration disposal facilities in the planned construction projects accounted for 37% of the total number of disposal facilities, while the number of non-incineration disposal facilities accounted

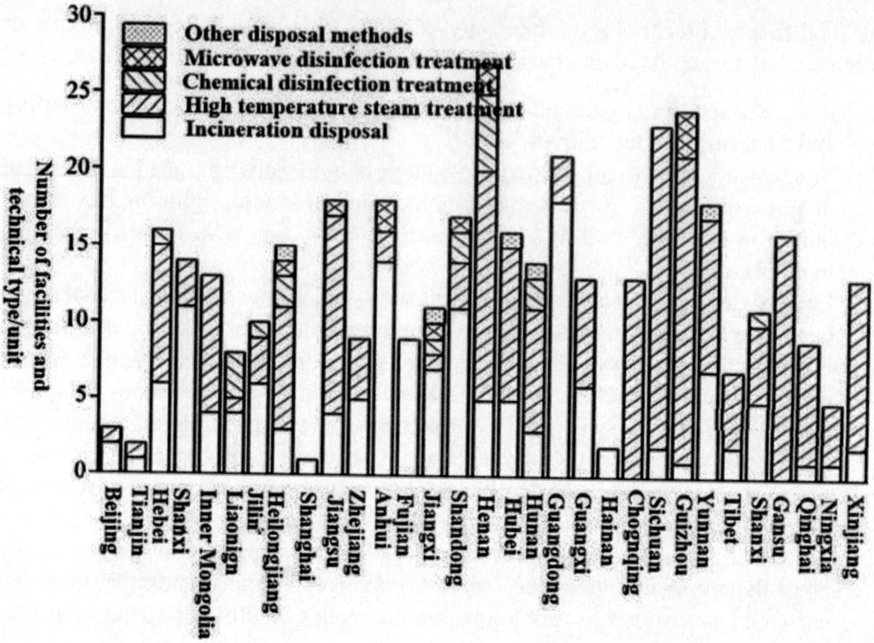

Fig. 1.1 Distribution of medical waste disposal facilities after the completion of the plan

for 63%. Incineration facilities are mainly pyrolysis incineration disposal facilities, non-incineration disposal facilities are mainly high-temperature steam treatment facilities. From the current status of domestic technology applications, the national medical waste incineration technology and the application of non-incineration technology are very common, where non-incineration treatment facilities in China has become a major medical waste treatment technology programs, and show a good momentum of development.

From the implementation of the plan, the problem of construction of medical waste disposal hardware facilities has basically been solved, while there is still a gap in the software supporting conditions to promote the standardization and safe operation of medical waste disposal facilities, especially in the specific context of China's economic policies and regulatory instruments related to the medical waste disposal industry is not in place, how to avoid and reduce the operation of hazardous waste incineration disposal facilities. The environmental risks in the operation of hazardous waste incineration and disposal facilities have become a scientific and practical problem that must be faced.

1.4.2 China's Medical Waste Disposal Management Model

The management of medical waste in China first began in the 1990s, and the main history includes the following stages:

(1) In 1995, the Ministry of Construction issued an industry standard for urban construction in the People's Republic of China, "Environmental Health Standards for Medical Waste Incineration", which sets out the corresponding requirements for the environmental health standards of the medical waste incineration process.
(2) In 1996, the State Technical Supervision Bureau and the Ministry of Health jointly issued the Sanitary Standard for Hospital Disinfection, which stipulates that contaminated items, whether recycled or discarded, must be disposed off in a harmless manner and no pathogenic microorganisms must be detected.
(3) In 1998, the State Environmental Protection Bureau and other ministries jointly promulgated and implemented the "National Hazardous Waste List" and revised it in 2008, which included medical waste.
(4) In 2000, the Ministry of Health promulgated and implemented the "Code of Practice for Hospital Infection Management (for Trial Implementation)", which provides for the disposal of hospital waste and clearly states that medical waste should be collected and treated separately; sharp instruments should be placed in leak-proof and puncture-resistant containers after use and be treated harmlessly; infectious waste should be placed in yellow plastic bags for airtight transportation and be treated harmlessly, etc.
(5) In 2001, the State Environmental Protection Administration issued the "Hazardous Waste Incineration Pollution Control Standard" (GB 18484-2001), "Hazardous Waste Storage Pollution Control Standard" (GB 18597-2001) and "Hazardous Waste Landfill Pollution Control Standard" (GB 18598-2001), the site selection of hazardous waste incineration plant, the technical indicators of the incinerator, hazardous waste storage and other aspects have made the corresponding provisions.
(6) In 2003, the State Council promulgated and implemented the "Regulations on Medical Waste Management", which is the first regulatory document on medical waste management in China, its introduction marked the management of medical waste in China from the generation, temporary storage, delivery, centralized disposal of the whole process into a standardized, legalized management track.
(7) In 2004, the State Council approved the "National Hazardous Waste and Medical Waste Disposal Facilities Construction Plan". The implementation of this plan has undoubtedly become a historic leap in China's efforts to promote the construction of hazardous and medical waste disposal capacity. Subsequently, the Ministry of Environmental Protection and the Ministry of Health also promulgated a series of regulations and standards, and the promulgation and implementation of the plan and related supporting policies, regulations and

standards marked a new stage of development in medical waste management and disposal in China.

(8) In October 2019, the Ministry of Ecology and Environment issued the "guidance on enhancing the capacity of environmental supervision of hazardous waste, the use of disposal capacity and environmental risk prevention capacity"; In February 2020, the Nation Health Commission of the People's Republic of China and Ministry of Ecology and Environment of the People's Republic of China and other ten ministries and commissions jointly issued the "comprehensive management of medical waste work program" which has clear requirements. By the end of 2020, each prefecture-level city x should build at least one standard medical waste disposal facilities. By the end of June 2022, the counties (cities) should build a more complete medical waste collection and transfer disposal system, and will be around to make up for the shortcomings, and comprehensively improve China's medical waste disposal capacity.

The management system related to medical waste in China is shown in Fig. 1.2.

As shown from the data in the figure above, in the application of medical waste incineration disposal technology, the state has also promulgated a series of standards, technical specifications and regulatory documents to regulate the treatment and disposal of medical waste and operation. In order to regulate the above three medical waste non-incineration treatment technology, in 2005 China has promulgated the "Centralized Medical Waste Disposal Technical Specification" (Huanfa [2003] No. 206), "Centralized Medical Waste Incineration Disposal Engineering Construction

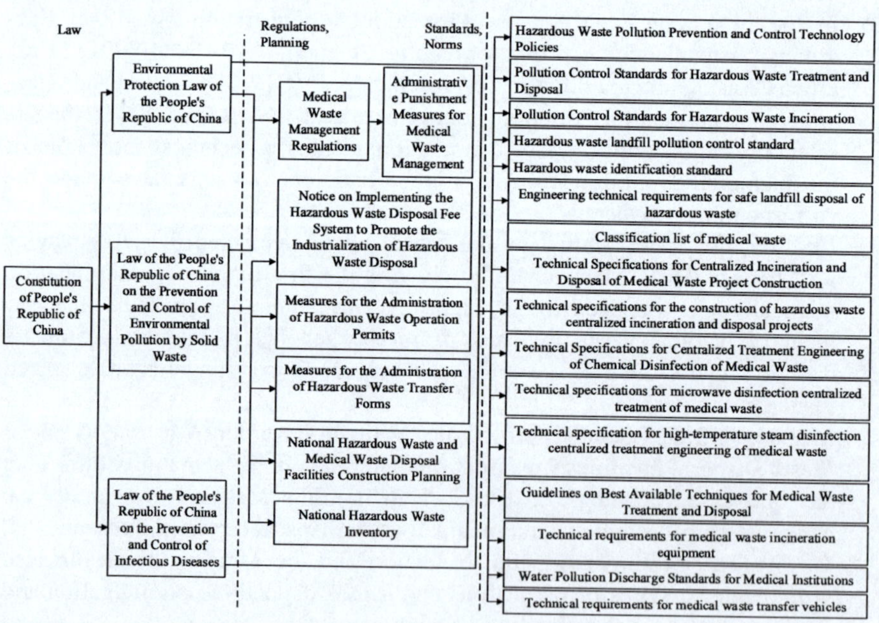

Fig. 1.2 The management system related to medical waste in China

Technical Specification" (HJ/T 177–2005) and "Medical Waste Incinerator Technical Requirements (for Trial Implementation)" (GB 19218-2003), the collection and packaging of medical waste incineration technology, collection and transportation, storage and transportation, disinfection treatment has been initially standardized. The three specifications are important for strengthening the safe management of medical waste, protecting the environment, preventing the spread of disease, protecting human health, achieving the goal of harmless and reduced treatment of medical waste, regulating the practical application of medical waste incineration disposal technology, guiding the planning, design, construction, acceptance and operational management of chemical disinfection of medical waste treatment projects. In the medical waste non-incineration treatment technology, in 2006, the "Medical waste chemical disinfection centralized treatment engineering technical specifications" (HJ/T 228-2006), "medical waste high-temperature steam centralized treatment engineering technical specifications" (HJ/T 276-2006) and "medical waste microwave disinfection centralized treatment engineering specifications (for trial implementation)" (HJ/T 229-2006). In June 2020, the Ministry of Ecology and Environment revised the above three non-incineration treatment technical specifications and publicly solicited comments. In addition, for the selection and application of medical waste treatment and disposal technologies, in 2009 the state issued the "Medical Waste Treatment and Disposal Best Available Techniques Guide" to guide the selection and application of medical waste treatment and disposal technologies. Hazardous waste (including medical waste) incineration and disposal facilities dioxin emissions monitoring technical specifications", "medical waste centralized incineration and disposal facilities operation supervision and management technical specifications", "hazardous waste (including medical waste) incineration and disposal facilities performance testing technical specifications", "medical waste treatment and disposal of pollution prevention and best available technology guide (for trial implementation)" and other technical specifications issued, is conducive to guiding and standardizing medical waste treatment disposal of pollution prevention, supervision and management and facility performance testing (Chen et al. 2012).

From the current trend of medical waste disposal management system, medical waste pollution prevention not only involves medical waste from the generation to the collection, storage, transportation, disposal of the entire life cycle of the implementation of the whole process management, but also involves medical waste disposal from the selection of technology to technology certification, technology application link of engineering construction, construction of disposal facilities after the operation of management, supervision and management of environmental protection departments and with the operation of facilities and supervision and management of the implementation of monitoring and management of the whole process management links (Shunze et al. 2005).

1.4.3 China's Medical Waste Disposal Technology and Management Model and Foreign Gap Analysis

Medical waste disposal technology and management model optimization is a continuous process of progress and development, and countries around the world, both developed and developing, are actively promoting their disposal technology applications and management capabilities, such as the United States, the United Kingdom, Jordan, Egypt, South Korea, Turkey, Brazil, Mongolia, Greece, and India (Limit et al. 2020).

Developed countries have gradually replaced incineration technology with non-incineration technology, such as dioxin emission standards for incineration flue gas, most international countries have adopted 0.1–0.5 ngTEQ/Nm^3 emission targets. Combined with the experience of Western countries, non-incineration treatment technology embodies the advantages of low construction and operating costs, the residue generated after disposal does not have hazardous characteristics and does not emit dioxins. However, from the medical waste non-incineration treatment technology application model, non-incineration treatment technology cannot dispose off all medical waste, such as high-temperature steam, microwave and chemical disinfection, etc., only infectious waste, injury waste and some pathologic waste, while chemical waste and pharmaceutical waste are not applicable to non-incineration treatment technology. Therefore, if the classification and disposal is not properly managed, the infectious problem of medical waste is not easy to eliminate completely. In addition, the non-incineration process of medical waste will also produce odor, VOCs and dust and other secondary pollutants, the disposal process will also produce residual liquid and wastewater, and the disposal process of pollution control problems cannot be ignored. Therefore, foreign production of medical waste non-incineration treatment facilities are equipped with a more complete odor and VOCs pollution control system, and through automated control, to achieve the safe management of medical waste disposal process. In addition, non-incineration treatment technology is not properly applied and managed, but also to human health and the surrounding environment brings potential risks, such as the generation of odor and VOCs and other air pollution, improper treatment will still produce the spread of disease.

For non-incineration medical waste treatment facilities, the chemical and thermodynamic properties of medical waste determine the system configuration and unit design of the technology used, the operating conditions of the facility and the subsequent configuration of the waste gas treatment system and the wastewater/waste liquid treatment system, which determine parameters such as temperature and disinfection time. Therefore, in the selection of non-incineration systems and design of medical waste to fully analyze and consider the type and nature of waste, a reasonable choice of efficient and economical treatment units, and according to the treatment capacity and treatment requirements to determine the appropriate design parameters and process parameters to ensure the sterilization efficiency of medical waste non-incineration. In turn, to determine whether a facility meets emission standards and ensures suitability for a particular waste disposal type, a systematic evaluation is required, taking into account the characteristics of the waste and the working conditions of the facility. Performance testing, as a comprehensive method of evaluating facility performance, has been more widely used abroad in the field of hazardous waste and medical waste incineration. In fact, to carry out a scientific and systematic test of the performance of a set of non-incineration facilities, it is necessary to combine the characteristics of non-incineration facilities, a comprehensive test of the safety performance of a set of facilities, from the waste characteristics, the performance of the facility working conditions, the performance of pollutant emissions and the operation of the facility parameters combined with the whole process of facility performance testing and evaluation methods.

Although China has basically established a system of regulations, policies and standards for effective medical waste management, and the construction of a large number of medical waste disposal facilities, but China's medical waste management and disposal has not yet established a whole process management system, the medical waste management is not strong, safety is not enough, the difficulty of controlling the release of pollutants and other problems (Yang 2011), mainly in the following areas:

(1) The existing medical waste management system needs to be further standardized and improved

The current management system does not reflect the actual needs of medical waste life cycle and the whole process of management, such as insufficient consideration of medical waste reduction at source, which is not conducive to reducing the discharge of pollutants in the process of medical waste disposal; medical waste classification within the medical institution failed to consider the interface with the end of the disposal technology; medical waste collection, storage, transportation and other aspects of the specification is not specific; the lack of scientific supervision and management of the operation of facilities and the lack of methods and means. The existence of the above-mentioned problems is subject to scientific analysis and diagnosis from the perspective of technical management system improvement to promote the establishment of a sustainable medical waste management system.

(2) Supervision and management of medical waste needs to be strengthened

Medical waste supervision and management system is still not perfect, such as medical waste generation institutions, supervision and management institutions, medical waste disposal institutions and related institutions of the division of powers and responsibilities is not clear enough, the legal responsibility is not clear enough; facility construction and completion of the trial run management procedures and the reliability of the corresponding monitoring, acceptance procedures are still lacking in operability; medical waste disposal facilities operation supervision and management still lacks sufficient basis, the lack of Complete law enforcement supervision system, the lack of continuous, comprehensive, strict, procedural means of supervision and management, it is difficult to ensure effective supervision and management. How to promote the application of medical waste life cycle and the whole process management system of medical waste supervision and management based on the characteristics of medical waste generation and subsequent disposal technology is essential.

(3) Medical waste disposal technology should promote the BAT/BEP concept

Medical waste disposal technology is difficult to meet the basic requirements of the future international community on medical waste treatment and disposal technology and the need to protect the environment, such as in the selection of disposal technology does not fully consider the BAT/BEP concept; domestic incineration of medical waste disposal technology should meet the requirements of compliance and there is a large gap, any technology selection and application and application of technology should be based on the application of technology, management, economics and social acceptability as a whole model to be effective.

(4) Medical waste incineration and disposal facilities emission standards need to be improved

At present, the dioxin emission standard for incineration flue gas in China is 0.5 ngTEQ/Nm^3, which is still somewhat different from the 0.1 ngTEQ/Nm emission target recommended by the international POPs Convention^3. In addition, due to the gap between the operational capacity and level of management and operators, it also causes the difference of emission from the disposal facilities, how to promote the operation and management of facilities from the perspective of operation and management capacity is the key to ensure the disposal standards.

(5) Classification and reduction of medical waste at source to be linked with post-disposal technologies

The guiding idea of medical waste classification in China is to achieve the purpose of medical waste reduction through classification and scientific differentiation between domestic waste and medical waste (Shao et al. 2001); after reasonable classification of medical waste, according to its material and the degree of pollution, the use of different harmless disposal methods for treatment, in order to minimize the pollution of the environment. However, although the classification of medical waste is divided into five categories according to its hygienic characteristics and hazards, due to the single regional medical waste disposal technology, resulting in a disconnect between classification and disposal, which does not reflect the practical significance of classification; and some areas of medical waste management and treatment and disposal institutions responsible for their own vague understanding of classification led to the requirement of medical institutions to classify too much, greatly increasing the amount of polymer waste for packaging. In the case of medical waste non-incineration treatment technology, only infectious waste (Huang and Jim 2008), sharps and part of the pathologic waste can be treated, such as chemical waste and pharmaceutical waste into non-incineration treatment facilities will pose environmental risks.

(6) Medical waste emergency disposal capacity to be further enhanced

The outbreak of the new crown epidemic in early 2020 exposed the shortcomings of China's medical waste emergency response capacity in many aspects. On the one hand, the national ecological and environmental management system is generally weak in emergency awareness and emergency thinking, and various government departments lack sufficient understanding of the importance, complexity, severity and persistence of a rapid and efficient fight against the epidemic. The general lack of practical emergency plans for medical waste at the grassroots level. Coupled with the lack of emergency facilities, emergency technology, transfer vehicles, emergency operators, resulting in a passive emergency response, emergency management "temporary", the obvious lack of continuity. On the other hand, China's environmental pollution management requirements in the process of emergency disposal of medical waste is still a gap. Compared with professional disposal facilities, the temporary opening of the medical waste emergency disposal facilities in the composition of facilities, functions, operating parameters control, automatic control and other aspects of some problems, the lack of supervision and monitoring of pollutant emissions, pollution hazards hold objective understanding and analysis. In addition, China's centralized medical waste disposal capacity is designed according to the amount of medical waste generated under normal conditions, without taking into account the rapid increase in the amount of medical waste generated during a major epidemic. Moreover, compared with the normal situation, the inorganic waste in medical waste increases during the epidemic, and the water content increases, which makes the subsequent disposal more difficult. Therefore, the lack of top-level design of medical waste emergency disposal, the serious lack of emergency disposal capacity, the shortage of emergency disposal technology, the lack of emergency disposal supervision capacity, etc., are the fundamental reasons why China has to make up for the shortcomings of medical waste emergency disposal capacity (Yang et al. 2017).

(7) Single mode of medical waste disposal

"SARS" epidemic, China's implementation of the centralized medical waste disposal model. In general, a prefecture-level city to establish a centralized medical waste disposal center, responsible for the collection and disposal of medical waste within its jurisdiction. However, due to the small number, location, small volume and other factors, in the urban infrastructure construction system, centralized medical waste disposal facilities are easily ignored. Moreover, in the centralized disposal mode, medical waste inevitably requires long-distance transportation, which is prone to scattering and leakage during transportation, resulting in a greater risk of disease transmission and environmental pollution. In the guidance document on personal and public health protection and medical waste management for the new crown epidemic, WHO recommends the local disposal of medical waste involved in the epidemic under the premise of ensuring safe collection and classification. Therefore, under the concept of segregated medical waste management, taking into account safety, health and environmental needs, medical waste during an epidemic can be divided into highly infectious medical waste, traditional medical waste and waste mixed with highly infectious waste (e.g. domestic waste from isolation wards, bedding sheets, etc.). In order to reduce the risk of medical waste in the smallest scope and shortest time, medical waste can be treated by inactivating it locally or nearby using non-incineration treatment technologies and then disposed off by centralized incineration. With the development of non-incineration disposal technology and equipment, the implementation of a combination of centralized and local disposal of medical waste disposal model is imperative.

1.5 China's Medical Waste Disposal Technology Application Management Needs

1.5.1 Establish a Sound Life Cycle and Whole Process Management System

Combined with national conventions and the experience of developed countries in Europe and the United States, the actual needs of the whole process management should be taken as a starting point to promote the construction of medical waste life cycle and whole process management system. The proposed framework for the construction of medical waste technology management system in China is shown in Fig. 1.3.

Figure 1.3 effectively reflects the concept of life cycle management and whole process management. The so-called life cycle management refers to the management of the entire process of medical waste from generation to collection, classification, packaging, temporary storage, transportation, treatment, disposal; whole process management is from the process of environmental management, that is, for a technology after the development of technology certification, technology application of engineering construction, construction of facilities after the operation management, supervision and management of environmental protection departments and monitoring management implemented with the operation and supervision of facilities. Through this process, we promote the establishment of a comprehensive medical waste management system that meets the environmental, economic, social and public hospital needs and promotes the sustainable management of medical waste (Guangming and Hongfang 2004).

Medical waste management and disposal should be a whole process from the source, until the end of safe disposal, the core content is to be based on the different characteristics of the incineration and non-incineration two types of technology, effectively from the point of view of technical applicability, from the source to carry out, do a good job of medical waste reduction, classification and packaging work, comprehensive promotion of source classification and late disposal technology application; in addition, also from the waste generation and In addition, the waste generation and disposal process, do a good job of its temporary storage and transportation, to eliminate the threat of infection, should be around the actual management

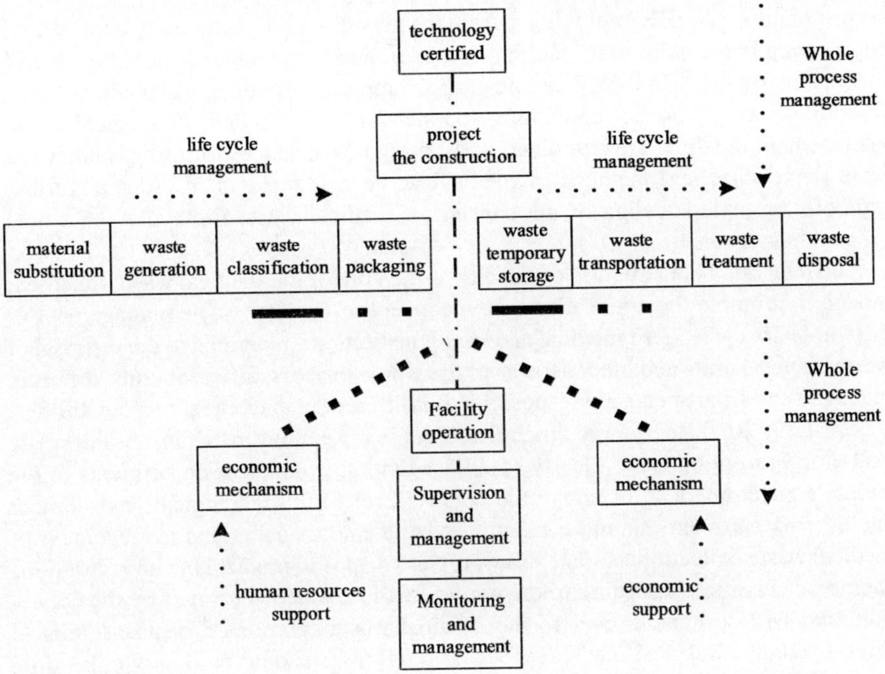

Fig. 1.3 Proposed framework for the construction of medical waste technology management system in China

of medical waste from generation to disposal of the entire life cycle of management needs, the state should be from the perspective of technology selection, engineering construction, facility operation, supervision and management, as well as monitoring and management, the establishment of rules and regulations, combined with the existing medical waste management system to develop technical standards and norms applicable to different aspects, so that there are rules to follow, there is a law to follow. In addition, medical waste disposal work to effectively address two key issues, one is capacity building, including medical waste collection, classification, packaging, transportation, disposal and other related personnel throughout the process of capacity enhancement; the second is the economic operation mechanism, should be done in the medical waste disposal facilities engineering, facilities operation and other processes have sufficient economic support to ensure that the entire process of medical waste management and disposal of stable operation.

In addition, according to China's compliance requirements, the BAT/BEP in the field of medical waste management and disposal with Chinese characteristics should be actively explored. the adoption of BAT/BEP is the way to China's future medical waste management and disposal, therefore, according to different medical waste disposal technologies, such as incineration, pyrolysis, high-temperature steam, chemical disinfection, microwave and other technologies of practical application, in the comprehensive development of BAT/BEP demonstration, the preparation of the corresponding BAT/BEP guidelines, from the perspective of technology application, engineering construction and facility operation and management, the definition of medical waste field BAT/BEP and push application to provide a management basis. In addition, the policies, regulations and standards in the field of medical waste management in China are combined with foreign experiences to further clarify the main approaches and methods of medical waste management in order to further promote the understanding of the concept of medical waste life cycle and whole process management.

The integrated construction of the city as the core of the medical waste treatment model, to promote the application of the concept of medical waste management of the whole life cycle, each medical and health institutions generated to the disposal of waste disposal units and other aspects of the whole process, different units, different management departments are responsible for different management responsibilities.

April 29, 2020, the newly amended "People's Republic of China Solid Waste Pollution Prevention Law" clearly stipulates that the competent departments of the people's government at or above the county level health, ecological environment and other departments should each strengthen the supervision and management of medical waste collection, storage, transport and disposal; medical health institutions should be classified according to the unit of medical waste generated by the collection, and will be handed over to the medical waste centralized disposal units to centralized disposal; and medical waste disposal units should be timely collection, transport and disposal of medical waste. Occurrence of major infectious disease epidemics and other emergencies, health, ecological environment, environmental health, transportation and other competent departments should strengthen interdepartmental collaboration, according to the emergency disposal work, the coordination of medical waste and other hazardous waste collection, storage, transportation, disposal, etc., to protect the required vehicles, sites, disposal facilities and protective materials.

It can be seen, clear medical waste management and disposal of each responsible body, to achieve an effective division of labor, decomposition of tasks, to achieve unified and coordinated management of urban medical waste, to avoid a variety of omissions caused by unclear responsibilities and rights, is to prevent medical waste is harmful to public health, pollution of the environment necessary to protect.

1.5.2 Establishment of a City-Centered Medical Waste Management and Disposal System

Based on the current construction planning of China's medical waste disposal facilities, as well as the use of different medical waste disposal technologies, medical waste disposal should be achieved in three articulation, namely, incineration and non-incineration disposal technologies, to promote the advantages of different disposal technologies complement each other, including incineration and non-incineration treatment technologies within the city, including the application of technology between neighboring cities to complement each other; medical waste generation unit and medical waste disposal unit management interface, should achieve the source of medical waste classification and reduction work and the subsequent application of medical waste disposal technology to ensure that the application of disposal technology and safety; promote the ecological and environmental administrative departments and health administrative departments in charge of supervision and management of medical waste interface to ensure the consistency and unity of medical waste management. Specific content includes the following aspects.

(1) Further advance the technology development and technology application of degree and depth

From the current international development trend, medical waste disposal technology shows a continuous progress and development trend. Incineration disposal technology is changing rapidly, and non-incineration treatment technology is constantly updated. As the world's largest developing Country, China should comprehensively promote the capacity building of medical waste treatment and disposal R&D institutions from the national level, in the basic R&D capabilities, engineering construction and operation of facilities and other aspects gradually close to the international advanced level, the establishment of scientific and reasonable technical support system, to continuously solve the problem of medical waste environmental management, the search suitable for China's national conditions BAT / BEP to provide technical support. As for the specific incineration and disposal technology of medical waste, the research on the incinerator process needs to be further deepened, the tail gas purification technology needs to be further developed, and the operation and management process needs to be further explored for practical experience. From the medical waste non-incineration technology, based on the current application and management of various non-incineration technology in China is still in its infancy, and should be for its disposal process of secondary pollutants, such as odor, VOCs pollution control should be strengthened, for its microbiological detection technology and methods should be further explored, for the operation and management of such facilities should be increased investment in technology to promote the beginning development of this type of technology.

(2) Implement source separation of medical waste and promote the optimization of medical waste disposal technology

The issue of medical waste management within medical waste facilities is crucial, It is related to the effective elimination of medical waste infectivity and the reduction of human and environmental hazards; on the other hand it is also closely related to the subsequent selection of disposal technologies and the application of disposal technologies. As Diaz L.F. et al. argue, when developing medical waste management plans, a thorough understanding of the quantity and characteristics of waste is essential for risk and cost analysis, but these issues have not been clearly identified in either developed or developing countries. The core issue of environmental management in China is still the management issue, which is the premise and basis for the solution of technical problems. In the field of medical waste management and disposal, the following aspects can be achieved:

(1) Further improve the internal medical waste management mode of medical institutions, to explore the classification of medical waste and its treatment in medical institutions, which is due to the current classification of medical waste in medical institutions and in China is relatively crude, and more consideration is given to the convenience of management, and less consideration is given to the characteristics of medical waste; at the same time, the disposal of some special medical waste such as mercury-containing waste, waste chemical reagents, disinfectants, etc., is not given at the same time, the disposal of some special medical waste such as mercury-containing waste, waste chemical reagents, disinfectants, etc., has not been given sufficient attention; and the characteristics of China's high polymer waste production have not been fully considered. An effective medical waste management plan is a key component of medical waste management, which plays a significant role not only for care itself but also for the occupational safety of managers. In addition, health care facilities generate a certain amount of chemical and hazardous substances that cannot be disposed off using treatment facilities such as high temperature steam, including waste from chemotherapy processes, mercury, volatile and semi-volatile substances, radioactive substances, and other hazardous chemical substances.
(2) In the use of single-use medical supplies, its standardized management should be strengthened. The trend of medical waste recycling is inevitable, non-incineration resource and energy technology gradually replace the traditional incineration technology, and has become the development trend of single-use medical waste disposal technology. China's gradual adoption of mandatory recycling of plastic products is to represent the most advanced and environmentally friendly technology trends, not only to maximize the recycling of resources, but also to completely eliminate the various environmental drawbacks brought about by landfills, and incineration. Unified recycling, centralized storage, unified disposal, to prevent the use of abnormal ways into society, resulting in the spread of pathogenic diseases, and to ensure medical safety and patient health.
(3) Medical institution waste and medical waste should be correctly distinguished. According to the research results of WHO, only 10–25% of the waste generated by medical institutions belongs to medical waste, while in practice, all the waste generated in hospitals is often understood as medical waste, and the level of classification and management is poor.

In February 2020, the Nation Health Commission of the People's Republic of China and other ministries and commissions jointly issued a document that divides the waste generated by medical and health institutions into medical waste, household waste and infusion bottles (bags), and clarifies their corresponding treatment and disposal methods and management requirements. However, the classification and management of medical waste is not standardized, and the phenomenon of mixing medical waste with other garbage still prevails. This has caused an increase in the disposal of medical waste, which not only increases the burden of environmentally sound disposal of medical waste, but also creates more potential secondary environmental pollution, especially in the process of incineration and disposal of medical waste will produce dioxins and other toxic and harmful substances emissions.

The integrated treatment of harmless, reduced and resourceful is the development goal for disposable medical waste. In order to promote the source management of medical waste, experts and scholars at home and abroad have carried out a lot of research work, hoping to promote the source management of medical waste through a combination of administrative management and technical methods, such as hierarchical analysis. Therefore, we should first start from the source and promote the classification management of medical waste comprehensively through measures such as application reduction, comprehensive utilization and classification and recycling. When considering the medical waste management plan, on the one hand, it is necessary to achieve harmlessness and reduction, with due consideration to resource recovery; on the other hand, it is also necessary to consider the types of medical waste that can be disposed off at a later stage, and to carry out standardized classification management. This requires an in-depth study of the classification of medical waste generated by Chinese medical institutions and the development of detailed rules for the management of medical waste at source to promote medical waste management.

(3) Strengthen the operation and management of incineration and disposal facilities, and promote the standardized application of incineration technology

Incineration technology is currently the mainstream technology for medical waste disposal in China. The National Plan for the Construction of Hazardous Waste and Medical Waste Disposal Facilities proposed more than 300 medical waste disposal facilities, a large part of the incineration disposal technology to be used. In order to comprehensively promote China's compliance process, first of all, we must fully implement the concept of whole process management in the field of medical waste, starting from the source of medical waste, the implementation of reduction, especially such as medical waste incineration and disposal of the source of chlorine source of reduction control with foreign developed Countries and there is a certain gap, which is also the main reason why China's current hazardous waste incineration disposal tail gas management is more difficult and cannot meet the standards. One of the main reasons for the difficulty and failure to meet the standards of China's current hazardous waste incineration disposal tail gas management. Although China has put forward certain requirements from the standard level on the material of medical waste bags, such as polyvinyl chloride material is strictly prohibited for the production of medical waste bags, but, on the one hand, the management of hazardous waste is not only a waste of medical waste. On the other hand, due to the slow progress in environmental law enforcement and material substitution in China, the lack of basic policy guidance, results in hazardous waste and medical waste. The source reduction is not effectively controlled, such as Europe and the United States and some other developed countries in medical products using raw materials mainly polyethylene, while China mainly uses polyvinyl chloride; Europe and the United States and some other developed Countries in the packaging of medical waste is used in cardboard boxes, and is incinerated together with medical waste disposal, while China is the use of cleaning and disinfection boxes and medical waste bags, and the bags are incinerated together with medical waste. The existence of this gap directly leads to the gap between China and developed countries in terms of dioxin production. Therefore, strengthening source control is a prerequisite for achieving pollution control of hazardous waste incineration tail gas in China, which should be done as follows:

(1) To reduce and limit the production and application of chlorine-containing medical supplies at source, strict implementation of the "medical waste special packaging, container standards and warning labels" (Huanfa [2003] No. 188), to eliminate the use of PVC-containing materials in the medical waste packaging process to make packaging, to reduce the overall content of PVC used in the incineration of medical waste disposal, to fundamentally solve the problem of dioxin generation, the problem of chlorine source, the source of waste generation is minimized from the solution of hazardous waste incineration and disposal process of dioxin generation of the fundamental approach. This point has been increasingly clear consensus around the world. Only by reducing the chlorine source of dioxins at the source can make the final emissions to be fundamentally reduced and the load on the subsequent tail gas treatment be reduced.
(2) From how to improve the management level of the operator, we will promote the standardization of facility operation, and promote the solution of the problem of compliance through perfect management.
(3) from the level of environmental supervision and enforcement to promote environmental protection departments in the process of supervision and enforcement of the operation of facilities to effectively achieve from the end of control to process control, from macro environmental law enforcement to micro law enforcement, and to fully realize the road from human management to the legal management of technical environmental management, and then from the administrative law enforcement perspective to regulate the industry behavior in the field of medical waste disposal.
(4) Further improve the level of incineration disposal technology, increase investment to promote the demonstration of BAT/BEP and push to achieve safe management of medical waste incineration and disposal process.
(4) Promote the application of non-incineration treatment technology and grasp the standardized application of non-incineration treatment technology

From the current situation of China's medical waste management (Ye et al. 2017), and for the current planning of construction projects, especially for small and medium-sized medical waste disposal facilities of 3–5 t, should encourage the use of high-temperature steam, microwave and chemical disinfection and other non-incineration treatment technologies, and supporting the corresponding management tools to fundamentally disinfect the generation and emission of dioxins, which is also the fundamental solution to the problem of medical waste disposal process compliance in the future. This is the fundamental way to solve the compliance problem of medical waste disposal process in the future.

However, the use of non-incineration treatment technology for medical waste should also pay attention to pollution prevention issues, not only to consider the type of medical waste treatment, bacterial inactivation efficiency, technical reliability, automation level, environmental emissions and environmental impact, occupational safety and health and other basic factors, but also to consider the degree of recognition of policies and regulations, public acceptance, treatment space requirements and site selection, supporting public facilities and installation requirements and other basic factors, in order to ensure the effectiveness of non-incineration treatment technology applications. Therefore, in the process of non-incineration treatment technology application must pay attention to the following aspects:

(1) Be sure to separate out chemical, pharmaceutical and pathologic wastes that cannot be disposed of by non-incineration treatment technologies in the source separation process.
(2) To strictly control the process of non-incineration treatment parameters control and treatment effect detection, to ensure that the non-incineration treatment of medical waste can reach the corresponding national standards below and no longer infectious.
(3) Non-incineration treatment of medical waste should also be properly managed, the implementation of standardized landfill or incineration disposal, not arbitrarily discarded, otherwise there will be the possibility of reverting to the original characteristics of infected waste.
(5) Strengthen supervision and monitoring technology system construction

The environmental supervision of medical waste is the core link to comprehensively promote the construction of the entire medical waste policy, regulations and standards system. In the supervision and management system construction, should first promote supervision and management policies, regulations and standards system construction, while focusing on promoting policies, regulations and standards from the perspective of implementation, around the medical waste collection, classification and packaging, storage, transportation, treatment, disposal and resource utilization of the whole process, from the technical and management point of view to strengthen the management of the technology and means needed to make environmental protection departments at all levels and other Related management departments to the technical management track. Monitoring is a necessary tool for supervision and management. Therefore, the construction of the system should be combined with the needs of environmental management of medical waste and the operation of facilities, focusing on strengthening the environmental monitoring capacity of environmental management departments, from the perspective of monitoring methods, monitoring technology and monitoring means to enhance the construction of environmental monitoring system to improve China's environmental monitoring capacity of medical waste, and thus enhance the ability of the whole process of environmental management of medical waste.

(6) Establishment of a standardized economic operation mechanism

The ultimate goal of the construction and improvement of the economic operation mechanism is to achieve sufficient economic support in the process of medical waste disposal facility construction, facility operation, etc., and to ensure the stable operation of the entire medical waste management and disposal process. The construction of economic operation mechanism is the necessary support means to promote sustainable environmental management of medical waste (Yang et al. 2009). Therefore, the actual needs of disposal technology research and development, engineering construction, facility operation and other aspects of the exploration and suitability of economic instruments, from the perspective of commercial collaboration, medical waste disposal pricing, financial subsidies, taxation and environmental taxes, etc., to regulate the standardized construction and operation of medical waste disposal behavior.

(7) Strengthen the management of centralized medical waste disposal units

Collect medical waste from medical and health institutions within the service area as required, packed and loaded according to specifications. Equipped with spare vehicles and emergency personnel and maintain good condition, to ensure that the collection of medical waste in health care institutions on schedule. Medical waste transport must implement the system of hazardous waste transfer coupons. Incineration of medical waste in strict accordance with the process specifications, regular maintenance of maintenance equipment, to maintain good condition. Waste water, waste gas treatment must meet the national emission standards; slag must be sent to landfill; fly ash, waste activated carbon and other hazardous waste must be stored according to the norms, sent to qualified units for treatment. The establishment of emergency response plans, for a drill every year, a large change in personnel, should be timely rehearsal, to ensure that the emergency response plans according to the skilled disposal of emergency events.

(8) Improve the construction of medical waste emergency disposal system

Since the beginning of 2020, with the gradual increase in the prevention and control of epidemics, China's medical waste emergency collection and disposal system has highlighted a series of problems and shortcomings. On February 21, 2020, the Political Bureau of the CPC Central Committee held a meeting where Xi Jinping, General Secretary of the CPC Central Committee, requested, "to fight the battle of pollution prevention and control, promote sustained improvement in the quality of the ecological environment, and accelerate the patching up of shortcomings in medical waste and hazardous waste collection and treatment facilities". Therefore, improve the awareness and attention to improve the medical waste emergency response system, the overall planning of emergency response capacity building, to make up for the shortcomings of the ecological and environmental emergency response system, to protect the sustainable development of the city is of great significance. Municipal people's governments as the main body of emergency response management, should organize the preparation and release of medical waste emergency disposal work under the jurisdiction of the implementation plan for the management of medical waste shortcomings, planning and construction of recycling and treatment and disposal facilities, to encourage the development of mobile medical waste treatment and disposal technology, improve the medical waste information management platform and enhance the emergency security capacity, to comprehensively improve the emergency management of medical waste ability to effectively protect medical waste in the event of a public health or environmental pollution incident to get rapid, efficient and harmless disposal, and to maintain local environmental and health safety.

(9) Increase publicity and education to promote public participation

Foreign experience and domestic lessons objectively require us to increase the publicity and education efforts, and promote public participation in a very important position. In order to effectively achieve the whole process of medical waste management. It is necessary to make the relevant management and technical personnel know how to achieve the corresponding environmental management objectives, what methods and measures are taken to meet the requirements. On this basis, it is also necessary to make the relevant government officials, environmental managers, managers of disposal facilities operating units, operators and workers to effectively understand the relevant national policies, regulations, standards, and effectively clarify the operating mode and operating procedures of various types of medical waste disposal facilities, and then achieve the management and operation process has rules to follow. Therefore, the establishment of the training system involved in the whole process of medical waste management should be explored, including the training of medical waste collection, classification and packaging knowledge within medical institutions, including medical waste storage, transportation and disposal management and technical training, in order to maximize the requirements of medical waste management and disposal, and promote the effective implementation of the concept of whole process management. In order to improve the level and effectiveness of medical waste management, it is necessary to change the current training methods, study the training methods and contents of medical waste management knowledge within Chinese medical institutions, and develop a set of medical waste management and training materials suitable for Chinese medical institutions and different disposal technology points. In addition, more funds should be invested in training institutions and trainers in order to provide the basis and conditions for the establishment of a standardized management and technical training system.

In addition, it should also actively promote management and technical training to improve the capacity and level of management and technical personnel, based on extensive publicity and education to expand public participation, in order to better promote the formation and development of good medical waste management awareness and behavioral habits.

1.5.3 Promote Diversification of Medical Waste Disposal Management

(1) Establish a combination of centralized and decentralized medical waste disposal model

At present, China's establishment is the core of the administrative division of medical waste management system. Medical waste disposal business license to the municipal environmental protection administrative department for approval, and generally only a centralized medical waste disposal facilities.

Remote areas or medical health institutions far from the local centralized medical waste disposal facilities due to their remote location, economic backwardness, distribution and lack of knowledge of medical waste management of medical and nursing staff, resulting in such areas of medical waste disposal classification isolation storage is not standardized, low frequency of transfer, long storage time, illegal self-disposal and other issues cannot be effectively improved for a long time.

Especially during a major epidemic, the above problems are more prominent. The resulting environmental risks, increased disposal costs and other aspects of the problem cannot be ignored. To explore the centralized and decentralized collection and disposal mode of medical waste with the city as the core, is the way to solve the medical waste management to meet the needs of urban development. On the basis of insisting on the implementation of centralized disposal, fully consider the characteristics of the generation and distribution of medical waste in remote areas, in accordance with the principle of local conditions, combined with the real needs, scientific supporting the construction of small medical waste treatment facilities to reduce the long-distance transfer, to carry out local, nearby, timely disposal, while serving as an appropriate complement to the centralized disposal facilities. These small medical waste treatment facilities should be simple and easy to use, such as, the use of non-incineration treatment technology or treatment after the safe landfill, etc., the use of existing local conditions to achieve the purpose of timely elimination of infectious medical waste, and reduce its disposal process on the secondary pollution of the environment.

In addition, reference can also be made to the international implementation of the management experience, infectious medical waste generated by medical institutions or infectious diseases after a certain scale of public health emergencies, the number of medical waste using mobile skid-mounted treatment facilities for local, efficient treatment and disposal, is also a strong operability, lower environmental risks, and easy to manage the regional centralized disposal method.

(2) Actively explore the experience of medical waste co-processing and disposal

The existing medical waste treatment and disposal technologies, especially non-incineration, have certain limitations in terms of their scope of application. Incineration disposal facilities can basically dispose off all types of medical waste harmlessly, while non-incineration treatment facilities, but not for chemical, pharmaceutical and some pathological medical waste for effective disposal. To this end, the city as a unit, the establishment of a regional collaborative disposal mechanism between cities, the use of different treatment and disposal process complementarity, the type of medical waste cannot be disposed off to the relatively close municipal medical waste disposal units for safe disposal, can provide the conditions for solving the problem of environmentally sound disposal of all types of local medical waste. At the same time, the establishment of cross-regional collaborative disposal mechanisms between neighboring cities or provinces with surplus medical waste disposal capacity is also a positive and effective initiative to effectively respond to the need for emergency medical waste disposal.

Medical waste disposal industry research literature shows that the national calorific value of medical waste is higher than 4500 kJ/kg, moisture >40%, combustible materials to paper, plastic, cotton yarn type, suitable for disposal by incineration technology.

Based on this, the state issued the "new coronavirus infection of pneumonia epidemic medical waste emergency disposal management and technical guidelines (for trial implementation)" proposed that hazardous waste incineration facilities, domestic waste incineration facilities, industrial kilns and other incineration facilities for emergency disposal of medical waste during an epidemic in turn. However, these non-professional disposal facilities should set up special unloading area, and the configuration of special feeding equipment, cleaning and disinfection equipment. The local government can integrate the above environmental infrastructure of similar environmental utility, integrated management, so that the first time in the emergency period corresponds to disposal needs. At the same time, should also be in the planning or implementation of medical waste emergency capacity building, give full consideration to medical waste disposal facilities and follow-up disposal facilities (such as landfills, etc.) and emergency disposal facilities (such as, co-disposal facilities) scientific and reasonable layout, so that all types of medical waste treatment and disposal facilities to maximize the effectiveness.

(3) Promote high level of medical waste management with intelligent management tools

In recent years, with the "Internet+" and other information technology in various fields of application, the Internet of Things (IoT) technology plays an active role in the practice of medical waste management. Cloud-based management of medical waste management system, the integrated application of Internet of things technology, mobile terminal technology, GIS, GPRS and other technologies, medical waste from the generation, measurement, handover, collection, bulk storage and other aspects of the implementation of dynamic supervision, the use of electronic induction and other means of Internet of things will be all the process of manual entry into the scan or automatic upload entry, to achieve the integration of Internet of things and information technology system linkage, for medical waste management to provide solid technical support.

Through such a means of Internet of things management, all kinds of medical waste generated by medical and health institutions, such as medical waste, household waste and disposable infusion bottles (bags) and other waste as a renewable resource to achieve the collection, storage, sales, transfer and treatment and disposal process re-engineering, the implementation of the whole process of scanning records, zero input, to minimize and prevent and control waste handover process of health and environmental risks. In the management of medical waste, and its traceability closed-loop supervision effect is more prominent. Not only can the generation of various types of medical waste, storage, transportation, disposal and other visual management, but also to achieve real-time data, historical data analysis, statistics and other functions, so that managers based on data trends, vertical and horizontal comparison, and to assist management practices and decision-making.

In addition, the construction of medical waste information management platform, medical health institutions, medical waste disposal units on the management of medical waste data interconnection and sharing, and set different management rights, according to the work needs of different functional departments and according to the different scope of work to enjoy the resources for access control, etc.

According to a large number of case studies, medical waste information management is now also in improving the efficiency of medical waste collection and handover, saving labor costs; increasing the transparency of medical waste disposal billing, avoiding inconsistencies in weight or quantity between disposal units and medical institutions; reducing all kinds of paper registration and archival records, saving a lot of material consumption, etc., is widely recognized.

The process of information management of medical waste is still in continuous improvement and development, which will certainly provide a more efficient and accurate operation platform for modern medical waste management to play a more broad role.

References

Abdulla F, Qdais H A, Rabi A. Site investigation on medical waste management practices in Northern Jordan [J]. Waste Management, 2008(28): 450–458.

Addink R, Altwicker E R. Formation of polychlorinated dibenzo-p-dioxins/dibenzofurans from soot of benzene ando-dichlorobenzene combustion [J]. Environ. Sci. Technol., 2004, 38(19): 5196–5200.

Alagöz A Z, Kocasoy G. Determination of the best appropriate management methods for the health-care wastes in Istanbul [J]. Waste Management, 2008, 28(7): 1227–1235.

Almuneef M, Memish Z. Effective medical waste management: it can be done [J]. American Journal of Infection Control, 2003, 31(3): 188–192.

Alvim Ferraz M C M, Cardoso J I B, Pontes S L R. Concentration of atmospheric pollutants in the gaseous emissions of medical waste incinerators [J]. Journal of the Air and Waste Management Association, 2000, 50(1): 131–136.

Alvim-Ferraz M C M, Afonso S A V. Incineration of healthcare wastes: management of atmospheric emissions through waste segregation [J]. Waste Management, 2005, 25(6): 638–648.

Aristizbal B, Cobo M, Montes C, et al. Dioxin emissions from thermal waste management in Medellín, Colombia: present regulation status and preliminary results [J]. status and preliminary results [J]. Waste Management, 2007(11): 1603–1610.

Bdour A, Altrabsheh B, Hadadin N, et al. Assessment of medical wastes management practice: a case study of the northern part of Jordan [J]. Waste Management, 2007, 27(6): 746–759.

Blenkharn J I. Medical wastes management in the south of Brazil [J]. Waste Management, 2006(26): 315–317.

Chartier Y, Emmanuel J, Pieper U, et al. Safe management of wastes from health-care activities [M]. Geneva: World Health Organization Publications, 2014.

Chen J, Zhiyuan R, Yajing T, et al. Application of best available technologies on medical wastes disposal/treatment in China (with case application of best available technologies on medical wastes disposal/treatment in China (with case study)[R]. Procedia Environmental Sciences, 2012(16): 257–265.

Chen Y, Li PJ, Sun YZ, et al. Study on the implementation of POPs convention in China's medical waste sector [J]. Environmental Science and Technology,2008a(3): 123–126, 157.

Coutinho M, Rodrigues R, Duwel U, et al. The DG European dioxin emission inventory-stage II: characterization of the emissions of 2 hospitals waste incinerators in Portugal [J]. Organohalogen Compounds, 2000(46): 287–290.

Diaz L F, Savage G M, Eggerth L L. Alternatives for the treatment and disposal of healthcare wastes in developing countries [J]. Waste Management, 2005(25): 626–637.

Diaz L F, et al. Characteristics of healthcare wastes [J]. Waste Management, 2008(7): 1219–1226.

Dominica. Alternative to incineration of biomedical waste: autoclaving-a report for the commonwealth of Dominica [C]. http://www.oecs.org/esdu/documents/WasteManagementAutoclavesinominica.pdf, 2001.

Guangming G, Hongfang Z. Practice and discussion of medical waste management in hospitals [J]. Jiangsu Health Management, 2004(6): 48–51.

Huang M C, Jim J M L. Characteristics and management of infectious industrial waste in Taiwan [J]. Waste Management, 2008(28): 2220–2228.

Insa E, Zamorano M, López R, Critical review of medical waste legislation in Spain [J]. Resources, Conservation and Recycling 2010(54): 1048–1059.

Jang Y C, Lee Cargro, Yoon Oh-Sub, et al. Medical waste management in Korea [J]. Journal of Environmental Management, 2006(80): 107–115.

Lee C C, Huffman G L. Review of federal/state medical waste management [N]. Paper No. 91-30.9, 84th Annual Meeting and Exhibition, Air and Waste Management Association, Vancouver, BC, 1991-06-16(21).

Lee B K, Ellenbecker M J, Rafael M E. Alternatives for the treatment and disposal of healthcare wastes in developing countries [J]. Waste Management, 2004(24): 143–151.

Li Q X, et al. Medical waste disposal system in Japan [J]. Shanghai Environmental Sciences, 2003(7): 508–511, 518.

Limit C, Yandong T, Liang Y, et al. Analysis and reference of international medical waste management and treatment and disposal system [J]. Environmental protection, 2020(8). 63–69.

Mbongwe B, Mmereki B T, Magashula A. Healthcare waste management: current practices in selected healthcare facilities, Botswana [J]. Waste Management, 2008(1): 226–233.

Miyazaki M, Une H. Infectious waste management in Japan: a revised regulation and a management process in medical institutions [J]. Healthcare Wastes Management. 2005, 25(6): 616–621.

Muhlich M, Scherrer M, Daschner F D. Comparison of infectious waste management in European hospitals [J]. J. Hosp. Infect., 2003(55): 260–268.

Patil A D, Shekdar A V. Healthcare waste management in India [J]. Journal of Environmental Management, 2001(63): 211–220.

Qinghua X. Exploration of using hierarchical analysis to determine the weight coefficients of medical waste management comprehensive evaluation indexes [J]. Chinese Journal of Disease Control, 2008(4): 365–367.

Shao F, Wang Q, Zhao Y. Cai. Discussion on domestic medical waste disposal and management [J]. Chongqing Environmental Science, 2001(5): 54–56.

Shunze W, Ning S, et al. Current situation and countermeasures of medical waste management and disposal in China [J]. Environmental Protection, 2005(1): 35–38.

Tsukiji M, Gamaralalage P J D, Pratomo I S Y, et al. Waste management during the COVID-19 pandemic from response to recovery [M]. Nairobi: United Nations Environment Programme, 2020.

Water, sanitation, hygiene and waste management for COVID-19: technical brief [R/OL]. Copenhagen: World Health Organization, 2020-03-03 [2020-04-14]. http://www.apps.who.int/iris/handle/10665/331305.

WHO. Safe healthcare waste management policy paper by the World Health Organization [J]. Waste Management, 2005(25): 567–574.

Windfeld S E, Brooks S M. Medical waste management – A review [J]. Journal of Environmental Management, 2015, 163.

Yang C, Peijun L, Lupi C, et al. Sustainable management measures for healthcare waste in China [J]. Waste Management, 2009, 29(6): 1996–2004.

Yang C, Liyuan L, Qinzhong F, et al. Key issues study on the operation management of medical waste incineration disposal facilities [J]. Procedia Environmental Sciences, 2012(16): 208–213.

Yang L. Problems in medical waste treatment and its countermeasures [J]. Journal of Liaoning Medical College (Social Science Edition), 2011, 9(1): 50–52.

Yang C, Anhua W, Qinzhong F, et al. Transmutation and development of medical waste management model in the new era [J]. Chinese Journal of Infection Control,2017,16(6): 493–496.

Ye QF, Miao FONYU, Shan SHJ. Introduction of medical waste management project practices and results in medical institutions [J]. Chinese Journal of Infection Control,2017,16(4): 346–350.

Zhao L Y, Zhang F S, Wang K S, et al. Chemical properties of heavy metals in typical hospital waste incinerator ashes in China [J]. Waste Management, 2008(29): 1114–1121.

Zhipeng B. Research on medical waste management and disposal methods [J]. Environmental, 2003.

Chapter 2
Selection and Optimization of Medical Waste Treatment and Disposal Technology

Abstract This chapter analyzes and evaluates various types of medical waste treatment and disposal technologies, clarifies the applicability of different types of medical waste treatment and disposal technologies, analyzes and summarizes the influencing factors involved in the selection of medical waste treatment and disposal technologies, establishes an index system for the selection of medical waste treatment and disposal technologies, and also analyzes the hardware of medical waste treatment and disposal facilities and measures for pollutant control, medical waste treatment and disposal facilities hardware and pollutant control measures, and to provide a basis for the optimization of medical waste treatment, disposal technology and application (Yaoqiang, Primary Medical Forum 17:924–926, 2013).

2.1 Analysis and Evaluation of Medical Waste Treatment and Disposal Technologies

2.1.1 Medical Waste Incineration Disposal Technology Analysis and Assessment (Aihong and Huilan 2005; Chen et al. 2007a, 2012; Wang et al. 2005; Zhou et al. 2005)

Medical waste incineration facilities typically include systems for waste feeding, incineration, flue gas cleaning, and residue treatment (EEA 1999). Waste preparation and supply, waste incineration, and flue gas cleaning facilities can vary from one medical waste disposal facility to another. Therefore, medical waste incineration disposal technologies present a combination of different forms. The hardware components of medical waste incineration and disposal facilities and pollutant control measures are shown in Fig. 2.1.

From Fig. 2.1, medical waste incineration and disposal is a system project, fully reflecting the different functions of each system and the interface between different systems. Therefore, the performance evaluation of facilities should be combined with

Y. Chen et al., *Innovative Disposal Technology and Management Practice for Medical Waste*, https://doi.org/10.1007/978-981-99-6786-5_2

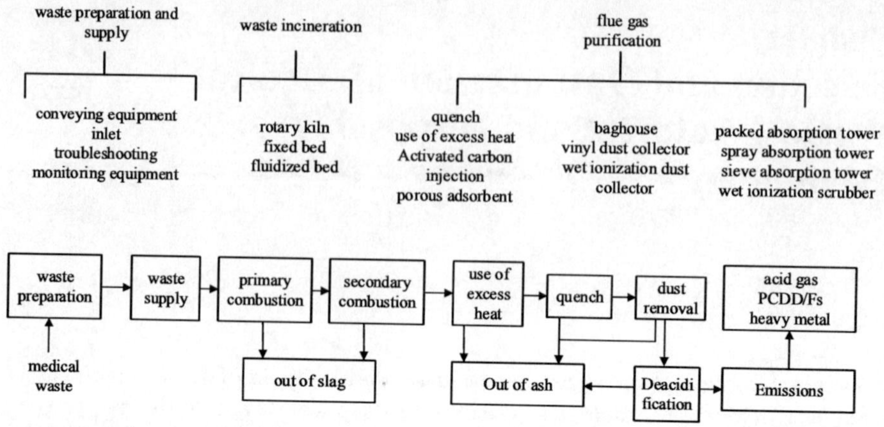

Fig. 2.1 Medical waste incineration and disposal facilities hardware composition and pollutant control measures

the overall system composition characteristics of the incineration process to verify. On the other hand, according to the configuration of different incineration facilities to meet local conditions. Different types of medical waste disposal technologies have different scopes of use.

Rotary kiln incinerator technology is mature, suitable for a variety of different physical (solid, liquid, semi-solid) and shape (granular, powder, lump) of the waste treatment, less secondary pollution, but because of its large one-time investment, high operating costs for the incineration of medical waste, mainly used in the scale of more than 10 to 15 t/d of medical waste disposal or hazardous waste and medical waste disposal projects. With strong adaptability, stable operation and other characteristics, suitable for larger cities and regions of centralized disposal of medical waste. In the incineration technology, rotary kiln technology is the best disposal effect, and more suitable for continuous operation, but the disposal costs are higher. Domestic rotary kiln type of medical waste incinerator problems is the ignition heating and cooling time is long, continuous operation time is short, high operating costs; in the incineration process, auxiliary fuel consumption is also large; refractory materials and other materials of low grade, and less operational experience (Zhang 2006).

Fixed-bed incinerators are suitable for infectious, injurious, chemical and pharmaceutical wastes. It is also suitable for pathologic wastes of small volume, but due to the low temperature of the first combustion chamber, incomplete combustion can occur for larger pathological or pharmaceutical wastes. Therefore, its applicability is however affected. The fixed bed incinerator is suitable for small and medium scale medical waste incineration with a disposal capacity of 1.5~8.0 t/d. It has the characteristics of low investment, simple operation, stable operation and low disposal cost, but lacks a perfect and mature flue gas cleaning system, which will produce secondary pollution to the surrounding environment.

Relatively speaking, pyrolysis incineration technology has been developed rapidly in recent years and is used in Canada, the United States, the United Kingdom and

Mexico, with good results. At present, pyrolysis incineration technology in foreign disposal of medical waste and other hazardous waste areas are more used. China's current production and operation of municipal medical waste incinerators more often use pyrolysis gasification incineration technology. Pyrolysis gasification incineration disposal technology in the disposal effect and disposal costs have greater advantages, with high combustion rate, low consumption of auxiliary fuels, the amount of flue gas produced, low concentration of pollutants in the flue gas, the load of post-disposal and dust entrainment are minimal. However, the technical equipment of pyrolysis incineration varies greatly, and it is difficult to achieve stable combustion and good self-control performance. The pyrolysis section, combustion section affect each other, and in actual operation, they often evolve into fixed-bed oxyfuel combustion because the feed state is far different from the design. The automatic control system is difficult to regulate to the ideal state, and the load of the tail gas system changes frequently, etc. The actual effect is not ideal, the threshold is low, and the technology market is confusing (Liu 2008; Qingjun 2007).

Grate incinerator due to the existence of incineration material, the burn rate is low, auxiliary fuel consumption and other shortcomings of the current application has been rare. In addition, taking into account the safety issues, should be prohibited for only the configuration of a single combustion chamber grate furnace used in medical waste disposal.

High-temperature plasma technology is effective in the disposal of medical waste, but its main feature is the incineration of non-degradable hazardous wastes (PCBs, etc.) at ultra-high temperatures, which is not yet fully exploited for medical waste. In addition, the high construction and operation costs also limit its application in the field of medical waste disposal (Huang and Tang 2007).

In addition to the main facilities involved in the above furnace type, should include other auxiliary equipment around the medical waste collection, transportation, disposal of the whole process. For a specific time, a specific location for some special medical waste incineration disposal and remote areas of the county cities using small incinerators, should be in the continuous feeding, flue gas purification, and automatic control technology.

2.1.2 *Analysis and Evaluation of Non-Incineration Technologies for Medical Waste (Fang-Yuan 2017; Chen et al. 2005, 2007b; Zheng et al. 2008; Shu-Wei et al. 2006)*

Medical waste non-incineration facilities usually consist of waste supply, waste treatment (high temperature steam, chemical, microwave treatment, etc.), tail gas purification, wastewater treatment, discharge and other systems. An examination of China's patent application trends over the past two decades re-veals that the patent application activity experienced a "slow growth period" in 2002–2010 and a "rapid development

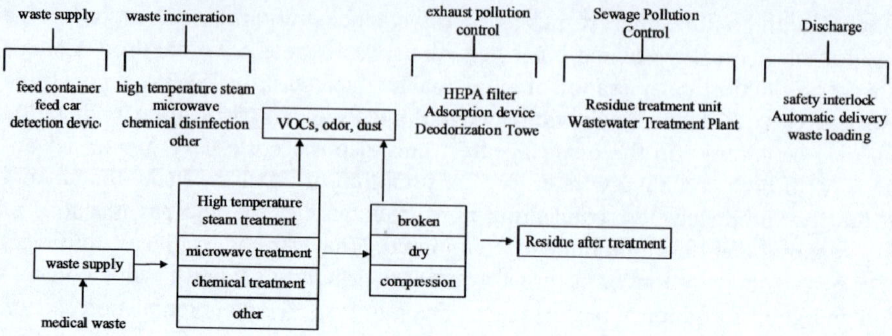

Fig. 2.2 Medical waste non-incineration treatment facilities hardware composition and pollutant control measures

period" in 2011–2019, mirroring global trends. Patent analysis in flue gas purification shows focus on high-temperature steam treatment, microwave and chemical disinfection process (30.4%), incineration process (64.5%), and wastewater treatment processes (5.1%).The hardware composition of medical waste non-incineration treatment facilities and pollutant control measures are shown in Fig. 2.2.

In 2001, the first high-temperature steam and microwave treatment equipment was put into operation in Tianjin. In 2005, the former State Environmental Protection Administration promulgated the implementation of non-incineration standards in the field of chemical, microwave and high-temperature steam treatment technologies issued engineering specifications. After that, the non-incineration treatment technology in medical waste treatment has made great progress.

2.1.3 *Analysis and Assessment of the Applicability of Medical Waste Treatment and Disposal Technologies (Huang 2004; Zhou et al. 2003)*

According to the 23 indicators identified in the medical waste optimal disposal technology index system, the key issues between the centralized technologies were firstly discussed and compared before the systematic evaluation and comparison.

(1) Comparison of Applicability of Disposal Objects

The scope of medical waste disposal technology is the basis for the application of technology, the scope of application of different medical waste treatment and disposal technology is shown in Table 2.1.

(2) Technical scale suitability comparison

Incineration process is more suitable for larger-scale medical waste disposal and hazardous waste disposal. 10 t/d or more medical waste disposal and hazardous

Table 2.1 Scope of application of different medical waste treatment and disposal technologies

Treatment and disposal technology		Infectious waste	Pathological waste	Injurious waste	Pharmaceutical waste	Cytotoxic waste	Chemical waste
Incineration	Incineration	√	√	√	√	√	√
	Pyrolysis	√	√	√	√	√	√
	Grate furnace	√	√	√	√	√	Allow one small part
	Plasma	√	√	√	√	√	√
Non-incineration	High temperature steam	√	×	√	×	×	×
	Microwave disinfection	√	×	√	×	×	×
	Chemistry disinfection	√	×	√	×	×	×
	High temperature dry heat	√	×	√	×	×	×
	Electronic irradiation	√	×	√	×	×	×

waste disposal often use technologies such as rotary kiln incineration. But for small-scale medical waste incineration facilities (such as 3 t/d, 5 t/d), all its process links (such as rapid cooling of the tail gas, de-acidification, bag filter, etc.) are similar to large-scale incineration facilities, but there is a greater difficulty in achieving that. In fact, only the configuration of the national standard requirements of the incineration tail gas online monitoring device will require an investment of about 1 million Yuan. The small-scale incineration facilities are not stable incoming material, 3 t/d scale of full collection is often actually only about 1 t/d, and it is difficult to achieve stable and continuous operation. In addition, the impact on the tail gas treatment process fluctuates greatly, the frequent start and stop furnace intervals lead to serious pollution, to maintain the combustion of auxiliary fuel costs are extremely high. Therefore, it is only theoretically possible for small-scale incineration facilities to achieve medical waste incineration and meet emission standards. In addition, China has acceded to the Stockholm Convention on Persistent Organic Pollutants, which will require future incineration dioxin emission standards such as the existing 0.5 ngTEQ/Nm^3 to 0.1 ngTEQ/Nm^3, which is a devastating pressure on small-scale incineration facilities. For small-scale medical waste disposal projects, high-temperature steam, microwave, chemical and high-temperature dry heat treatment technologies are more suitable for small-scale medical waste disposal because of their intermittent operation, low operating costs, adaptability, less secondary pollution, no dioxins and other pollutants, easy operation and management, and stable process operation (Chengzhi et al. 2007; Yang et al. 2009; Shi and Shao 2006).

(3) Technical reliability comparison

The incineration disposal technology has good adaptability to different wastes, and has been widely used because it can make medical waste treatment achieve harmless, reduction, stabilization and complete destruction of the treatment purpose. However, the current level of domestic pyrolysis technology equipment in operation can hardly support the reliability of its equipment. China's medical waste pyrolysis incineration threshold is low, the technology is immature, the market is confusing, and the actual operation often deviates from the original design of the theoretical incineration process, poor operating results. According to future development needs, most of these technologies are facing transformation or elimination range. For non-incineration treatment technology, different technologies for the disposal of medical waste have different adaptability. Therefore, the technical equipment should be considered in accordance with the degree of reliability (operation level, classification level, technical level, disinfection effect).

(4) Comparison of technical pollutant emissions

Incineration disposal technology in the process of disposal of medical waste dioxins and heavy metals and other pollutants, especially in the case of unstable waste material, will cause many problems in the tail gas purification, environmental risks. Non-incineration treatment technology is a positive complement to the incineration technology, its intermittent operation and process characteristics make the technology has a flexible operation, simple operation, low cost of treatment, more suitable for the generation of small amounts, unstable incoming material, small-scale medical waste treatment. Compared with the incineration technology, the temperature of the non-incineration treatment process does not exceed 200 °C at most. Medical waste containing chlorinated polymers such as plastics do not decompose, so no dioxin-like carcinogenic substances are produced, and dioxin "zero emissions" can be achieved (Huafeng et al. 2005).

(5) Comparison of technology construction cost and operation cost (Lu et al. 2005)

Compared with incineration technology, non-incineration treatment technology has more significant advantages in both construction and operation costs, from the point of view of construction costs, non-incineration treatment facilities do not have a similar incineration disposal technology and should have a complex tail gas purification system, so its construction costs are lower, compared with the same treatment scale, the construction costs are only 1/2 of the incineration disposal facilities. The cost of fuel, power, raw and auxiliary materials consumed by non-incineration technology is 1/4~1/3 of that of incineration technology, so the lower operating cost makes the technology more attractive and competitive, and the amount of wastewater and waste gas generated is small and easy to handle, and the disposal effect is guaranteed to be higher. Non-incineration treatment technology does not change the existing classification and packaging collection system within the hospital, and the proportion of waste drugs, chemical waste, pathological tissue waste that cannot be included in its disposal system is often small, and there is generally a corresponding disposal system, and is not included in the incineration system. The technology is particularly suitable for small-scale medical waste disposal projects such as 3 t/d, 5 t/d.

(6) Technology Management Matching Comparison (Chen 2004)

Both incineration disposal facilities and non-incineration treatment facilities, in terms of management are in a continuous process of progress and development. In order to regulate the construction of incineration disposal facilities, the Ministry of Environmental Protection has promulgated and implemented in 2004 and 2006 respectively for the centralized incineration of medical waste disposal and non-incineration centralized treatment engineering specifications, from the construction and operation of the two aspects of the strict requirements. Relatively speaking, incineration facilities require a high level of operation due to the complexity of the process, while the limitations of non-incineration technology is that it is not a spectrum of disposal technology, for pharmaceutical waste, pathologic waste, chemical waste is not applicable, so the application of non-incineration technology requires strict classification management procedures within the hospital. The health and environmental protection departments have promulgated a number of industry regulations that strictly govern the segregated management and collection of medical waste, reducing the amount of waste and hazardous components at the source. Therefore, non-incineration technologies require a high level of internal management and whole-process supervision in medical institutions in order to address the applicability of different disposal technologies.

(7) Comprehensive comparative analysis of different disposal technologies

After summarizing the assessment and analysis of incineration and non-incineration treatment technologies in the previous chapters, the main medical waste disposal technologies currently applied and potentially applied in China are systematically analyzed and evaluated by combining the 23 indicators of technology optimization, as shown in Table 2.2.

2.2 Medical Waste Disposal Technology Selection Index and Optimization

2.2.1 Medical Waste Disposal Technology Selection Indicators (Xiao et al. 2005)

The essence of the implementation of the POPs Convention in the field of medical waste in China is to reduce the dioxins generated in the process of their incineration and disposal (Chengzhi et al. 2007). At present, the ultimate goal of China's medical waste management and disposal is to establish a medical disposal technology and management model that meets the requirements of international conventions and is consistent with national conditions, and ultimately achieve sustainable environmental management. There are various technical routes for medical waste disposal, and many technologies have their own characteristics and applicability (Chen and Hao 2020). To promote the application of a medical waste disposal technology in a region, the selection criteria of international conventions and international organizations on the application of medical waste disposal technologies should be combined with the actual situation of the country to determine. After studying and analyzing the application of technologies at home and abroad, the selection of a technology should consider the following factors:

(1) Environmental desirability: this refers to the adoption of waste disposal technologies and management capabilities that ensure public health and environmental safety (Zhou et al. 2006).
(2) Management continuity (administrative diligence): this refers to the ability of management to ensure that the policies and measures adopted are implemented and effective over time, with emphasis on the environmental impact.
(3) Economic effectiveness: this refers to the cost effectiveness of the disposal technology and management tools adopted, and at the same time take into account the economic value of the waste itself.
(4) Social acceptability and equity: this refers to the adoption of disposal technology and management means can be supported and accepted by the local community, including the effectiveness of waste management methods.

For the selection of technologies, the US and European health care waste elimination organizations consider factors such as; waste disposal capacity, type of waste to

Table 2.2 Comparison of medical waste treatment and disposal methods

Item	Pyrolysis incineration	Rotary kiln incineration	High temperature steam treatment	Microwave processing	Chemical treatment	High temperature dry heat processing
Scope of application	Infectious, pathological, injurious, pharmaceutical and chemical medical waste	Infectious, pathological, injurious, pharmaceutical and chemical medical waste	Infectious and injurious medical waste	Infectious and injurious medical waste	Infectious and injurious medical waste	Infectious and injurious medical waste
Suitable treatment size	5~10 t	10t or more	<10 t	<10 t	<10 t	<10 t
Technical Reliability	Meet the burning reduction, disinfection requirements	Meet the burning reduction, disinfection requirements	Meet disinfection requirements	Meet disinfection requirements	Meet disinfection requirements	Meet disinfection requirements
Technology Maturity	Nationalized equipment is mature	Nationalized equipment is mature	Nationalized equipment is mature	Nationalized equipment is mature	Nationalized equipment is mature	Nationalized equipment is mature
Equipment requirements	High temperature and corrosion resistance	High temperature and corrosion resistance	Airtight, insulated, resistant to high temperature and pressure	Hermetically sealed, high temperature resistant, electromagnetic protection	Negative pressure operation, corrosion resistance	Hermetically sealed, high temperature resistant, corrosion resistant
Technology Benefits	Low flue gas volume and High heat utilization	Good disposal effect, strong adaptability, large treatment capacity, complete combustion, stable operation effect	Low operating cost, strong adaptability, less secondary pollution, no dioxins and other pollutants, easy to operate and manage, stable operation effect			

(continued)

Table 2.2 (continued)

Item	Pyrolysis incineration	Rotary kiln incineration	High temperature steam treatment	Microwave processing	Chemical treatment	High temperature dry heat processing
Technical disadvantages	Not easy to achieve stable combustion, frequent changes in tail gas system load, easy to produce dioxins	High operating costs, poor energy efficiency, easy to produce dioxins	Condensate and steam boiler exhaust must be treated	Waste first crushed to increase safety risk, electromagnetic radiation must be protected	Easy to produce secondary contamination of disinfectant	Must achieve a certain disinfection effect
Operation method	Continuous/ intermittent operation	Continuous operation	Intermittent operation	Intermittent operation	Intermittent operation	Intermittent operation
Operation requirements	General operation difficulty and labor intensity	Difficult and labor-intensive operation	General operation difficulty and labor intensity	General operation difficulty and labor intensity	General operation difficulty and low labor intensity	General operation difficulty and labor intensity
Pollutant Emissions	Acid gases, heavy metals, dioxins	Acid gases, heavy metals, dioxins	VOCs, bad odor	VOCs, microwave radiation	VOCs, waste disinfectants	Odor, VOCs, pathogenic organisms, noise
Floor space	Relatively large	Big	Relatively small	Relatively large	Relatively small	Relatively small
Operation and maintenance	Higher operation and maintenance requirements, higher costs	High operation and maintenance requirements, high cost	Higher operation and maintenance requirements, higher costs	Average operation and maintenance requirements, low cost	High operation and maintenance requirements, medium cost	High operation and maintenance requirements, medium cost

be disposed, microbial inactivation efficiency, pollutant emissions, policy acceptance level, space requirements, ancillary facility requirements, waste reduction, occupational safety and health, noise and odor, automation, technology reliability, commercialization level, technology and equipment provider background, cost, social and public acceptance, etc. (Qinghua 2008).

Based on the above considerations, the influencing factors involved in the selection process of medical waste disposal technology were analyzed and summarized, and the index system for the selection of medical waste disposal technology was derived, as shown in Table 2.3.

Table 2.3 lists the main factors that should be considered when making technology choices, and the details are described below.

(1) Technical performance index A

A1—appropriateness of disposal scale: this refers to the disposal of medical waste disposal system in the specified time according to the amount of waste that can be disposed off.

A2—the effectiveness of the disposal effect: this implies that the disposal of medical waste can achieve the efficiency of killing pathogenic microorganisms, the degree of reduction in volume and shape destruction and subsequent disposal requirements.

A3—disposal of waste applicability: this refers to the applicability of the disposal technology to various types of medical waste.

A4—system configuration completeness: this implies that the disposal facility should have a full set of equipment and unit situation.

A5—Unit design sophistication: this refers to the degree of sophistication of each disposal unit.

A6—Automation control level: this refers to the degree of braking control of the relevant working parameters and operating parameters of the disposal facility.

A7—disposal facility security: medical waste disposal system should have a complete emergency protection program, in the event of a sudden emergency, and can be set up by the work area monitoring system, graded alarm display, interlocking self-locking devices, emergency power supply and equipment and other components of the emergency security system to ensure the safety of the system.

A8—supported infrastructure: this mainly refers to the main equipment and ancillary equipment, in addition to what other facilities, such as water supply, power supply, secondary pollutant disposal, etc.

A9—energy saving performance: the energy consumption of the disposal system is usually measured by the amount of energy consumed per unit of waste disposed.

A10—the ease of operation of the disposal facilities: mainly in the ease of disposal operations, the size of the intensity of the operation and the length of operation time.

A11—regulatory means achievable: mainly combined with the level of disposal technology, considering the degree of implementation of local regulatory hardware and software conditions can be achieved.

(2) Environmental Impact Indicator B

Table 2.3 Indicator system for consideration of medical waste disposal technology selection

Serial number	Scope of consideration	Number	Select indicator	Key considerations
A	Technical performance	A1	Disposal scale appropriateness	Adaptability to the scale of local planning
		A2	Effectiveness of Disposal	Matching with local supporting facilities
		A3	Disposal of waste applicability	Complementarity with local waste types and supporting facilities
		A4	System configuration completeness	Conformity with standards and norms
		A5	Advanced unit design	Conformity with standards and norms
		A6	Automation control level	Conformity with standards and specifications and advanced automation
		A7	Disposal facility safety	Consider the configuration of security measures for facilities
		A8	Supporting infrastructure required	Consider the ease of supporting facilities
		A9	Energy saving performance	Consider the comparative advantages of energy efficiency
		A10	Ease of operation of disposal facilities	Consider the complexity and operability of the facility
		A11	Realizability of regulatory instruments	Consider the corresponding local regulatory and monitoring capabilities
B	Environmental impact	B1	Generate toxic and harmful pollutants risk	Consider the generation of toxic and hazardous substances such as POPs
		B2	Risk of secondary contaminants	Consider the environmental factors such as exhaust gas, waste water, waste residue, noise, etc
		B3	Occupational safety and health risks	Consider the health risks of technology applications to operators
		B4	Risk of environmental impact on surrounding residents	Consider the degree of impact of pollutant emissions on surrounding residents

(continued)

Table 2.3 (continued)

Serial number	Scope of consideration	Number	Select indicator	Key considerations
		B5	Ecological impact risk	Consider the degree of ecological impact of pollutant emissions
C	Economic performance	C1	Construction costs	Consider the cost of the main equipment and supporting facilities
		C2	Running costs	Consider waste disposal costs and equipment depreciation
		C3	Earnings level	Consider profit levels and operational sustainability
D	Social conditions	D1	Public acceptability	Comprehensive evaluation in terms of pollution level and public response
		D2	The extent to which the policy allows	Evaluation from a policy perspective on the application of technology
		D3	Ease of site selection	Evaluation from the perspective of health and safety protection distance
		D4	Ease of technology acquisition	Evaluation from the ease of technology acquisition

B1—the risk of toxic and harmful pollutants: various medical waste disposal technologies will produce more or less toxic and harmful gases, liquid or solid pollutants, etc., to the environmental hazards of the size of the possibility.

B2—the risk of secondary pollutants: this refers to the disposal process of toxic and hazardous substances generated by the disposal of pollution control devices after the discharge, the possibility of secondary pollution of the environment from the discharge size. Usually reflected by the type and concentration of emissions from the pollution control device.

B3—occupational safety and health risks: refers to the possibility of unsafe factors and the consequences of hazards caused to staff in the disposal process, unsafe factors include toxic and hazardous hazards and hazards of hazardous operations, such as mechanical injuries, thermal surface burns, radiation, chemical injuries and pathogenic infections, etc.

B4—risk of environmental impact on the surrounding residents: The size of the possibility of health hazards to the surrounding residents from disposal activities.

B5—ecological impact risk: this refers to the disposal of medical waste activities generated or emitted harmful substances on the disposal unit location of the soil, water bodies, the atmosphere and other ecological conditions caused by the negative impact.

(3) Economic performance index C

C1—construction cost: The total capital invested in the project, including construction investment, working capital and annual capital requirements during the construction period.

C2—operating costs: refers to the various costs incurred in the production and operation of the project. It is usually reflected by the disposal cost per unit of waste volume and the total cost.

C3—revenue level: The level of revenue reflects the profitability of the project. Financial internal rate of return, financial net present value and payback period are the main profitability indicators.

(4) Social Conditions Indicator D

D1—public acceptability: the degree of rejection or acceptance shown by the public according to their perception of medical waste disposal methods and technological solutions.

D2—policy allowance: the degree to which the selected disposal technology is recognized or biased by national and local standards and regulations.

D3—ease of site selection: for alternative medical waste disposal technologies, the project site conditions available at the site match and adapt to the site selection requirements.

D4—ease of technology acquisition: the ease of technology acquisition is related to the background of technology suppliers, the degree of technology commercialization, the way of technology introduction and the degree of localization, etc.

Each aspect contains a number of index factors, both quantitative and qualitative. All these factors reflect the overall, comprehensive, relevant and hierarchical characteristics of technology program evaluation. In fact, the successful application of a technology will ultimately be reflected in the project. At the same time, the comprehensive evaluation of medical waste disposal engineering technology program is a complex system composed of many interrelated and mutually constraining factors, involving technical, economic, environmental and social constraints and other aspects. Based on the problems of medical waste disposal facilities in China, as mentioned earlier, there is a large gap between the existing disposal facilities and their applications, especially in the specific context of China's implementation of the Stockholm Convention and strict control of dioxin emissions, it is particularly important to take appropriate technical optimization measures for existing and new sources.

2.2.2 *Optimization of Medical Waste Treatment and Disposal Technology (Guozhu et al. 2006)*

2.2.2.1 Medical Waste Treatment and Disposal Technology Selection Model

China's vast territory, the economic level of different regions, the level of health care and environmental awareness vary greatly, the technical and economic conditions of medical waste disposal vary. Therefore, how to better exploit the strengths and avoid the weaknesses, and effectively play the application of the corresponding disposal technology in the regional medical waste disposal, should become the basis for determining the choice of medical waste disposal technology in China. The choice of medical waste disposal technology in China should follow the principle of multiple technologies and safe disposal (Huang et al. 2008). Summarize the lessons learned from past medical waste management (Shinee et al. 2008), drawing on the successful experience of international medical waste management in developed countries, combined with the reality of China's national conditions and implementation needs, the following recommendations are made:

(1) For the disposal scale (>10 t/d) of the provincial hazardous waste disposal center (and disposal of medical waste), the use of rotary kiln-based incineration disposal technology; for the completed construction of municipal medical waste disposal center using pyrolysis incineration disposal technology, the focus should be on two aspects: on the one hand, to ensure that such facilities in the process of future renovation and improvement. On the one hand, to ensure that such facilities in the future renovation and improvement process, in accordance with the previously mentioned disposal technology optimization indicators for comparison and analysis, to ensure that the construction level of disposal facilities; on the other hand, the focus should be on management to effectively promote the standardized operation and management of such facilities, to achieve safe disposal of medical waste at the same time, and to achieve stable emission of pollutants to standards.

(2) For the disposal scale not large (<10 t/d), especially 3 t/d, 5 t/d and other small-scale medical waste treatment projects, in the project construction conditions allow the use of non-incineration technology as far as possible to ensure the flexibility and economy of operation after the completion of the facility. At the same time, we must also regulate the non-incineration technology market, to avoid repeated pyrolysis incineration equipment market chaos and disorderly competition, and strengthen the strict supervision and management of the equipment bidding process.

(3) for remote areas, should be fully integrated with the characteristics of medical waste generation and distribution, in accordance with the principle of local conditions, in order to promote centralized disposal, taking into account the construction of simple and easy medical waste disposal facilities (transport distance of more than a hundred kilometers can be considered before building

medical waste disposal facilities), such as the use of non-incineration treatment technology, the use of chemical disinfection treatment after landfill, etc., aimed at using existing local conditions to eliminate the infectious nature of medical waste in a timely manner, and reduce its disposal process on the secondary pollution of the environment (Li et al. 2019).

(4) For remote areas, emergency situations and regional medical waste disposal facilities are not yet completed, you can consider the use of mobile disposal technology for disposal, and the type of technology can be used non-incineration technology.

Of course, as mentioned in the previous sections, no technology is a panacea. Therefore, in the implementation of a technology selection, must be fully integrated with the local technical, management, economic and social factors, and effectively provide the background and basis for the final choice of disposal technology. In addition, for high-temperature steam, chemical disinfection and other non-incineration treatment technology, China's medical waste classification catalog of chemical waste, pharmaceutical waste and part of the pathologic waste cannot be treated by such methods. Therefore, the level of medical waste classification within medical institutions determines the final disposal technology choice of medical waste.

2.2.2.2 Medical Waste Treatment and Disposal Technology Application Model

China's medical waste treatment and disposal technology application model should take the principle of centralized disposal and reasonable layout. As stipulated in the "medical waste management regulations," in China, should be implemented in the centralized disposal of medical waste. Therefore, should continue to implement the planning proposed in principle to the city as a planning unit, the construction of centralized medical waste disposal facilities, in a reasonable transportation radius to accept the disposal of medical waste within the jurisdiction of all counties; in general, do not advocate, do not allow hospitals to disperse; local people's governments at all levels should plan to build a centralized medical waste disposal facilities, centralized disposal of medical waste; and in accordance with the relevant national provisions to medical and health institutions to charge medical waste disposal costs (Yang et al. 2004).

Centralized medical waste disposal facilities at the prefecture-level cities as a unit for the construction of the basis, to encourage developed transportation, dense urban areas of the city to jointly build and share centralized medical waste disposal facilities. At the same time, hazardous waste facilities and medical waste facilities should be integrated construction, hazardous waste centralized disposal facilities to deal with medical waste generated by the city.

Medical waste is infectious and contagious, cytotoxic, radioactive hazards and other characteristics, medical waste from the source to the final disposal should be carried out in a fully enclosed state, and the implementation of the isolation

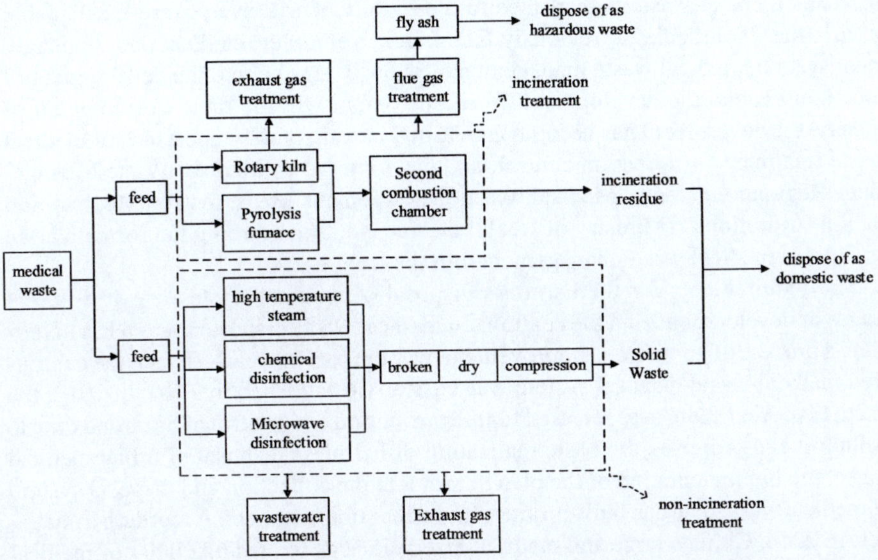

Fig. 2.3 Best feasible technology combination for medical waste treatment and disposal process (Chen et al. 2010)

of people and the environment. Medical waste from the generation, classification and collection, warning labels, closed packaging and transport, storage, harmless disposal of the entire process of implementation of the whole management, that is, from the "cradle" to the "grave" of each link to implement the whole process of strict management and control (Huang et al. 2008).

For a city, the overall process technology selection for medical waste pollution prevention is shown in Fig. 2.3.

2.3 Review of Medical Waste Disposal Technology System and Development

2.3.1 Construction and Development of Centralized Medical Waste Disposal Model (Ministry of Ecology and Environment 2004, 2020; Liu et al. 2005)

Since the 1950s, medical waste disposal has attracted widespread attention from all over the world. Medical waste generated by hospitals contains a large number of pathogenic microorganisms, parasites and other harmful substances. Therefore, if medical waste is not managed properly, it is bound to become a source of disease transmission. Not only will medical staff and patients face the risk of infection, but once this medical waste is dispersed to the society, it will even cause incalculable harm to the whole society. Previously, China lacks a complete medical waste management system, medical waste management obviously lags behind the development of social and economic development. In recent years, with the rapid development of society, the government has become increasingly aware of the importance of medical waste treatment for infection control, and has issued the "Medical Waste Management Regulations" and "Medical Waste Management Measures for Medical and Health Institutions" (Ministry of Health Decree No. 36), marking the formalization of China's medical waste management into the legal track.

The requirements for the harmless disposal of medical waste vary at different stages of development, and the centralized medical waste disposal system has undergone a process of continuous improvement over the past 10 years. In fact, the centralized medical waste disposal system was basically established in 2010. In 2012, the "12th Five-Year Plan" was released for the prevention and control of hazardous waste pollution and proposed the basic realization of harmless disposal of urban medical waste, the implementation of the plan to promote the collection and disposal system of medical waste in urban built-up areas for further improvement. According to statistics, in 2014, China's large and medium-sized cities generated 622,000 t of medical waste, disposal volume of 607,000 t, disposal rate of 97.5%, and most cities have established a centralized medical waste treatment facility.

The "Thirteenth Five-Year Plan" ecological environmental protection plan requires the extension of the scope of centralized medical waste disposal to rural areas, while requiring the strengthening of standardized medical waste management; "Guidance on enhancing the ability to supervise the environment of hazardous waste, the ability to use and dispose it off, and the ability to prevent environmental risks" (Environmental Solids [2019] No. 92) proposed that each County (city) should build medical waste collection and transfer disposal system requirements, the "Thirteenth Five-Year Plan" period, the harmless disposal of rural medical waste has received more and more attention. After the New Crown epidemic, the centralized disposal of medical waste has been vigorously promoted, and the epidemic issued during the "comprehensive management of medical institutions waste work program", "medical waste centralized disposal facilities capacity building implementation program" and other documents to further clarify the "14th Five-Year" period of medical waste management goals, and China's centralized disposal of medical waste. The system will usher in a new round of upgrading.

2.3.2 *Medical Waste to Fill the Shortage of Technology Needs*

At present, in the treatment of medical waste, there are mainly high-temperature incineration disposal technology, non-incineration disposal technology and other technologies. Among them, high-temperature incineration technology mainly has the following shortcomings: first, the incineration flue gas contains a variety of harmful substances; second, the disposal system is a relatively large one-time investment, high operating costs; again, the incineration disposal project has a protective distance requirements, site selection is difficult; finally, the incineration residue is very easy to cause secondary pollution. For non-incineration disposal technology, there are limitations to its use. Other technologies such as plasma method, there are high investment and operating costs, specific operational difficulties, technical, and the overall maturity of the disposal technology has yet to be studied.

In the management of medical waste disposal there is a lack of emergency disposal capacity, and collaborative disposal mechanism is not sound. During the outbreak of the new crown epidemic, the national medical waste generation grew rapidly, the epidemic in serious areas of medical waste generation, transfer vehicles, transfer boxes, emergency disposal facilities and other reserves are clearly insufficient. The Ministry of Ecology and Environment mobilized the entire system to mobilize facilities and equipment nationwide to assist Hubei, only to solve the problem of medical waste backlog in Hubei Province and Wuhan City. At present, medical waste disposal centers are planned and built in the city as a unit. In the collection and transportation process, medical waste needs to be transported across long distances across counties, the distance from dozens to hundreds of kilometers, easily causing medical waste scattering, leakage, causing environmental pollution accidents and the spread of infectious diseases. For medical waste after disinfection of the residue with the sanitary landfill of domestic waste. This method does not only achieve the resource treatment of waste, but also occupies a large amount of land, and it is a dangerous practice of leaving the source of pollution to future generations, and does not comply with the principle of waste reduction and resource disposal. Compared with the rapid increase in the amount of medical waste generated, the centralized disposal capacity of some cities is slow to improve. At the same time, medical waste in rural and remote areas is not properly disposed off. China's centralized medical waste disposal system basically covers the urban built-up areas and most of the eastern rural areas, the central and western rural and remote areas of medical waste due to the scattered sources, the generation of small quantities, long transport distances, inconvenient transportation and other factors, has not been included in the centralized medical waste disposal system.

With the continuous development of information technology, the level of urban information applications continue to improve, the construction of smart cities came into being. In the medical waste disposal, intelligent treatment also has a lot of room for development. More and more diversified intelligent medical waste disposal technologies have been widely used in the field of medical waste disposal at home and abroad, and continue to play a role in intelligent waste disposal control and guidance, facility operation and maintenance management, etc. How to further use diversified technologies related to intelligent waste disposal to solve the practical problems in the process of medical waste disposal, to promote the application of emerging technologies on the ground, to improve the safety and efficiency of medical waste disposal still needs further consideration.

2.3.3 "Fourteen Five" Medical Waste Disposal Technology Development (Cheng et al. 2021)

Centralized medical waste disposal system in the new crown epidemic, the existing centralized disposal system is facing a new round of quality upgrade, and should make up for the shortcomings of the disposal capacity, coverage, emergency response capacity, management system, etc., as soon as possible.

In view of the current shortcomings, the following are the main recommendations to optimize the centralized medical waste disposal system: First, will be orderly to enhance the level of centralized urban medical waste disposal capacity, comprehensive medical waste collection and disposal of various factors affecting the amount of a comprehensive assessment of the previous facilities medical waste disposal capacity, to the disposal facilities operating load rate higher than normal level of the city as the focus, the orderly promotion of medical waste Centralized disposal capacity; secondly, focus on promoting the collection and disposal of medical waste in rural and remote areas, to further expand the coverage of the existing centralized medical waste collection and transfer disposal system can consider centralized disposal facilities, neighboring counties (cities) can take a shared mode of construction of centralized medical waste disposal facilities. Centralized disposal facilities collection and transfer system is difficult to cover rural and remote areas, through the local construction of non-incineration facilities or equipped with mobile disposal facilities for local decentralized disposal; again, to accelerate the construction of centralized disposal system emergency response capacity to urban areas with a resident population of more than 3 million large cities as the focus, to enhance the centralized medical waste disposal system emergency response capacity, each city can be determined according to its own situation Centralized disposal system capacity configuration form, to ensure that the emergency disposal capacity of each million people is not less than 10 t/d; Finally, relying on centralized medical waste disposal facilities, the formation of medical waste emergency disposal of professional team reserves and material reserves, the establishment of a sound emergency dispatch mechanism. Continuously improve the centralized medical waste disposal management system, start the revision of the "Medical Waste Management Regulations", accelerate the introduction of the "Medical Waste Classification Catalog", dynamically update the "National Hazardous Waste List", accelerate the "Medical Waste Treatment and Disposal Pollution Control Standards", "Medical Waste Centralized Disposal Technical Specifications" and a variety of centralized disposal engineering specifications for the preparation and revision of the work, accelerate the designation of mobile and small Medical waste disposal engineering specifications to ensure coordination and unity between the relevant requirements of the medical waste management system.

The occurrence of the epidemic accelerated the upgrading of the centralized medical waste disposal system, in order to meet the "14th Five-Year" period of medical waste management goals, medical waste management and disposal of the short board must be made up as soon as possible.

References

Pan Aihong, Xue Huilan. Exploration of problems related to medical waste disposal process[J]. Agricultural Mechanization Research, 2005(2): 51-53.

Chen X D, Hao L W. Exploration on the technical route of cement kiln co-disposal of domestic waste and medical waste hazardous waste under the epidemic[J]. China Cement, 2020(4): 71-75.

Chen Y, Li PJ, et al. Analysis of barriers and countermeasures for the application of non-incineration treatment technology for medical waste[J]. Non-ferrous metallurgy design and research, 2007b(3): 27-29.

Chen Y, Ding Q, Jiang C, et al. Exploring the best available technologies and best environmental practices for medical waste disposal in the context of compliance[J]. China Nursing Management,2010,10(4): 75-79.

Chen Y, Wu A-H, Feng Q-Z, et al. Medical waste treatment and disposal technology and source separation countermeasures[J]. Chinese Journal of Infection Control,2012,11(6): 401-404.

Chen Y, Wang K Y, Liu F Q, et al. Application and development trend of non-incineration treatment technology for medical waste[J]. Environmental Protection, 2005a(7): 57–58,63.

Chen Y, Li PJ, Sun YZ, et al. Study on compliance requirements and technical measures for hazardous waste incineration and disposal facilities in China [G]. Beijing: Proceedings of the Annual Academic Conference of the Chinese Society of Environmental Sciences (Vol. 1), 2007a(8): 563–567.

Cheng L, Zhang Z, Tao Y, et al. Optimization ideas of centralized medical waste disposal system in China's 14th Five-Year Plan[J]. Journal of Environmental Engineering, 2021, 15(2). doi: https://doi.org/10.12030/j.cjee.202008110.

Wang Chengzhi, Hu Xiaomin, Shi R. Process control safety technology of dioxins in medical waste incineration and disposal[J]. Chinese Journal of Safety Science,2007: 17(1): 157–161.

EEA, 1999. Incineration of medical wastes. Emission inventory guidebook, Environmental European Agency, Oxford [S]. [6] Integrated pollution prevention and control reference document on the best available techniques for waste incineration[R], 2006.

Jim Fang-Yuan. Status and development trend of medical waste treatment and disposal technology[J]. Heilongjiang Environmental Bulletin,2017,41(2): 92-94.

QI Guozhu, WANG Chengzhi, LI Xiongyong. Technical points of waste pyrolysis incineration process equipment selection[J]. Environmental Protection Science,2006(6): 28-30.

Liu Huafeng, et al. Environmental risk assessment of hazardous waste incineration facilities[J]. Environmental Science Research,2005,18(Suppl.): 48–52.

Huang W-P. A few suggestions for hazardous waste incineration monitoring [J]. Environmental Monitoring Management and Technology, 2004(2): 35-36.

Huang H, Tang L. Treatment of organic waste using thermal plasma pyrolysis technology [J]. Energy Conversion and Management, 2007(48): 1331-1337.

Huang MC, Jim JML. Characteristics and management of infectious industrial waste in Taiwan [J]. Waste Management, 2008(28): 2220-2228

Huang J, Zhang B, Ai N Shan, et al. Comparative analysis of eight medical waste treatment methods[J]. Chinese Journal of Disinfection,2008(3): 313–315.

Li Y., Chen Y., Wu A. H., et al. Exploration of medical waste disposal technology and management model in remote areas of China[J]. Chinese Journal of Infection Control,2019,18(1): 83-88.

Liu Z. S. Process introduction and environmental analysis of medical waste pyrolysis gasification incinerator system [J]. Chemical Engineering and Equipment, 2008(8): 124-127.

Liu F, Wang D M, Li W Q. Construction program of centralized medical waste disposal facilities system in Tianjin [J]. Urban Environment and Urban Ecology,2005,18(2): 44-46.

Lu C, Xue J, Pei D-B. Review of non-incineration treatment technologies for medical waste [J]. Urban Environment and Urban Ecology,2005(8): 27-29.

Ministry of Ecology and Environment. Notice on the release of "Technical Requirements for the Construction of Centralized Incineration and Disposal of Hazardous Wastes (for Trial Implementation)" and "Technical Requirements for the Construction of Centralized Incineration and Disposal of Medical Wastes (for Trial Implementation)" [Z/OL]. [2004–01–19].

Ministry of Ecology and Environment. Announcement on the release of "technical specifications for the construction of hazardous waste centralized incineration and disposal projects" and two other environmental protection industry standards [Z/OL]. [2005–05–24].

Ministry of Ecology and Environment. Letter on soliciting comments on the national environmental protection standard "Pollution Control Standard for Hazardous Waste Incineration" (draft for comments) [Z/OL]. [2014–10–24].

Ministry of Ecology and Environment. Questions and answers related to the work of emergency disposal of medical waste at domestic waste incineration facilities [R/OL]. (2020–02–27) [2020–04–14]. http://sthjt.jl.gov.cn/zwzx/qghb/202002/t20200227_6862133.html.

Xu Qinghua. Exploration of using hierarchical analysis to determine the weight coefficients of medical waste management comprehensive evaluation indexes[J]. Chinese Journal of Disease Control, 2008(4): 365-367.

Zou Qingjun. Brief discussion on the treatment of medical waste by pyrolysis gasification incineration [J]. Environmental Health Engineering, 2007(3): 47-49.

Shi M-F, Shao K-Z. Waste incineration flue gas purification and control technology of dioxin pollutants[J]. Environmental Science and Technology,2006,28(9): 78-79.

Shinee. E, Gombojav. E, Nishimura. A, et al. Healthcare waste management in the capital city of Mongolia, Waste Management (2008) (2): 435-441.

Li Shu-Wei, Zhang, Jing-Yan, Chen, Zhi-Hong. "Microwave-high-temperature and high-pressure steam" technology for the environmentally sound disposal of medical waste [J]. China Environmental Protection Industry,2006(1): 22–24.

Wang W, Lu H L, Li M. A review of flue gas cleaning technologies for medical waste incineration[J]. Environmental Science and Technology,2005(2): 101-103.

Xiao W, Nie X, Feng B. Research on the comprehensive evaluation system of multi-technical options for medical waste disposal project and its auxiliary software design [J]. Environmental Science Research,2005(Suppl.): 57–62.

Chen Yang, Wen Wu, Shao Chunyan. Study on technical requirements for the design of centralized hazardous waste incineration and disposal facilities [J]. Environmental Protection Science,2004(8): 26-28.

Chen Yang, Shao Chunyan, Xu Dundou, et al. Study on the performance evaluation method of hazardous waste incineration and disposal facilities[C]. Dalian: The 5th Global Chemistry Conference, 2009(5): 9–12.

Huang Yaoqiang.Research progress in medical waste treatment [J]. Primary Medical Forum,2013(17): 924-926.

Zhang H, Yang LQ. Exploration of medical waste disposal methods [J]. Environmental Health Engineering,2005(6): 19-21.

Zheng L, Yang YN, Wu SHUNZE. Screening and management study of applicable technologies for medical waste incineration treatment in China[J]. Environmental Protection,2008(11B): 63–65.

Zhou F, Liu Y, Guo H. Cheng. Study of key parameters in the incineration of medical waste [J]. Environmental Science Research,2003(3): 24–28,38.

Zhou F, Liu Y, Guo H. Cheng. Study of key parameters in the incineration and disposal of medical waste[J]. Environmental Science,2005,18(3): 24–28.

Zhou F, Liu Y, Guo Huai-Cheng, et al. Preferred method for safe medical waste treatment technology[J]. Environmental Science,2006(6): 1552–1556.

Chapter 3
Source Separation and Reduction of Medical Waste

Medical institutions are the source of medical waste generation, the starting point of medical waste life cycle management, and an important link in the total life cycle management of medical waste. In February 2020, the National Health and Health Commission and the Ministry of Ecology and the Environment and other 10 departments jointly issued a "Notice on the Issuance of a Comprehensive Management of Medical Waste Program". The notice proposed that by standardizing the classification and clear process, the formation of a waste management system in each medical institution of sorting, collecting, storing, handing over and transferring, to ensure that medical institution waste should be divided and traceable. This chapter discusses the management measures of medical waste reduction and segregation in the context of relevant regulations, standards and policies in the field of medical waste management, drawing on domestic and international experiences, and briefly analyzes the economic benefits.

3.1 Medical Waste Source Reduction and Classification Management Basis

3.1.1 Source Reduction of Medical Waste

About 80% of the waste generated by medical institutions belongs to non-hazardous general household waste, mainly including various wastes generated in office and living areas, daily wastes generated in medical and nursing living areas, doctors' offices in non-communicable disease treatment areas, nurses' stations and general patients, packaging wastes generated in pharmacy and logistic support warehouses, etc. Only about 20% of the waste generated by medical institutions is medical waste

Y. Chen et al., *Innovative Disposal Technology and Management Practice for Medical Waste*, https://doi.org/10.1007/978-981-99-6786-5_3

and waste that is included in medical waste management. A person without professional training, it is generally difficult to clearly distinguish the boundaries of ordinary waste and medical waste. For example, in the "Medical Waste Classification Catalogue" issued in 2003, infectious waste is characterized as "medical waste carrying pathogenic microorganisms with the risk of spreading infectious diseases". How to determine whether the waste that has been in contact with patients is medical waste or other, requires professional training. Therefore, medical institutions in the pre-service training of medical staff, most of them include medical waste classification and collection and other related content. According to the "National Hazardous Waste List" (2016 version), Article 3, medical waste is a hazardous waste, once the domestic waste mixed with medical waste, according to the principles of medical waste management, this ordinary garbage will also be treated as medical waste management and disposal. Whether it is a medical institution, or medical waste transfer, disposal units and other institutions or personnel, if medical waste mixed with domestic waste, will be determined in accordance with whether there is a subjective intention and the amount of waste to determine whether the criminal penalty. Therefore, if the source separation of medical waste management is not implemented, the amount of medical waste to be disposed of will be very large. Strict separation of domestic waste and medical waste is not only necessary to fulfill the requirements of laws, regulations and relevant provisions, but also the focus of source separation of medical waste management. Good management of medical waste at source is a very effective way to reduce the amount of medical waste management (Xu and Wu 2006).

In recent years, medical institutions, while maintaining and promoting the health of the people, are using more and more items for patient treatment with the increase of inpatients and outpatients, and the types and quantities of medical waste generated by medical institutions have increased significantly. With the progress of scientific research and technology, as well as the improvement of management and processing capabilities, those medical wastes that belong to the "gray area", such as infusion bottles (bags) have been clearly classified as household waste, but still must comply with the principles of targeted and closed-loop management; there are also a small amount of non-infectious disease patient blood, body fluids, cotton balls, dressings, etc. The future management may leave the category of medical waste as in Hong Kong, China. Therefore, medical waste classification methods are time-sensitive, and the "Medical Waste Classification Catalogue" needs to be adjusted regularly. Medical institutions clearly distinguish medical waste from domestic waste at the source, and reduce the use of disposable devices and supplies at the source is fundamental to the reduction of medical waste at source (Li et al. 2006).

3.1.2 *Medical Waste Classification Management Basis*

Medical waste generated by medical institutions is divided into five categories according to the "Medical Waste Classification Catalogue": infectious waste, pathologic waste, injury waste, pharmaceutical waste and chemical waste. Medical waste generated by medical institutions can be divided into the following types according to the material: polymer waste, glass waste, fabric waste, metal waste, pathologic waste, pharmaceutical waste, chemical waste, radioactive waste and other waste.

(1) Polymer waste

Polymer wastes mainly include plastic, latex and rubber wastes. Common sources of polymer waste generated within medical institutions include disposable syringes, disposable infusion sets, disposable blood transfusion sets, disposable speculums, disposable gloves, latex gloves, drainage tubes, blood lines, dialyzers, suction tubes, oxygen tubes, ventilator and anesthesia machine lines, urinary catheters, disposable masks, caps, disposable surgical gowns, disposable isolation gowns, disposable protective gowns, disposable layettes and padding sheets, medical waste special bags and sharps boxes, etc.

Polymer waste plastics mainly include polyethylene, polystyrene, polyurethane and polyvinyl chloride four chemical substances, including polyethylene material plastic waste accounted for the largest proportion. Four major medical plastic chemical composition and properties are shown in Table 3.1. The main common plastic medical waste and the corresponding raw material components are shown in Table 3.2.

(2) Glass waste

The common glass wastes in medical waste are mainly slides, cover slips, glass test tubes, glass ampoules, etc.

(3) Fabric waste

Table 3.1 Chemical composition and properties of 4 main medical plastics/wt%

Composition and properties	Polyethylene	Polystyrene	Polyurethane	Polyvinyl chloride
Moisture	0.2	0.2	0.2	0.2
Carbon	84.38	86.91	63.14	45.04
Hydrogen	14.14	8.42	6.25	5.6
Oxygen	0	3.96	11.61	1.56
Nitrogen	0.06	0.21	5.98	0.08
Sulfur	0.03	0.02	0.02	0.14
Chlorine	0	0	2.42	45.32
Ash	1.19	0.45	4.38	2.06

Table 3.2 Common plastic medical waste and the corresponding raw material components

Raw material composition	Common plastic medical waste
Polyethylene	Syringes, catheters, cannulae, catheters, blood transfusers, infusion sets, etc.
Polypropylene	Syringes, non-woven masks, gloves, surgical gowns, infusion bottles, etc.
Polyvinyl chloride	Catheters, intubation tubes, urinary catheters, blood transfusion sets, infusion sets, infusion bottles, infusion bags, plasma bags, examination tools, treatment tools, etc.
Polyethylene terephthalate	Non-woven fabrics, hemodialysis products, etc.

Common components of fabric waste include cotton balls, swabs, drainage tampons, gauze, urine pads, bandages, cotton fiber-based dressings contaminated with patient blood or body fluids; discarded contaminated surgical gowns, isolation gowns, etc.

(4) Metal waste

Common components of metal waste include medical needles, suture needles, scalpels, scalpel knives, surgical saws, skin preparation knives, dentistry forceps, probes, guide wires, steel nails, and discarded surgical instruments.

(5) Pathologic waste

Common components of pathologic waste include: surgical and other treatment processes generated by the waste human tissue, organs; medical laboratory animal tissues, cadavers; pathological sections after the waste of human tissue, pathological wax blocks, etc.

(6) Pharmacological waste

Common components of pharmaceutical waste include: bulk discarded drugs; discarded cytotoxic and genotoxic drugs; vials, infusion bottles and gloves with cytotoxic and genotoxic drug residues, etc. (Rao and Wankhede 2012).

(7) Chemical waste

Chemical wastes include waste solid, liquid and gaseous chemicals, which mainly originate from the diagnosis, cleaning, sterilization and maintenance of medical institutions. Chemical wastes include the following:

(1) Formaldehyde. Formaldehyde is an important source of chemical waste in medical institutions and is often used to clean or disinfect equipment and preserve specimens in pathology, anatomy, diagnostic, embalming, and nursing departments. Formaldehyde is an infectious waste in liquid form.
(2) Waste developer and fixer. Fixing solution mainly contains 5% to 10% of hydroquinone, 1% to 5% of potassium hydroxide and less than 1% of silver; developer mainly contains 45% of glutaric aldehyde; acetic acid in the developer, fixing solution are used. Because of the development of imaging technology, the current use of such chemicals is relatively small.

(3) Solvents. Wastes containing solvents originate from several departments of the hospital such as pathology rooms, autopsy rooms, and tissue laboratories. Solvents used in hospitals include halogenated compounds such as methylene chloride, chloride, trichloroethylene, refrigerants, and non-halogenated compounds such as xylene, methanol, acetone, isopropyl alcohol, toluene, and cyanogenic methane.
(4) Ethylene oxide. It is a gas waste, because it does not damage the sterilized items and has strong penetrating power, so most of the items that should not be sterilized by high temperature and pressure, immersion type sterilization can be sterilized and sterilized by ethylene oxide. Such as some medical devices, documents, lumpectomy, etc. It is mainly used in sterilization supply centers (rooms) of medical institutions.
(5) Chemical disinfectants. Waste peroxyacetic acid, glutaraldehyde, o-phthalaldehyde, etc.
(8) Heavy metal waste

Medical wastes containing high concentrations of heavy metal wastes are hazardous chemical wastes, and such wastes are usually highly toxic. Common heavy metals in medical waste is mainly mercury, from the spillage of broken medical equipment such as mercury sphygmomanometers, mercury thermometers and residues from the use of silver amalgam or silver mercury capsules in dentistry. In addition, it also includes cadmium-containing waste from discarded batteries in medical institutions.

(9) Radioactive waste

Radioactive waste is the medical waste generated in the application of radionuclides in the practice of medicine, the radioactivity exceeds the value of the national regulations, the waste includes the radioactive substances used in medical services and the release of medical waste.

Medical waste generated by medical institutions and related medical laboratories is a unidirectional waste stream from source to final disposal. After medical waste enters the waste stream, it irreversibly passes through the various links to reach the final disposal end and leaves the waste stream only after appropriate treatment. In this process, the earlier the waste is disposed of, the more likely it is to reduce the cost of management; at the same time, the less medical waste enters the waste stream, the lower the cost of management and treatment of disposal and environmental hazards. In a narrow sense, reducing waste generation at the source of medical waste generation; and reducing toxic and hazardous substances discharged into the environment at the end of disposal are effective reduction measures. From the analysis of, choosing to use recyclable items is the key to quantification, and using safe and non-toxic items instead is also an effective quantification measure.

3.2 Medical Waste Source Reduction and Sorting Management Measures

3.2.1 The General Idea of Classification and Reduction of Medical Waste and the Principle of Source Classification

3.2.1.1 The General Idea of Medical Waste Classification and Reduction

At present, China's medical waste classification and collection management is not standardized. Due to the lack of regulatory measures in place, there is a mix of medical waste and domestic garbage storage, different types of medical waste mixed phenomenon, causing greater environmental risks to the subsequent disposal. Rural and remote areas of medical waste collection, transfer, treatment capacity is weak, there is no classification and segregation of storage, low frequency of transfer, long storage time, illegal treatment and disposal of their own problems.

After a comprehensive analysis, it can be seen that the current medical waste disposal and management model in China reflects the following characteristics.

(1) In the planning and construction of the implementation of centralized disposal, rational layout of the general idea. China's implementation of centralized disposal of medical waste, the implementation of the city as a planning unit construction ideas. Each district city should build a centralized medical waste disposal facilities, taking into account the geographical distribution, service population and other factors set up regional collection, transfer or treatment and disposal of medical waste facilities, to achieve each county (city) are built medical waste collection and transfer disposal system. Encourage the development of medical waste mobile disposal facilities and pre-treatment facilities, to provide local disposal services in remote areas. Through the introduction of new technologies, updated equipment and facilities and other measures to optimize the disposal method, make up for the shortcomings, significantly improve the disposal capacity of existing centralized medical waste disposal facilities, the disposal of all types of medical waste standardized. Explore the establishment of a collaborative mechanism and interest compensation mechanism for the centralized disposal of medical waste across regions.

(2) Reflect the basic concept of risk management in risk control. Medical waste has a variety of characteristics such as infectious, cytotoxic and genotoxic. Medical waste from generation to final disposal should be carried out in a fully enclosed state, and the implementation of human and environmental isolation; implementation of medical waste from generation, classification and collection, warning labels, sealed transport, temporary storage to the transport unit or disposal unit handover, transfer, storage, harmless disposal of the overall process, that is,

from the "cradle" to "grave" of each link to implement the whole process of strict management and control.

3.2.1.2 The Principle of Source Classification of Medical Waste

The issue of medical waste management within the health care institutions is crucial: on the one hand, it is related to the ability to effectively eliminate the infectious nature of medical waste and reduce the harm to humans and the environment; on the other hand, different medical waste treatment and disposal technologies require the implementation of the corresponding scientific classification within the medical institutions. The source classification of medical waste should adhere to the following principles.

(1) Adhere to the principle of risk control

Medical waste has a variety of hazardous waste characteristics, the implementation of scientific classification is conducive to better eliminate or cut off the spread of disease-causing microorganisms. Prohibit medical waste mixed with household waste, cut off the pathway of transmission of pathogenic microorganisms, standardized classification and collection, tight sealing can avoid the health and environmental risks caused by the spread of pathogenic microorganisms in medical waste. When the bag containing medical waste is contaminated or suspected of contamination, the contaminated area can be disinfected or add another layer of medical waste bags. It can be said that the core task of medical waste management and disposal is: on the one hand, to implement effective infection control at the source of medical waste; on the other hand, to take into account the applicability of subsequent treatment and disposal technologies, to provide conditions for the treatment and disposal of medical waste, especially toxic and hazardous substances.

(2) Adhere to the principle of convergence with medical waste disposal technology

Different medical waste disposal technologies have different scope of application, so the source classification of medical waste should take into account the advantages and shortcomings of subsequent medical waste disposal technologies to ensure that the source classification of medical waste is compatible with subsequent disposal technologies. Classification is important if there are multiple means of waste disposal such as incineration, non-incineration, landfill, etc. Waste from chemotherapy processes, mercury, volatile and semi-volatile substances and other hazardous chemical substances cannot be treated by non-incineration technologies. Only well-segregated medical waste can be treated harmlessly using the appropriate treatment technology to achieve good results, cost savings and environmental friendliness.

(3) Adhere to the principle of medical waste reduction at source

Medical waste and domestic waste should be correctly distinguished, and the waste that belongs to domestic waste should not be mixed into medical waste, nor should medical waste be mixed into domestic waste. According to WHO research results, only 10–25% of the waste generated by medical institutions is medical waste. In fact, there may be a problem that all waste generated by medical and health institutions is often understood as medical waste, which directly causes a significant increase in the amount of medical waste that needs to be treated and disposed of, which not only increases the burden of subsequent medical waste treatment and disposal, but also causes secondary pollution, especially the incineration process provides a source for the generation and emission of more pollutants such as dioxins/furans.

(4) Adhere to the principle of adaptation of classification methods and corresponding treatment techniques for emergency management during major infectious disease outbreaks

During an epidemic, the number of pathogen carriers increases significantly, and medical and quasi-medical activities for epidemic prevention are strengthened accordingly, resulting in an increase in medical waste generated by existing medical institutions; moreover, temporary medical institutions (e.g., out-of-hospital medical sites), centralized isolation sites, and other quasi-medical sites also add a large amount of medical care waste, resulting in a surge in medical waste volume. The redundancy level of medical waste treatment capacity in most regions of China cannot withstand the impact load during this period, and it is necessary to use emergency treatment methods to dispose of medical waste generated during the epidemic in a timely manner.

In order to promote the source management of medical waste, experts and scholars at home and abroad have carried out a lot of research work, hoping to promote the source management of medical waste through a combination of administrative management and technical methods. Therefore, we should first start from the source, and comprehensively promote the management of medical waste by reducing the application, comprehensive utilization, classification and recycling measures. When considering the management of medical waste, on the one hand to achieve harmless and reduction, with due consideration to resource; on the other hand, we must also consider the type of medical waste that can be disposed of at a later stage, to regulate the management of classification (Malkan 2005). This requires the classification of medical waste generated by medical institutions, and the development of medical waste management methods at source, starting from fine management. As shown in Fig. 3.1, the source management of medical waste is the starting point of medical waste lifecycle management, and its importance is beyond doubt. Combining the policies, regulations and standards in the field of medical waste management in China, and drawing on foreign experience, the analysis of medical waste classification, treatment and disposal and quantification loci in China is shown in Fig. 3.2.

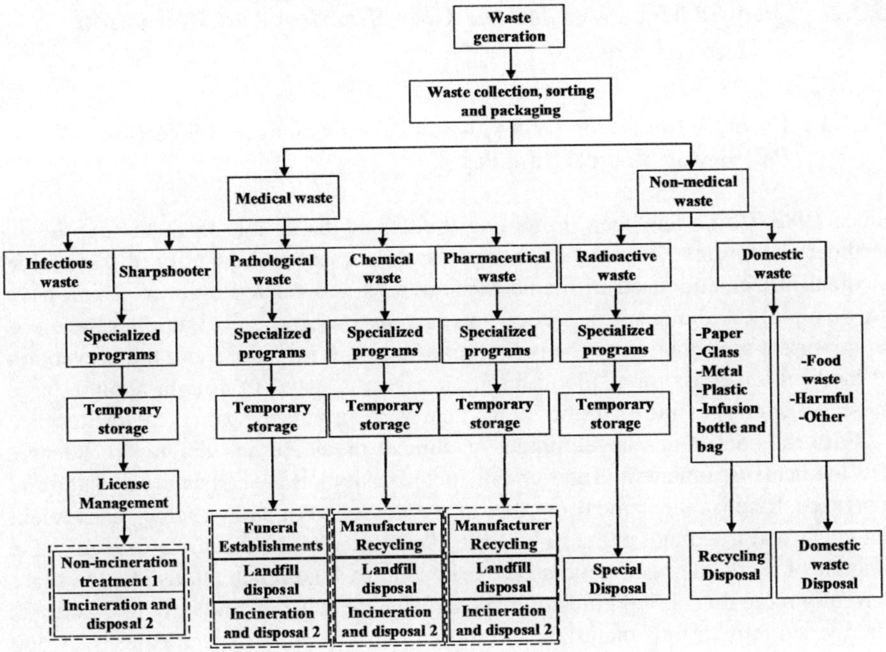

Fig. 3.1 Medical waste source management process

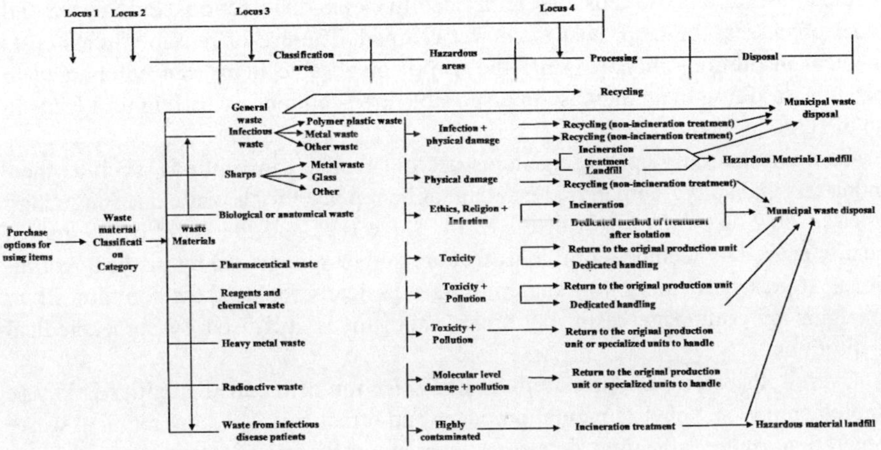

Fig. 3.2 Separation of medical waste, disposal and reduction site analysis

3.2.2 *Specific Measures for the Classification and Reduction of Medical Waste*

3.2.2.1 Promote Reduction by Adjusting Procurement and Selection Policies for Medical Supplies

Since 1987, China has been promoting the use of disposable medical supplies in medical institutions. Over the past 20 years, the variety and quantity of disposable medical instruments used in medical institutions and related units in China have grown rapidly. Although the use of various new disposable medical supplies has made an important contribution to the reduction of hospital infections and the prevention of medical transmission of blood-borne diseases, it has also brought about a large increase in the types and quantities of medical waste produced by medical institutions.

With the continuous development of clinical medicine and diagnostic science, new medical instruments and equipment, medical supplies are constantly updated and increased. Disposable medical supplies in the protection of patient safety, convenient diagnosis and treatment operations at the same time, but also caused a surge in the amount of medical waste generated. According to incomplete statistics, there are currently more than 1,500 kinds of disposable medical items, and only infusion sets are divided into various materials and more than 20 kinds of specifications. At the same time, there is the phenomenon of patients actively asking and requesting the use of disposable medical items. In this regard, there are reasons such as charging policies, medical institutions use reusable items need to clean, disinfection, sterilization and other processes, the labor cost is not yet charged. The use of reusable items costs medical institutions money, while the use of disposable items can be charged to patients, resulting in an increase in disposable medical items, which in turn leads to an increase in medical waste.

Common mercury-containing items used in medical institutions, such as thermometers and sphygmomanometers, cannot be replaced for a while due to the high price of new electronic instruments of the same type and the stability of product quality caused by technical problems that are not yet recognized by medical institutions. At present, dental filling materials have basically replaced the common silver amalgam by composite resin, but silver amalgam is still used by some medical institutions.

Overall, therefore, the procurement and selection of medical supplies is key to implementing best environmental practices and effectively reducing medical waste generation. At the same time, decisions on procurement and selection methods come with both pros and cons, and such trade-offs need to be justified and tested, and therefore should be carefully considered. To the extent that knowledge and experience are available, the following areas are of primary concern for medical supply procurement and selection policies:

(1) Non-invasive external examination tools can be reused items, such as stainless steel tongue depressor;
(2) Non-invasive in vivo examination equipment can be reused a limited number of times under the premise of strict sterilization and ensuring safety;
(3) In the premise of strict disinfection and safety, choose the feasible type of reuse of metal instruments;
(4) Use more paper materials, less polymer materials and chlorine-containing products;
(5) Items that are only slightly contaminated should be made of reusable materials as much as possible and reused after disinfection;
(6) Reduce unnecessary obsolescence and excessive replacement frequency of medical instruments and reagents;
(7) Try to use disinfectants that are highly effective and less toxic and can be easily decomposed in the environment;
(8) Purchasing drugs according to actual needs and reducing unnecessary stockpiling.

3.2.2.2 Promote the Reduction of Medical Waste and Domestic Waste Through the Development of Differentiation Standards

The clear classification of waste generated by medical institutions and the scientific distinction between domestic waste and medical waste is one of the key links in the implementation of medical waste reduction, and is also a prerequisite for the environmentally sound treatment of medical waste and the recycling of domestic waste resources. The criteria for differentiating medical waste and domestic waste should be considered from the following four aspects:

(1) The amount of patient blood, body fluids and secretions contained in the waste, after giving the measurement standard, the waste below this amount is not considered as medical waste; waste contaminated with blood, body fluids and secretions of patients with non-infectious diseases, basically no risk of disease transmission, theoretically can not be counted as medical waste.
(2) The waste contains drugs, chemicals and reagents and other items, according to whether the drugs and chemicals are toxic and harmful substances, as the basis for measuring whether they are medical waste. In order to improve operability, the common and clear understanding of whether the items have harmful effects can be judged first, while the items that have harmful effects or do not understand their harmful effects are treated as medical waste.
(3) Judgment is made based on the location of the waste generation site. For example, disposable mattresses, shoe covers, gloves, etc. generated in general wards can be treated as ordinary household waste.

(4) Judgment based on the situation of patients who produce waste. The waste generated by ordinary patients, only stained with blood, body fluids, secretions below the prescribed standard, can be considered according to ordinary garbage treatment; medical institutions admitted to the isolated infectious patients generated by domestic waste, should be managed and treated in accordance with medical waste.

3.2.2.3 Reduce the Amount of Packaging Supplies, Promote the Reduction of

According to the relevant regulations on medical waste management, medical waste generated in the place, medical waste must be collected in yellow medical waste bags or sharps boxes according to the category of "Medical Waste Special Bags, Containers and Warning Signs Standards" (HJ 421–2008) before in-hospital transportation and temporary storage. Due to the vast territory of China, the level of economic development varies around the country, and the treatment and disposal of medical waste is relatively single, resulting in a disconnect between the classification of medical waste within medical institutions and the later treatment and disposal, which has caused an unreasonable increase in medical waste packaging.

1. Specific bad performance

(1) According to the "Medical Waste Management Regulations" and related regulations, medical institutions should collect medical waste in five categories: infectious, pathological, injurious, pharmaceutical and chemical. A sample survey of medical institutions in some provinces and cities in China shows that some hospitals have increased the classification to more than 5 categories, such as infectious waste collected by infusion sets, syringes, and other categories. The later treatment and disposal for incineration and high-temperature steam, chemical disinfection, microwave disinfection and other disposal methods, there is no resource recycling methods and regulations, so that some of the early classification is invalid, while the required packaging containers can not be reduced due to the existence of classification. In particular, some regions require the use of sharps box time, so that the sharps container has not yet reached 3/4 full to be charged, not only increased medical costs, but also increased the amount of medical waste generated and processed.

(2) "Medical waste management in medical and health institutions" stipulates that the temporary storage of medical waste should not exceed 2d, so the collection and transportation personnel must be transported from the place of generation of medical waste to the medical waste temporary storage in medical institutions every day. Because most of the medical waste collection buckets are fixed location, fixed specifications, bag specifications are relatively single, medical waste is collected and transported once a day at the place of generation, medical waste sometimes does not reach the filling volume to 3/4, that is, cleared away. The survey results show that most of the wards each bag of medical waste weighs

less than 4 kg, the volume is less than 1/2 of the volume of the bag, resulting in the obvious waste of bags.

(3) The sharps boxes currently used in medical institutions are all made of plastic and are single-use. If the sharps boxes for the damaged waste can be opened and closed professionally by the disposal unit, disinfected and cleaned, they can be reused; for the glass ampoules and other waste, the use of cartons and plastic bags can be sealed to reduce the use of sharps boxes, which can not only reduce medical costs, but also effectively reduce the emission of polymer medical waste.

(2) Measures

(1) According to the local post-processing and disposal methods, determine the corresponding classification mode, merge similar items, and reduce unnecessary packaging.

(2) Use different sizes of medical waste collection buckets with different sizes of medical waste packaging bags to match the amount of medical waste generated each day, reduce the volume of ineffective and reduce the amount of packaging waste.

(3) Replace plastic sharps boxes with reusable plastic or metal sharps boxes, and use cardboard boxes for non-infectious sharps such as glass ampoules while ensuring a tight seal.

3.2.2.4 Reduce Inappropriate Handling and Disposal of Chlorine, Mercury, Moisture, Metallic Materials, and Organic Polymer Materials

Some studies have shown that the presence of chlorine and metallic elements such as copper, iron and aluminum during incineration is one of the important factors in the generation of POPs such as dioxins (Chen et al. 2008). Medical waste containing chlorine in medical institutions mainly comes from polyvinyl chloride (PVC) plastics, chlorine-containing disinfectants and chlorine-containing pharmaceuticals. Therefore, the implementation of strict classification management measures to exclude such items from the incineration of waste is an operationally feasible and effective way to reduce the generation of dioxins (Pulles et al. 2005). In addition, reducing the amount of mercury-containing medical waste disposed of at the source is one of the most important ways to ensure that this type of waste does not pollute or harm the environment and humans. Effective measures include the following:

(1) Establish a procurement policy to reduce the purchase of medical supplies made from PVC, and use non-incineration methods for medical waste containing PVC;

(2) Chlorinated disinfectants, whether they are expired products or used residues, should not enter solid medical waste as much as possible and can be disposed of in sewage treatment ponds;

(3) Strictly implement separate collection, according to the list of chlorine-containing drugs and mercury-containing medical devices provided by the relevant departments, such waste drugs will be collected separately and handed over to the corresponding qualified disposal institutions;
(4) After pressure steam sterilization of metal waste, recycling, reuse or landfill disposal.

Some of the water in solid medical waste is generated during the use of medical supplies, some are unconsciously brought in during the operation, and some are artificially put in due to the lack of knowledge of medical waste management by health care workers and operators. The use of incineration of medical waste, requiring low water content of waste, the calorific value of waste greater than 18,600 kJ/kg, can maintain their own combustion, more than 6,000 kJ/kg, there is a high value of heat recovery, less than 3,000 kJ/kg, there is no value of heat recovery (Jangsawang et al. 2005). A study shows that the comprehensive calorific value of medical waste in China, due to the high water content, most of which is less than 2500 kJ/kg, the need to add fuel oil or coal briquettes to fuel combustion. This situation actually indirectly increases the amount of waste treatment, aggravates the possibility of dioxin emissions, and also increases the cost of treatment. Therefore, effective measures should be developed to reduce the introduction of moisture in solid medical waste.

3.2.2.5 Recycling of Renewable Resources to Achieve Reduction

The substances with potential recycling value in medical waste are shown in Table 3.3.

At present, for the use of non-incineration and other technologies for the disinfection of recyclable plastic medical waste, according to foreign practices (Kasane 2007), after disinfection and environmentally sound treatment as a recycling resource can be reused. Through recycling, reduce the consumption of raw materials, in order

Table 3.3 Substances with potential recycling value in medical waste

Material	Examples of common species	Total weight percentage/wt%	Potential recycling value
Polymers	Infusion sets, syringes, etc.	30–35	Wire sheathing, shoe soles, door curtains, agricultural storage baskets, etc.
Metal Materials	Needles, blades, caps, etc.	2–3	Melting into raw metal
Glass Products	Infusion bottles, medicine bottles, etc.	7–9	Melting into raw materials for glass products
Biomass	Cotton balls, gauze, dressings, etc.	30–35	Energy Recovery

to fundamentally solve the problem of dioxins generated by the source of chlorine, which has reached a clear consensus in the world.

Before recycling polymeric materials, metallic materials and glass products, such wastes must be rendered harmless using non-incineration treatment technologies. The objectives of the treatment should meet the following requirements:

(1) To achieve the disinfection effect, i.e. to achieve the effect of killing pathogenic microorganisms;
(2) No aerosol leakage or emission of germ-carrying gas that exceeds the standard or contains pathogenic microorganisms at the time of treatment;
(3) No excessive VOC emissions;
(4) No significant physical and chemical property changes in the material;
(5) The material can be purified and homogenized to meet the requirements of recycling;
(6) Processing costs are within acceptable limits.

Medical waste management and treatment and disposal using the aforementioned methods, 39 to 47% of the waste materials can be recycled, and only 53 to 61% of the waste requires incineration or other technologies for final disposal, which can greatly reduce the terminal emissions of medical waste.

3.3 Source Reduction Economic Benefit Analysis

3.3.1 Expected Effects of Medical Waste Minimization

The potential contribution of different components of medical waste to dioxin production when incinerated is different. On a unit weight basis, polymeric materials contribute the most to dioxins and biomass the least. The use of non-incineration technologies for the treatment of medical waste without resource recycling can also result in the waste of potential resources by encroaching on a large amount of agricultural land or other land in landfills.

There are four possible reduction sites in the total medical waste process, through the implementation of source separation and reduction of medical waste, the expected effect of 65% reduction in the total amount of medical waste can be achieved. Four points of medical waste source reduction analysis is shown in Table 3.4.

Combined with Table 3.4, according to the national medical waste total of 506,000t accounting, site one emission reduction involves production, pharmacy, clinical, price and other departments, although it is a potential emission reduction site, but requires the coordination of multiple national authorities, so the amount of emission reduction can not be precisely estimated, the expected emission reduction target of 3%, 15,200 t reduction. Site two reduction can be achieved when the internal use of BEP management is adopted in medical institutions, by strictly separating domestic waste and medical waste, which can contribute 6.5% reduction, and liquid into the

Table 3.4 Source reduction analysis of medical waste

Volume reduction sites	Reduction category	Current Situation	Optimization measures	Reduction/%
Locus I	1. Choose items that can be reused after disposal and reduce disposable items 2. Reduce the items that are prone to dioxin generation during incineration	No consideration	Give full consideration to	
Locus II	1. Strict separation of domestic waste and medical waste 2. Control of liquids into solid waste	There are shortcomings in the classification	1. Strictly distinguish between domestic waste and medical waste 2. Liquid enters the wastewater treatment system	15
Locus III	1. Medical waste packaging 2. polymer waste, metal waste, glass waste	Incineration	1. Reduce ineffective packaging 2. recycling after non-incineration treatment	50
Locus IV	Various types of waste suitable for non-incineration treatment	Incineration	Non-incineration treatment	

Note The total amount of national medical waste is calculated as 506,000 t. The total reduction of 65% in the table, the total weight reduction of 328,900 tons

wastewater treatment system, which can contribute 8.5% reduction, totaling 15%, 75,900t.

The reduction at site three requires national policy support and appropriate treatment equipment; if implemented, the reduction effect is significant, up to 50%, 25.3 million t. According to Table 3.4 According to the data shown in Tables 3.4, the percentages of polymer materials, metal materials and glass products in medical waste are different, and the experiments in several hospitals in China show that the reduction effect of site IV is fully achievable (Chaerul et al. 2008). The reduction effect of locus IV is limited to the reduction of the proportion of medical waste incineration and the resulting reduction of dioxin emissions, but not on the total emissions of medical waste reduction effect. The reduction percentage of the four loci is greater than 65%, and the total reduction is not less than 328,900 t.

3.3.2 *Expected Reduction Effect of Medical Waste Dioxins*

Based on the total annual medical waste volume of 506,000 t, if 80% of the final treatment is incinerated, the emission factor is 3000 μg TEQ/t, and the annual emission is 1214.4 g TEQ. The reduction of dioxin emission can be promoted by adopting standardized medical waste classification and reduction measures. The expected reduction effect of medical waste dioxins is shown in Table 3.5.

As can be seen from Tables 3.5, medical waste accounted for 33% of the total amount of polymer materials, 166.98 million t, with a reduction contribution of 450.846 gTEQ; 2.5% of metal materials, 12.65 million t, with a reduction contribution of 37.950 gTEQ; 8% of glass materials, 40.48 million t, with a reduction contribution of 121.440 gTEQ; 33% of biomass, 166.98 million t The study showed that the emission factor of dioxins when incinerated is more than 300 times lower than that of polymer materials under the same conditions, at 10 μgTEQ/t, with an annual emission of 1.336 gTEQ; other wastes are mainly residual liquid as the main component, most of which are discharged from the sewage system, calculated at 40% into the incineration treatment, a total of 47,560,000t need to be treated, the dioxin emission factor of such substances is unknown, It can be seen that the BAT/BEP scheme can reduce 1162.494gTEQ per year with an annual treatment capacity of 506,000t, and the reduction rate reaches 95.23%, which is a scientific and advanced environment-friendly medical waste It is a scientific and advanced environment-friendly medical waste source separation and reduction program.

Table 3.5 Expected reduction effect of medical waste dioxins

Category	Item	Quantity/ million t	Emission factors (TEFs)/ (μgTEQ/t)	Incineration ratio /%	Total emissions/ gTEQ
Current Situation	Total medical waste in 2005	50.60	3000	80	1214.400
Implementation of BAT/BEP Program	Polymers	16.698	3000	10	50.094
	Metal Materials	1.265	None	10	0
	Glass material	4.048	None	10	0
	Fabric	16.698	10	80	1.336
	Other waste (residual liquid)	11.891	10	40	0.476
	Total	50.60			51.906

Note ① In the total amount of medical waste, the polymer material is 33%, metal material 2.5%, glass material 8%, fabric 33%, other waste (residual liquid) 23.5%; where the emission factor of other waste is unknown, temporarily calculated by the emission factor of fuel combustion 10μgTEQ/t. ② In the table, the reduction was 1162.494 g TEQ, and the percentage reduction was 95.23%

From Table 3.5, it can be seen that the medical waste of polymer materials accounted for 33% of the total, 166.98 million t, with a reduction contribution of 450.846 g TEQ; metal materials 2.5%, 12.65 million t, with a reduction contribution of 37.950 g TEQ; glass materials 8%, 40.48 million t, with a reduction contribution of 121.440 g TEQ; biomass 33%, 166.98 million t. The study shows that the emission factor of dioxins in incineration is more than 300 times lower than that of polymer materials under the same conditions, which is 10ug TEQ/t, with an annual emission of 1.336gTEQ; other wastes are mainly residual liquid as the main component, most of them are discharged from the sewage system, calculated by 40% into the incineration treatment, a total of 47.564 million t need to be treated, the dioxin emission factor of such substances is unknown, tentatively calculated by fuel emission factor 10ug TEQ/t, the annual emission of 0.476 g TEQ, the above total 51.906 g TEQ. It can be seen that with the adoption of BAT/BEP program, the annual treatment capacity of 506,000t can reduce 1162.494 g TEQ per year, with an emission reduction rate of 95.23%, which is a set of scientific and advanced environment-friendly medical waste source separation and reduction program.

3.4 Source Classification of Medical Waste Information Technology and Applications

3.4.1 Medical Waste Source Separation Technology Needs

Source separation of medical waste is often not a single, in consideration of source separation at the same time, but also give full consideration to the applicability of the classification technology. The implementation of practical source separation in order to avoid risks, reduce secondary pollution, and achieve safe management and disposal of medical waste without harm (Council and Ebrary 2007).

The whole process of medical waste treatment and disposal mainly includes the generation of medical waste in medical institutions, collection, storage processes and centralized medical waste disposal units for medical waste collection and disposal process two major parts.

Medical institutions within the medical waste collection from the source of classification, transfer and temporary storage, the need to register the source of medical waste, type, quantity or weight, handover time, the final destination, and the signature of the operator. Centralized medical waste disposal units to medical institutions to collect and transfer, also need to register the source of medical waste, type, quantity or weight, handover time, treatment and disposal methods, the final destination, and the signature of the operator.

(1) Information technology needs of medical institutions

At present, China's medical institutions as the main source of medical waste generation, most of the purely manual operation to achieve internal medical waste handover, transfer, registration and other aspects of management, the traditional form of handover registration is usually unable to meet the requirements of modern hospital management, but also not conducive to the timely management of medical waste collection and treatment and disposal information. If the personnel operation is not standardized, it is difficult to trace the responsible person, once the loss of medical waste is more difficult to achieve the source of traceability, specifically in the following five areas:

(1) Packaging. Label randomly without barcode, packaging containers are not tightly closed.
(2) Weight. Each department is equipped with a spring scale, but the weighing error is large.
(3) Recording. Manual records on paper forms, incomplete records, illegible handwriting, and difficult to keep on paper.
(4) Traceability. Statistical analysis of manually recorded data is cumbersome and difficult to trace.
(5) Personnel. The operation of medical waste collection and temporary storage is basically undertaken by outsourced property staff, and the standardized operation relies on the staff's consciousness, which is difficult for the hospital management to supervise.

(2) Medical waste disposal institutions information technology needs

The chain of medical disposal institutions includes medical waste collection units, medical waste transportation units and medical waste disposal units. The level of information management of different institutions varies widely. Its information technology needs mainly lies in the management of medical waste handover, transfer, transport, in and out of storage and other aspects, similar to the existing management model of medical institutions, the vast majority of purely manual operation and paper file management, the same problems exist if the operator is not standardized, there are errors in the weighing of waste, in and out of storage ledger confusion, resulting in the loss of medical waste traceability difficulties, the bottom of the disposal ledger is unclear. In addition, medical disposal institutions will be poor or delayed due to poor information channels, resulting in medical waste can not be collected and transported in a timely manner, the low frequency of transfer, long storage time, illegal self-disposal and other issues.

3.4.2 *Principle of Medical Waste Sorting and Collection Technology*

The front-end management of medical waste classification and collection mainly achieves the goal of classification management and standardized storage. For medical waste, it is subdivided into individual wards and departments, and individual wards or departments are identified by two-dimensional code or radio frequency identification (RFID) to effectively identify the source of medical waste generation, and the

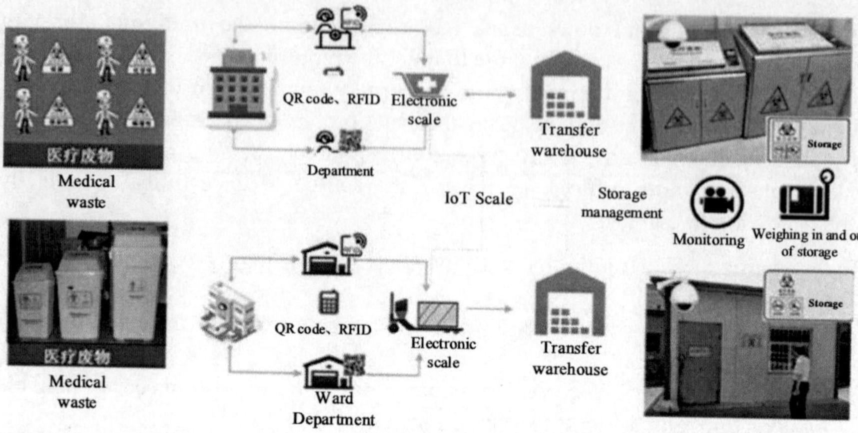

Fig. 3.3 Principle of new technology of front-end of medical waste sorting and collection

weighing and measurement of waste generated by individual wards or departments is realized by medical waste IOT scales. Medical waste IoT scale is an application of IoT weighing sensor in the field of medical waste measurement, which enables users to monitor the scale equipment, and can send the information data of the equipment itself and the generated overweight information to the background in time, providing a more efficient method than the traditional way for all kinds of users to manage the equipment and understand the production process in time. The principle of the new front-end technology of medical waste sorting and collection is shown in Fig. 3.3.

3.4.3 *Medical Waste Source Separation Technology Applications*

(1) Grid-based production and waste section

The production waste department is the starting point of the whole process of information technology supervision of medical waste, medical waste transfer boxes can be placed in each production waste department, from the production waste department on the start of the classification of sealed boxes transferred to the temporary storage room. According to the size of the hospital, the waste production department can be a department or ward, or can contain several departments. The ward with independent temporary storage room or medical waste generated by a large number of departments can be set up as a waste production department. Not set up a separate temporary storage room or a small amount of waste production departments, you can combine a number of departments into a waste production department. Outside the production waste section set up outside the production waste section intelligent signage.

Fig. 3.4 Medical waste identification code

(2) Source classification collection

Different types of medical waste collection equipment are used according to the gridded waste production departments for sorting, and electronic tags or QR codes are established for each waste placed. With the medical waste transfer box or bag as the object, establish a unique electronic label or two-dimensional code, each barcode identification number corresponds to each transfer box or bag. Using mobile intelligent data terminals to scan the code collection and identification of medical waste data, including the type of waste, waste weight, the production department, the time and personnel, etc., each link to scan the medical waste identification code as the basis for data entry, the identification code as the basis for traceability, and at the same time the relevant information real-time transmission to the background management system. Medical waste identification code is shown in Fig. 3.4.

(3) Mobile data collection

Using mobile intelligent data terminals to scan and sense the handover cards, departmental plates, etc., to collect the source, type, quantity, weight, handover time, handover personnel, etc. of medical waste. Instead of the original manual counting method, the collected data is more accurate and reliable.

For medical waste transfer, storage links, through the backstage management system of medical waste in the hospital to put early warning, notify the removal of personnel to collect in a timely manner, through the mobile terminal APP to achieve the transfer of medical waste in the hospital storage, medical waste can be traced in the hospital flow, medical waste transfer and disposal time frame of more than 48 h, issued an early warning and notify the regulatory authorities.

For out-of-hospital disposal of medical waste transfer process, transfer personnel arrive, first scan the handover information, lock the responsible person, and then scan the medical waste information, lock the type of medical waste. IOT metering system will automatically count the weight, and finally scan the information of the transfer personnel. After the medical waste arrives at the treatment and disposal enterprises through the QR code information for storage, out of storage, the system automatically

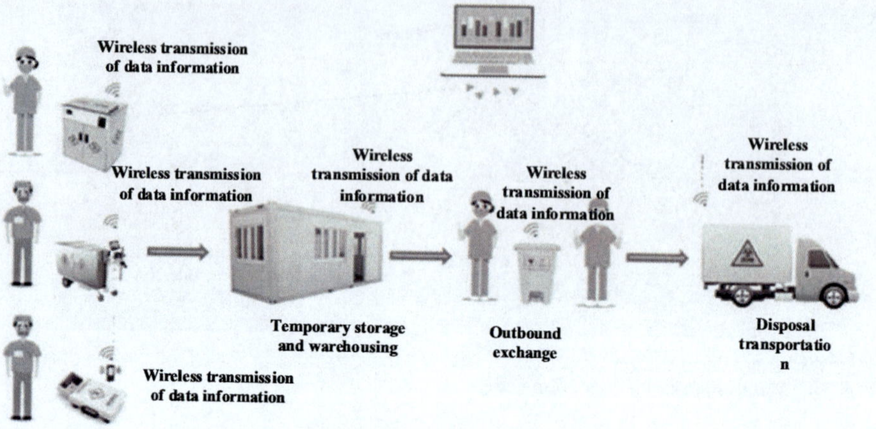

Fig. 3.5 Medical waste sorting and collection IOT application process

binds all the personnel, medical waste types, weights, time and other information of this operation together to achieve the traceability of out-of-hospital transfer disposal management. Medical waste classification and collection of the Internet of things application process is shown in Fig. 3.5.

References

Chaerul M, Tanaka M, Shekdar A V. A system dynamics approach for hospital waste management [J]. Waste Management, 2008(28): 442-449.

Chen Y, Sun YZ, Li PJ, et al. Study on compliance requirements and countermeasures for emission reduction of POPs as by-products in China [J]. Environmental Science and Technology, 2008b (12): 575-591.

Council N , Ebrary I .Review of Chemical Agent Secondary Waste Disposal and Regulatory Requirements[J]. 2007. DOI: https://doi.org/10.17226/11881.

Jangsawang W, Fungtammasan B, Kerdsuwan S. Effects of operating parameters on the combustion of medical waste in a controlled air incinerator [J]. Energy Conversion and Management, 2005(46): 3137-3149.

Kasane B. Best environmental practices and alternative technologies for medical waste management [R]. Eighth International Waste Management Congress and Exhibition, 2007.

Li R D, Nie Y F, Bernhard R. Options for healthcare waste management and treatment in China [J]. The Chinese Journal of Process Engineering, 2006(2): 261-266.

Malkan J N. Global trends in responsible healthcare waste management - a perspective from health care without harm stacy [J]. Waste Management, 2005(25): 567-574.

Pulles T , Kok H , Quass U ,et al.Dioxin Emissions in Candidate Countries[J].Tno Environment Energy & Process Innovation Tno Repoer, 2005.

Rao M S, Wankhede S. Management of hospital wastes with potential pathogenic microbes [J]. Microorganisms in Environmental Management, 2012: 365–401.

Xu Hongyu, Wu Jiuju. Problems and countermeasures of medical waste classification management [J]. Contemporary Nurses, 2006(4): 38-39.

Chapter 4
Optimized Incineration Disposal Technology of Medical Waste

Medical waste incineration disposal technology is a commonly used technology in the United States, WHO, organizations and other countries (Malkan and Nelson 2005; Xie et al. 2009; Windfeld et al. 2015; Uysal and Tinmaz 2004). China also takes the incineration method as the main disposal method, and has formulated relevant standards and technical specifications (Chen et al. 2012; Zhang et al. 2009). This chapter will describe the principle, process flow and pollution generated node of medical waste incineration disposal in detail, and analyse the medical waste incinerator and tail gas control technology. Based on comprehensive analysis, the optimization measures of medical waste incineration and disposal technology are proposed.

4.1 Principle of Medical Waste Incineration Disposal Technology

4.1.1 Principle of Pyrolysis Gasification Technology

The pyrolysis and gasification technology of medical waste is a thermochemical reaction technology (Science Times 2003; Noyes Data Corporation 1986; Zhu et al. 2008). According to the different ways of medical waste feeding, medical waste pyrolysis gasification technology can be divided into continuous pyrolysis gasification technology and intermittent pyrolysis gasification technology.

Continuous pyrolysis gasification technology means that the waste feed system continuously inputs the processed materials into the pyrolysis furnace in a certain interval and in batches, so as to maintain a continuous and stable pyrolysis process in the pyrolysis furnace. During the whole working process, the pyrolysis products generated at the exit of the pyrolysis furnace fluctuate less or are basically stable. Intermittent pyrolysis gasification technology means that the waste feeding

Y. Chen et al., *Innovative Disposal Technology and Management Practice for Medical Waste*, https://doi.org/10.1007/978-981-99-6786-5_4

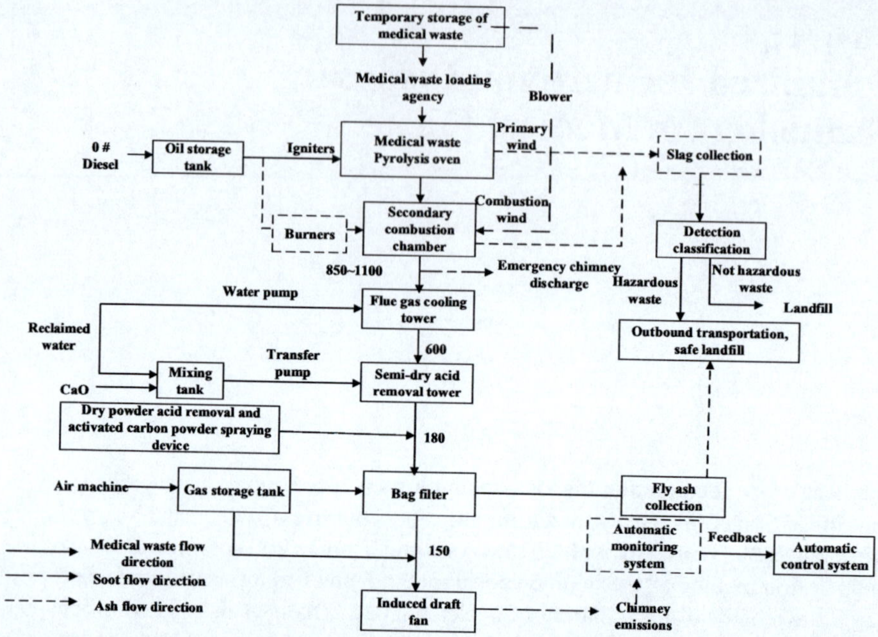

Fig. 4.1 General process of pyrolysis and gasification of medical waste

system takes a single feeding method for the processed materials, and the feeding and pyrolysis process of the pyrolysis furnace are in batch and intermittent working mode.

The feed system and the pyrolysis furnace operate in a cycle of feed → pyrolysis → ash discharge → feed → pyrolysis → ash discharge. During the whole working process, the temperature in the pyrolysis furnace and the amount of pyrolysis products produced at the outlet are fluctuating in a wave-like cycle. The general process flow of pyrolysis gasification is shown in Fig. 4.1.

(1) Medical waste pyrolysis gasification process

(1) The medical waste feeding mechanism introduce the medical waste into the pyrolysis furnace, where the decomposition process is completed. The pyrolysis gasifier uses the principle of anoxic pyrolysis to supply insufficient combustion air, so that the medical waste can be pyrolysed within a certain temperature range. The medical waste in the pyrolysis gasifier is 600–900 °C. Under the condition of oxygen, the medical waste is decomposed into short-chain organic gas, methane, hydrogen, carbon monoxide and other combustible gases. The pyrolysis gasifier also adopts the continuous operation of subsequent combustion furnace.

(2) The combustible gas after pyrolysis gasification is incinerated at 850–1100 °C in the second combustion chamber to reach the complete combustion state. The

cracked gas produced by the pyrolysis gasifier continues to be fully combusted in the secondary combustion chamber. The secondary combustion chamber is equipped with a diesel burner, the opening range or closing of which is automatically controlled by the set temperature. In the starting stage, due to the low furnace temperature, the burner needs to be turned on to heat up the secondary combustion chamber, and when the pyrolysis gas spontaneously combusts, the furnace temperature will rise rapidly and reach the set temperature (about 850 °C), and the combustion device will be automatically turned off. As the temperature in the secondary combustion chamber reaches above 850 °C, the flue gas stays at this temperature for more than 2 s, so the various harmful components in the flue gas (including dioxins) are fully decomposed and eliminated in the secondary combustion chamber.

(3) In the tail gas purification, the process of cooling, semi-dry emergency cooling and acid removal, activated carbon adsorption, bag dust removal and ash treatment are adopted. The function of the flue gas cooling system is to cool the high temperature flue gas (the second combustion chamber outlet temperature up to 1100 °C or more) to the temperature at which the subsequent flue gas treatment equipment can work normally. The subsequent semi-dry quench deacidification tower requires the flue gas temperature of the tower to be below 600 °C. The semi-dry cooling deacidification device is both used to rapidly reduce the flue gas from 600 to 180 °C, and remove the acid at the same time when the dioxin is generated at low temperature to prevent subsequent equipment corrosion. The activated carbon powder is evenly sprayed into the flue gas using a venturi device to adsorb residual dioxins in the flue gas. The bag-type dust collector intercepts the activated carbon powder, residual smoke and dust adsorbed with dioxins on the surface of the filter bag.

(4) The slag formed after burning solid waste in the pyrolysis furnace can be directly landfilled if it is not a hazardous waste after the toxicity leaching test due to high temperature treatment. The residue of the pyrolysis furnace and the soot (fly ash) collected by the flue gas purification system are hazardous waste, which can be collected in a closed area and transported to the hazardous waste disposal centre for solidification and disposal on a regular basis (Niessen 2002).

(2) Medical waste pyrolysis gasification technology features

(1) Medical waste is first decomposed to produce combustible gas under reduction conditions. The metal in the waste is not oxidized, and the copper, iron and other metals in the waste are not easy to generate catalysts in order to promote the formation of dioxin.

(2) The low air coefficient of pyrolysis gas combustion can greatly reduce the exhaust volume, improve the energy utilization, reduce NO_x emissions, and reduce the investment and operating costs of flue gas treatment equipment.

(3) The ash containing carbon is incinerated at a high temperature of about 1000 °C, which can inhibit the generation of dioxin-like toxic substances and make the generated dioxin decompose, and the slag is disinfected at high temperature to realize its recycling, which can maximize the waste reduction and resource utilization.

4.1.2 Principle of Incineration Technology

Medical waste rotary kiln incineration technology is a high-temperature thermo-chemical reaction technology and its incineration system is composed of rotary kiln and two combustion chamber (Pei and Wu 2007; Tudor et al. 2005). The rotary kiln is slightly inclined, and the kiln head is slightly higher than the tail. Oxygen-enriched burning and pouring mode is adopted in the rotary kiln, and the combustion temperature is controlled at 850–1000 °C. Medical waste enters the kiln from the kiln head. With the rotation of the kiln, the medical waste moves downward along the inner wall of the kiln, thus completing the drying, burning and cooling process. The cooled ash is discharged at the end of the kiln, and the boiled steam together with the combustion gases enter the second combustion chamber. The temperature of the second combustion chamber is maintained at 900–1200 °C. The flue gas residence time is 2 s or more to ensure that the combustible components in the flue gas reach a complete combustion state and a high degree of dioxin decomposition (Kilgroe 1996). Typical process flow of a rotary kiln is shown in Fig. 4.2.

(1) Medical waste rotary kiln disposal process
(1) The measured and tested medical waste is fed into the sorted storage room, and refrigerated in the storage room.
(2) The rotary kiln uses supplementary fuel to automatically ignite and heat up the kiln body. The medical waste is sent to the incineration plant by the closed waste transfer trolley and put into the automatic feeder of the rotary kiln. The automatic feeder is sent to the furnace mouth and sent into the chute inside the

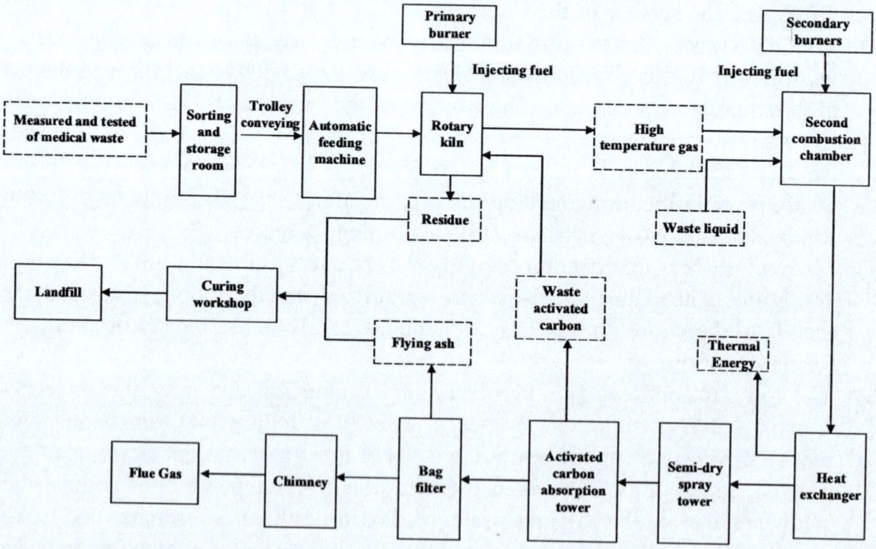

Fig. 4.2 Typical process flow rotation

furnace body. The furnace body rotates at a speed of 0.5–3 r/min (adjustable speed), and the medical waste is automatically turned over and stirred, so that the incinerator material moves slowly forward to complete the process of drying and boiling. During the process of drying, boiling, high-temperature oxidation, main combustion, burning out, slagging, and production of volatile combustible high-temperature gas, the combustion temperature in the furnace can reach about 700–850 °C.

(3) The high temperature gas generated by the boiling oxidation of the rotary kiln enter the secondary combustion chamber. The secondary combustion system uses supplementary fuel to further burn to further burn the flue gas at high temperature in the secondary combustion chamber. The combustion temperature reaches 850–1300 °C, so that the organic waste gas is fully decomposed at high temperature.

(4) The high-temperature flue gas generated from the two high-temperature incineration passes through a heat exchanger to reduce the flue gas temperature to about 600 °C.

(5) The flue gas then passes through the semi-dry spray tower to further rapidly reduce the temperature to below 200 °C, to control the flue gas in the temperature range of 200–600 °C with residence time of less than 1 s, thereby removing most of the acidic harmful gases at the same time, and also to inhibit the production of dioxins in large quantities. Thereafter, it is sent into the activated carbon absorption tower where activated carbon adsorption of dioxins and other harmful substances in the flue gas is carried out. Furthermore, it is sent into the bag filter to remove very small dust, and dioxins and other harmful substances for further interception and filtration such as waste activated carbon incineration treatment and bag fly ash curing landfill treatment.

(6) The generated clean flue gas is sent into the chimney by the exhaust system for up to standard discharge, and the incinerated ash is discharged from the lower part of the furnace body for solidification and landfill disposal.

(2) Medical waste rotary kiln incineration technology features

(1) It has a simple structure, stable control, mature technology, strong adaptability to medical waste, and can handle a variety of different shapes of solid and liquid waste.

(2) The medical waste can be fully stirred and rolled in the furnace, with good mixing effect with air and good turbulence. There is no dead corner in the furnace due to uneven distribution or too thick material layer.

(3) The medical waste tumbles forward in the kiln. Three types of heat transfer (radiation, convection, conduction) coexist in one furnace. The heat utilization rate is high, the kiln rotation speed can be adjusted, the residence time of medical waste can be controlled, and the rotary kiln and the second combustion chamber have enough space to make the medical waste burn completely.

4.1.3 Medical Waste Incineration Disposal Process and Nodes of Pollution Production

Whether it is pyrolysis gasification or rotary kiln incineration, its process generally includes functional units such as; feed unit, decomposition unit (or primary incineration), secondary combustion unit, waste heat recovery unit, residue collection unit, purification unit, water treatment unit, automatic control unit and other auxiliary units (Vesilind et al. 2002). The process flow and pollution production nodes are shown in Fig. 4.3.

It can be seen from Fig. 4.3 that secondary pollutants will be generated in the external process of medical waste pyrolysis and incineration, mainly including flue gas pollutants, dust, dioxin, SO_2, HCL, heavy metals, wastewater, noise, etc. (Coutinho et al. 2000).

The main materials and energy consumption of medical waste pyrolysis incineration disposal process are diesel, electricity and water. The specific consumption is shown in Table 4.1.

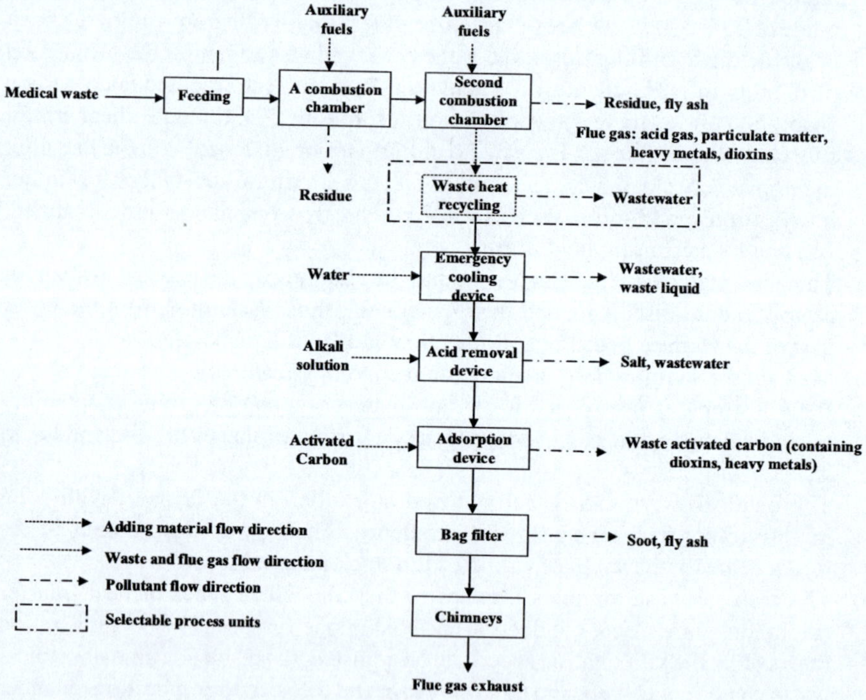

Fig. 4.3 Process Flow and Pollutant Generation Nodes of Medical Waste Pyrolysis and Incineration Treatment

Table 4.1 Material and energy consumption (in t waste)

Serial number	Name	Unit	Consumption		
			Continuous pyrolysis gasification	Intermittent pyrolysis gasification	Rotary kiln incineration
1	Diesel	kg	20–50	16.2–23.3	398–520
2	Electricity	kW h	390–520	394–456.7	300–400
3	Water	T	3.5–8	8–11.2	10–14

4.1.3.1 Medical Waste Incineration Process Pollutant Emissions

(1) Air pollutant emissions

The types and contents of pollutants in the flue gas generated by the pyrolysis of medical waste disposal have a wide range of variation (Alvim Ferraz et al. 2000; Sun et al. 2005). The flue gas usually contains pollutants such as particulate matter, carbon monoxide, sulfur dioxide, nitrogen oxides, hydrogen chloride, hydrogen fluoride, heavy metals and dioxins. The specific concentrations are shown in Table 4.2.

(2) Wastewater pollutant discharge

Wastewater pollutants refer to the wastewater discharged from the process engineering and a small amount of domestic wastewater, mainly including disinfection and rinsing wastewater from transfer vehicles, disinfection and rinsing water from crates, softening water discharge wastewater, rinsing wastewater from temporary storage places at unloading sites and refrigerated storage rooms, and domestic wastewater, etc. The specific values of the main indicators of water quality are shown in Table 4.3.

(3) Solid residue discharge

Table 4.2 Medical waste pyrolysis disposal air pollutant emissions

Serial number	Pollutant control projects	Unit	Value
1	Particulate matter	mg/Nm^3	40–120
2	Carbon monoxide	mg/Nm^3	40–120
3	Sulfur dioxide	mg/Nm^3	50–300
4	Nitrogen oxides	mg/Nm^3	30–300
5	Hydrogen chloride	mg/Nm^3	10–105
6	Hydrogen fluoride	mg/Nm^3	3.5–15
7	Heavy metals	mg/Nm^3	0.05–0.15
8	Dioxins	$ngTEQ/Nm^3$	0.1–1.5

Table 4.3 Discharge of pollutants from medical waste pyrolysis disposal wastewater

Serial number	Pollutant control projects	Unit	Value
1	pH	mg/L	6–9
2	BOD_5	mg/L	10–20
3	COD_{Cr}	mg/L	30–60
4	SS	mg/L	15–30
5	Ammonia nitrogen	mg/L	5–15
6	Number of coliform bacteria	MPN/L	100

Solid residues can be divided into two categories: incineration residue and fly ash. The incineration residue is the slag discharged from the bottom of the pyrolysis incinerator and secondary combustion furnace and its components are mainly inorganic substances such as glass. Incineration slag is harmless and can be sent directly to the municipal waste disposal site as domestic waste for treatment. The fly ash mainly includes the purification slag collected by the waste heat boiler and bag filter, which contains heavy metals and dioxins, and other pollutants which must be managed and disposed of as hazardous (waste Stockholm convention on persistent organic pollutants (POPs) [R], 2001).

(4) Noise emission

The noise is mainly concentrated in the plant and auxiliary workshops of various mechanical equipment and power facilities, such as; blowers, induced draft fans, generator sets, various pumps, air compressors and boiler safety valves, with a minimum of more than 85 dB.

4.1.3.2 Analysis on Pollution Control Technology of Medical Waste Pyrolysis

4.1.4 Pyrolysis Incinerator Type and Control Parameter Selection

In various medical waste incineration technology, according to their different working principles and combustion methods, can be divided into small single combustion chamber incinerator, mechanical grate incinerator, rotary kiln incinerator, controlled gas incinerator (CAO), two-stage pyrolysis gasification incineration (batch type), vertical pyrolysis gasification incinerator, electric arc furnace method, or a combination of technologies. There are many combustion modes and types of medical waste incinerators. According to the incineration method, there are two types of combustion methods, such as peroxide combustion and pyrolysis gasification method. The second method is the furnace type and it consist of rotary kiln type, reciprocating grate furnace, chain furnace, vertical rotary furnace, etc. Generally, there is a tail gas purification system, but the feeding system and monitoring system are relatively simple (WHO 2004).

Due to the low degree of automation and lack of a complete flue gas cooling system, the small single-combustion chamber incinerator produces secondary pollution to the environment and cannot meet the new environmental protection requirements, so it is no longer suitable for the treatment of medical waste and has been basically eliminated.

Rotary kiln incinerator technology is mature and less in secondary pollution. However, due to its large one-time investment, the operating costs for incineration of medical waste is high, and generally needs to be combined with other incineration purposes. This technology has good treatment effect, strong adaptability and stable operation, and it is suitable for the centralized treatment of medical waste in larger cities and regions.

Electric arc furnaces and plasma are more effective in treating medical waste. Its main feature is the incineration of hard-to-degrade hazardous waste (PCBs, etc.) at ultra-high temperatures, which cannot be fully exploited for medical waste yet. In addition, the high construction and operation costs also hinder their application in the field of medical waste treatment and disposal.

Relatively speaking, pyrolysis technology has developed rapidly in recent years and has been used with good results in Canada, the United States, the United Kingdom and Mexico. At present, pyrolysis incineration technology has been used more in the field of treatment of medical waste and other hazardous waste abroad (Canadian Council of Ministers of the Environment 1992).

At present, most of the municipal medical waste incinerators produced and put into operation in China use pyrolysis gasification incineration technology. Pyrolysis gasification incineration treatment technology has great advantages in terms of treatment effect and treatment cost. It is an incinerator mainly used to treat hazardous wastes such as medical wastes. The incinerator has the advantages of high burn-out rate, low consumption of auxiliary fuels, less flue gas, low concentration of pollutants in the flue gas, low post-treatment load, dust entrainment and so on. However, the technical equipment of pyrolysis incineration is quite different and it is difficult to achieve stable combustion and good self-control. The pyrolysis section, combustion section and burnout section interact with each other. In actual operation, often due to the large difference between the feed state and the design, the self-control system is difficult to regulate to the ideal state, and the tail gas system load changes frequently. It also evolved into a fixed bed of over-oxygen combustion, so the actual effect is not ideal. Due to its low technical threshold, the technology market is chaotic.

Grate incinerator is also widely used. In China, grate incinerators have been used in Tianjin, Anshan, Jilin, Dalian, Guiyang, Shunde and other cities. In practical application, there is a low burnout rate of incineration, auxiliary fuel consumption and other shortcomings. At present, the domestic has banned the configuration of only a single combustion chamber grate furnace used in medical waste treatment.

Fixed bed incinerator has the characteristics of low investment, simple operation, stable operation, low treatment cost but it lacks a complete and matured flue gas cleaning system, thereby producing secondary pollution to the surrounding environment.

In the selection of the incinerator type of medical waste, the appropriate incinerator type should be selected according to the characteristics of the medical waste and the treatment scale of the incineration plant. It is strictly prohibited to select the incineration device that cannot meet the pollutant emission standards. Incinerators with mature technology, high automation level and stable operation should be selected. It is strictly prohibited to use single combustion chamber incinerator and incineration devices without automatic control system and tail gas treatment system (Liu 2008). The choice of medical waste incinerator should meet the following requirements:

(1) The incinerator structure consists of a primary combustion chamber and a secondary combustion chamber. The primary combustion chamber is for combustion or pyrolysis, and the secondary combustion chamber is for complete combustion.
(2) The design of the incinerator bed should prevent the leakage of liquid or incompletely burned waste. Ensure that the incompletely burned medical waste does not leak into the slag through the bed, and make the air evenly distributed along the bottom of the bed. The air supply hole should be designed to avoid clogging due to ash accumulation or scaling.
(3) The incineration control conditions should comply with the "Medical Waste Treatment and Disposal Pollution Control Standards" (GB 39,707–2020), "Medical Waste Management Regulations" and "Medical Waste Incinerator Technical Requirements (for Trial Implementation)" and other relevant provisions.
(4) There should be appropriate overload processing capacity, and the waste feed should be adjustable.
(5) During normal operation, the incinerator should be in a slightly negative pressure combustion state.
(6) The temperature of the accessible shell exterior of the furnace body should be $\leq$50 °C.
(7) Control the secondary combustion chamber flue gas temperature $\geq$850 °C, and flue gas residence time $\geq$2.0 s.
(8) The combustion efficiency of the equipment should be $\geq$99.9%, and the thermal scorching rate of the incineration residue should be <5%.
(9) The oxygen content in the flue gas exiting the incinerator should be controlled at 6–10% (dry gas).
(10) The incinerator can be operated, monitored and managed from a central console, including continuous display of operating parameters and conditions (temperature, pressure, oxygen content, air volume, fuel volume, etc.), and feedback control can be implemented.
(11) The control of pyrolysis and combustion process should be realized to prevent incomplete combustion or furnace burn down.
(12) The second combustion chamber of the incinerator should be equipped with emergency discharge chimney while that of the first combustion chamber of

the pyrolysis incinerator should be equipped with explosion-proof pressure-exhausting equipment.

4.1.5 *Pyrolysis Incineration Tail Gas Control Technology*

4.1.5.1 Acid Removal Technology

(1) Semi-dry deacidification control technology

The semi-dry de-acidification process is a spray drying system (the equipment is usually called a spray drying absorber tower). The flue gas enters from the upper part of the absorber tower and exits from the lower part. The slurry of slaked lime (calcium hydroxide slurry with a concentration of 5% to 10%) is sprayed into the absorption tower using a high-efficiency atomizer, and the flue gas is in full contact with the sprayed slurry and neutralization occurs. Due to the good atomization effect and large contact surface in the system, the semi-dry method can not only effectively reduce the flue gas temperature and neutralize the acidic gas in the flue gas, but also the sprayed slurry can be completely evaporated in the absorption tower.

The advantages of the semi-dry process are higher deacidification efficiency, less investment in equipment than the wet process, no secondary pollution of wastewater, less land area, low operating costs while the disadvantages are more complex pulping equipment, nozzles are easy to wear, scaling, lime slurry transport pipeline is prone to failure, and system maintenance requirements are high. The process has high requirements on the residence time of flue gas in the semi-dry tower, the type, particle size and configuration concentration of absorbent in the absorbent slurry, and the nozzle.

In the semi-dry purification process, there are two ways of feeding the flue gas: one is the upper intake, where the flue gas flows downward with the absorber slurry and then enters the subsequent de dusting equipment. The other is the lower intake, where the flue gas flows upward with the absorber slurry. In the down-flow method, the absorbed flue gas usually enters the cyclone separator for solid–gas separation, and some of the resulting solids are returned to the deacidification tower, so that the incomplete reaction of the absorber can be fully utilized, thus improving the efficiency of the use of the absorber, and the flue gas passes through the cyclone separator before entering the dust removal facility. Usually the former process equipment is called spray drying absorption tower, and the latter process equipment is called gaseous suspension absorption tower. In actual engineering applications, the former process is more commonly used.

(2) Dry deacidification control technology

Usually the slaked lime neutralizer is used and the slaked lime injection device is set between the emergency cooling tower and the baghouse. Through the mixer on the flue, the slaked lime is mixed evenly in the flue gas under the action of compressed air. The slaked lime is deposited on the bag wall of the dust collector to form a reaction layer, so that the slaked lime can neutralize the gaseous acid in the flue gas to remove the acid. In order to enhance the reaction rate, the actual amount of alkaline solids is 3–4 times of the reaction requirement, and the solids residence time should be at least 1.0 s.

Slaked lime (hydrated lime), which is mainly composed of $Ca(OH)_2$, is white powder, slightly soluble in water, and its clarified aqueous solution is a colourless and odourless alkaline transparent liquid. In general, $Ca(OH)_2$ is required to have a sieving rate of more than 90% at 200 mesh. In recent years, in order to improve the removal efficiency of some pollutants that are difficult to remove by dry gas washing method, sodium sulfide (Na_2S) and activated carbon powder mixed with lime powder is sprayed in, which can effectively absorb gaseous mercury and dioxins.

Dry gas washing tower and bag filter process is a common method of tail gas pollution control in incineration plants. The advantages are simple equipment, easy maintenance, low cost, and the slaked lime pipeline is not easy to block. The disadvantage is that due to the limited contact time between the solid and gas phases and poor mass transfer, it often require excessive dosing, the consumption of reagents is large, the overall removal efficiency is also lower than the other two methods, and the amount of reactants and unreacted materials produced is also more, which requires proper final disposal. At present, although some manufacturers have used the recycling system to mix the fly ash, reactants and unreactants collected by the dust collector with fresh slaked lime powder in a certain ratio for reuse, in order to save the consumption of chemicals. However, its effectiveness is not significant, and it will make the whole preparation and injection system complicated. Furthermore, the pipeline system also increases the frequency of wear and tear or blockage due to the intervention of fly ash and reactants, but will lose the advantages of simple operation and easy maintenance of the original system equipment.

(3) Wet deacidification control technology

The most commonly used wet deacidification control technology in incineration flue gas treatment systems is the wet scrubber tower. The system is a convection-operated packed absorption tower, where the tail gas and the downward flowing alkaline solution are in constant contact and reaction in the packing void and surface, so that the polluted gas in the tail gas is effectively absorbed. Commonly used absorption chemicals (alkaline chemicals) are mainly NaOH solution (0–15%, mass concentration) or Ca $(OH)_2$ solution (0–10%, mass concentration). The alkaline scrubbing solution of the gas washing tower is recycled, and when the pH or salinity of the recycled solution exceeds a certain standard the discharge part is replenished with fresh NaOH solution to maintain a certain acid gas removal efficiency. There are usually many dissolved heavy metal salts (such as $HgCl_2$, $PbCl_2$, etc.) in the discharge solution, and the chlorine salt concentration is as high as 3%, which must be properly treated. The biggest advantage of the wet scrubber is the high efficiency of acid gas removal, HCl removal rate of 98%, SO_x removal rate of more than 90%, and can remove highly volatile heavy metal substances (such as mercury). The disadvantages are higher cost, higher electricity and water consumption, and the need to install additional exhaust gas re-heater to avoid the phenomenon of white smoke after exhaust gas emission. In addition, the wastewater produced by the wet scrubbing method, which contains metals and high concentration of chlorine salts, needs to be treated.

4.1.5.2 Bag Type Dust Pollution Prevention and Control Technology

Tail gas de-dusting is very critical for the operation of all incinerators. Dust collectors can be divided into mechanical dust collectors, wet dust collectors, bag-type dust collectors and electrostatic precipitators according to the purification mechanism. Various technologies are used around the world. Incinerator de-dusting device should be selected for bag type dust collector, and the use of cloth bag dust collector is recommended in China. Bag filter is suitable for removing dust particles above 0.1 μm particle size, with 99% dust removal efficiency.

Among the dust removal equipment for medical waste incineration, baghouse dust collector has more advantages than other dust removal equipment, especially for dry or semi-dry de-acidification process. The baghouse is not only a dust collector, but also a reaction device for removing other harmful substances from the flue gas, and is one of the most critical pieces of equipment for exhaust gas treatment. Hazardous waste/medical waste incineration flue gas purification should be given priority to, and the use of bag-type dust collector, do not use electrostatic de-dusting and mechanical de-dusting device. If wet dust removal device is chosen, then complete wastewater treatment facilities must be utilized.

A bag filter usually consists of several bags between 16–20 cm in diameter and 2–3 m in length, which are made of glass fiber or PTFE material and arranged in sequence. The porous nature of the fabric and the filter cake formed on the surface of the bag filter, forms the filter layer of the bag filter, which can remove particles in the size of 0.05–20 μm, pressure drop in the range of 1–2 kPa, and efficiently remove dust-containing substances from the flue gas, as well as heavy metals (especially Hg) and dioxins adsorbed on the dust particles in the exhaust gas. The dust removal efficiency is generally over 99%.

According to the different cleaning methods, bag filter can be divided into mechanical vibration bag filter, reverse airflow blowing bag filter, pulse blowing bag filter and so on. In the pulse blowing cleaning method, the exhaust gas flows from the outside of the filter bag to the inside, and the granular pollutants accumulate in the outer layer of the filter bag, which is only fixed at the upper end. During cleaning, the filter bad is expanded with the help of high-pressure gas from the inside to the outside. This method is more rapid, and can be used in a continuous online operation. It is recommended to give priority to the bag filter with pulse blowing cleaning method. The structure of this type of bag filter mainly includes box, ash hopper, pillar, stairs (climbing ladder), railing, platform, filter bag frame, filter bag, lifting valve, ash cleaning air system, ash cleaning controller, ash discharge equipment, ash discharge device, pulse valve sub-air box, pulse valve, air bag, etc. The length of the filter bag design is limited by the limit of the sprayed gas pressure. To maintain the cleaning effect, it is generally are less than 5 m.

As the bag is more sensitive to acid, it is usually set up in front of the de-acidification device. It is recommended to use the semi-dry de-acidification process, which plays the role of de-acidification, and on the other hand reduces the temperature of the flue gas to the appropriate range of the bag filter bag, which plays a role in protecting the bag. In addition, special attention needs to be paid to the water content of the flue gas. For incineration facilities with high difficulties in reaching the discharge standard, wet deacidification process can be used.

Particulate matter emission concentration in the flue gas outlet of the bag filter can, to a certain extent, indirectly reflect the dioxin emission concentration. At present, the "medical waste treatment and disposal pollution control standards" being developed for particulate matter emission limits of 30 mg/m^3 (1 h average) and 20 mg/m^3 (24 h average or daily average), dioxin emission concentration is specified as 0.5ngTEQ/Nm^3. Improving the bag filter efficiency and reducing the outlet particulate matter emission concentration, has a positive effect on the control of dioxin.

4.1.5.3 NO_x Processing Technology

Currently, two NO_x treatment technologies, SCR and SNCR, are generally used in China. Selective catalytic reduction (SCR): under the action of catalysts (such as V_2O_5/TiO_2 and $V_2O_5–WO_3/TiO_2$), the reducing agent NH_3 reduces NO and NO_2 to N_2 at 290–400 °C, while almost no oxidation of NH_3 occurs, thus improving the selectivity of N_2 and reducing the consumption of NH_3. Selective non-catalytic reduction (SNCR) refers to the reduction of NO_x in flue gas to harmless nitrogen and water by spraying a reducing agent in the temperature range suitable for denitrification reaction without a catalyst. This technology generally uses ammonia, urea or hydrogen ammonia as the reducing agent to reduce NO_x. The reducing agent only reacts with NO_x in the flue gas, and generally does not react with oxygen, and this technology does not use catalysts. SNCR is now commonly used with ammonia and urea as reducing agents. In terms of safety, urea is easy to store and safe to use, while ammonia has strict requirements for preparation and use. From the economic point of view, the cost of urea and transportation cost is much higher than ammonia. From the consideration of the use effect, the NO_x removal rate of ammonia can be as high as 70–80%, which is much higher than the NO_x removal rate of urea 30–45%. In conclusion, if safety is a priority, it is recommended to use urea as denitrification reductant; if economic efficiency is a priority, it is recommended to give priority to ammonia as denitrification reductant.

4.1.5.4 Dioxins and Major Heavy Metals Process Control Technology

(1) Flue gas high temperature combustion technology

The flue gas produced by the primary combustion of medical waste contains a large number of incomplete combustion products, the flue gas must be burnt at high temperatures. The mixed flue gas from the first combustion chamber enters the second combustion chamber for combustion. The second combustion chamber is a gas combustion chamber. Its main function is to burn out the flammable pyrolysis gas generated by the first combustion chamber, to minimize harmful substances. From the point of view of inhibiting dioxin production, high temperature combustion is very important and necessary. Most of the dioxins generated by chlorine-containing precursors (such as PVC, chlorobenzene, etc.) in the combustion process can be decomposed under high temperature conditions.

Whether the secondary combustion is complete, can be judged according to the concentration of carbon monoxide in the outlet flue gas. Generally, in the case of complete combustion, the concentration of carbon monoxide in the outlet flue gas should be below 50×10^{-6}. Usually, the combustion efficiency can be measured by the index, and the required combustion efficiency should be above 99.99%.

To achieve full and complete combustion, the flue gas residence time, combustion temperature, turbulence and enough air supply are the main indicators that affect the combustion efficiency (Walker and Cooper 1992). High temperature combustion technology is mainly designed and operated around these indicators.

The combustion furnace adopts vertical cylindrical structure, including head, straight section and auxiliary devices, diesel burner fixtures, blast port, explosion-proof door, emergency discharge port and refractory lining.

(2) Emergency flue gas cooling technology

Many medical wastes contain heavy metal compounds and heavy metal components that are volatile or evaporate easily (Zhao 2008). When medical waste is incinerated, these heavy metals enter the flue gas with or adhere to the fly ash. Among them, some will volatilize or evaporate to form a gaseous state into the flue gas, with the flue gas out. In the flow process, with the change of flue gas temperature, these heavy metal gas components will be condensed and agglomerated; if the flue directly into the environment, it will cause great harm.

The flue gas temperature from the second combustion chamber is over 850 °C, so it must be cooled down by cooling before flue gas cleaning. Flue gas cooling is generally carried out in two stages: the first stage is to reduce the flue gas temperature from 850 to 600 °C; the second stage is to reduce the flue gas temperature from 600 °C to about 200 °C. In the process of flue gas cooling, some of the evaporated heavy metal gases will re-condense or agglomerate on the dust particles, and then collect the dust through the dust collector to remove heavy metals. The lower the temperature, the better the removal effect.

(3) Activated carbon adsorption technology

To reduce dioxin emissions, activated carbon powder or porous adsorbent (mostly activated carbon) is usually sprayed into the flue before the bag filter to form a retention layer on the surface of the bag filter to adsorb dioxin substances and heavy metal substances in the flue gas. This technology is called activated carbon injection (ACI) adsorption technology. After the bag filter, an absorption tower (bed) with activated carbon or porous adsorbent is set up, and this technology is called activated carbon fixed bed (FCB) adsorption technology.

Activated carbon is a type of microcrystalline carbon material with black appearance, well-developed internal pore structure, large specific surface area and strong adsorption capacity made mainly of carbon-containing materials. Increasing the amount of activated carbon injected can significantly reduce the number of dioxins emitted to the atmosphere. A lime powder and activated carbon injection reactor can be installed in the flue gas pipeline before the baghouse in the system. Activated carbon is delivered to the flue using compressed air. As the activated carbon is easy to absorb moisture and caking, the traditional feeding equipment is not reliable. The activated carbon is manually poured into the loading bin of the gasification chamber. The compressed air is connected to the top of the gasification chamber, and the activated carbon powder of a certain concentration in the gasification chamber is sent into the flue by the pressure line air. This device overcomes the shortcomings of conventional devices, which are easy to block and poorly control the powder spraying volume. Alternatively, a spiral plate feeder is used to feed the activated carbon in the spiral baffle and draw it in through the negative pressure in the pipe.

After the flue gas has passed through the rapid cooling tower and the semi-dry de-acidification tower, most of the acidic substances and dust have been removed from the flue gas. As the flue gas composition changes with the working conditions, the acidic substances are still present in the flue gas after de-acidification to a greater or lesser extent. To ensure the purification effect of flue gas, lime powder is sprayed into the flue gas as a deacidifier before entering the baghouse, and then sprayed into the flue gas pipe before the baghouse, together with activated carbon powder. This can further remove the acid in the flue gas and remove most of the harmful substances such as dioxins, and facilitate the absorption of water in the flue gas to ensure the effectiveness of the subsequent operation.

Powdered activated carbon is added quantitatively to the flue gas. At low temperature (200 °C), dioxins are easily adsorbed by activated carbon. The activated carbon is sprayed into the flue gas and mixed with the flue gas for initial adsorption; then the mixed flue gas enters the baghouse; the activated carbon particles are adsorbed onto the surface of the filter bag and continue to be adsorbed on the surface of the filter bag, thus improving the efficiency of dioxin removal.

In addition, adding activated carbon to the flue gas is also very effective in removing mercury from the flue gas. The purchased activated carbon is stored in a closed storage tank and fed into the reactor through a small rotary feeder and mixed with compressed air, and the amount of activated carbon injected can be adjusted by adjusting the rotary feed speed.

The storage bin is located above the flue and is connected to the activated carbon dry powder and lime dry powder pipes by connecting pipes. The pipe is open in the downstream direction of the flue gas. Behind the dry powder spraying pipeline, the flue gas pipeline is partially indented to increase the flow rate of flue gas, and the sprayed activated carbon dry powder and lime dry powder flow in the same direction as the flue gas, to reduce the system resistance. In the process of flue gas speed up, the sprayed activated carbon dry powder and lime dry powder are mixed with flue gas evenly to achieve the purpose of absorption.

(4) Low temperature plasma decomposition technology

The technique consists of a high voltage impulse current generation device discharging in the gas phase. In this process, strong currents rush into the discharge channel in a very short time (hundred nanoseconds), forming an electron avalanche and causing electron heating (10^4–10^5 K). The discharge channel is completely filled with dense plasma, and hydroxyl radicals, ozone and ultraviolet light are generated; at the same time, due to the very short time of the rising edge of the narrow pulse, the plasma channel expands outward at a speed of 10^2–10^3 m/s, completing the entire breakdown, and using the very high electron energy to break the bonds of dioxin molecules and reorganize them into single atoms or harmless molecules directly to achieve the precipitation and removal effect and the degradation of dioxins is accomplished. The principle is that, the inelastic collision of high-energy electrons bombards the dioxin molecules, causing the breakage of the bonds of the molecular ring structure, and the dioxin is decomposed into CO_2, H_2O, HCl and other inorganic products. As an innovative technology option for deep purification, this technology can be applied in combination with other technologies to achieve the desired removal effect.

4.1.5.5 Fly Ash and Residue Pollution Control Technology

The disposal of these forms of waste is important because they are discharged into the environment in the form of fly ash, bottom ash or filter residue. For example, they can be pretreated or landfilled in specially designed and operated landfills based on best available technology. Confined transport and specialized landfills are common methods for managing these incineration residues.

The residue treatment system should include slag treatment system and fly ash treatment system. Slag handling system should include de-slagging cooling, conveying, storage, slag crushing and other facilities. Fly ash treatment system should include fly ash collection, conveying, storage and other facilities.

Bag dust generated by fly ash and other facilities to trap the dust, due to the considerable number of dioxins and heavy metals, are hazardous waste, and should be implemented in accordance with the relevant regulations and requirements for environmentally sound treatment. Due to the relatively small amount of generation, the centralized disposal facilities for medical waste should not be configured to cure stabilization and other environmentally sound treatment facilities. It is recommended that hazardous waste should be stored safely and disposed by the local centralized disposal units.

The residue treatment system should have stable and reliable mechanical performance and easy maintenance characteristics. The choice of slag handling device should meet the following requirements: the slag removal connected with the incinerator should have reliable mechanical properties and measures to ensure the sealing of the furnace; preferably recommended with a water seal chain plate slag machine, running under the water seal sprocket and drive parts should be made of stainless steel; the slag conveying equipment should have sufficient width and headroom height.

Slag and fly ash treatment system devices should be kept airtight. When the flue gas purification system adopts the semi-dry method, the fly ash treatment system should take mechanical ash removal or pneumatic ash removal method. Pneumatic ash removal system should take measures to prevent the air from entering and prevent the ash from clumping. When using wet flue gas purification method, effective dewatering measures should be taken.

Fly ash collection should be used to avoid the scattering of fly ash sealed containers. Collection of fly ash with ash storage tank capacity should be determined by the rated amount of fly ash generated. Ash storage tank should be equipped with level indication, dust removal and prevention of ash slating facilities, and it is desirable to set up humidification facilities near the ash discharge port.

4.1.5.6 Automatic Control System Integration and Optimization Technology

An important requirement for the technical optimization of the medical waste incineration system is to have a complete automatic control system, with complete process control functions, safety functions, easy operation, greatly simplifying the intensity of human operations. Medical waste incineration system using the PC + PLC (programmable logic controller) automatic control method. At present, PLC control technology is quite mature, strong control function, can make the whole incineration process more smooth, the control variables of each process is easier to coordinate. Incineration plant automation control system must be applicable and reliable, should be designed according to the characteristics of hazardous waste incineration facilities, and should meet the requirements of safe and economic operation of the facilities and prevent secondary pollution of the environment. Incineration plant automation system should use mature control technology and high reliability, performance to price ratio of appropriate equipment and components. The design of new products and new technologies should be used in the relevant areas of successful operation of the experience. The main elements generally include:

(1) Medical waste incineration line monitoring system
(1) Video surveillance of medical waste unloading process.
(2) Video monitoring of medical waste loading process.
(3) Video monitoring of the flame in the kiln.
(4) Video monitoring of the chimney vent.
(5) Solids feed weight per batch and cumulative weight display.
(2) Incinerator kiln system automatic monitoring project
(1) Pyrolysis gasification outlet temperature.
(2) The outlet temperature of the second combustion chamber.
(3) Pyrolysis gasifier air valve opening.

(4) Second combustion chamber air valve opening.
(5) Negative pressure in the gasifier and second combustion chamber.
(6) Detection and automatic display of O_2 in the flue gas at the outlet of the second Combustion chamber.
(3) Self-flue gas cleaning and smoke exhaust system dynamic monitoring project
(1) Flue gas temperature and pressure at the inlet and outlet of the emergency cooling tower.
(2) Emergency cooler cooling water supply pressure and flow rate.
(3) Dust collector inlet and outlet flue gas temperature and pressure.
(4) Differential pressure of the dust collector.
(5) Machine inlet temperature and airflow.
(6) Lime powder and activated carbon powder flow rate and pipeline pressure.
(4) Online exhaust gas monitoring system

Should be burnt in the flue gas of soot, carbon monoxide, sulfur oxides, nitrogen oxides, hydrogen chloride, carbon dioxide, oxygen content to achieve automatic continuous online monitoring. Flue gas blackness, hydrogen fluoride, heavy metals and their compounds should be sampled at least once a quarter to monitor. Dioxin sampling and testing frequency of not less than 1 time/year. Should be the main process parameters of the incineration system and characterize the operational performance of the incineration system indicators including flue gas CO, CO_2, NO_x, SO_2, soot, O_2, HCl concentration to implement online monitoring.

(5) Automatic chain control project
(1) The feeding system of the upper and lower gate chain are interlocked and controlled. The charging operation of pyrolysis furnace requires starting and stopping the hoist, horizontal conveyor and upper and lower gates in sequence. In order to prevent the leakage of harmful gas in the process of feeding, the upper and lower gates are required to be controlled by chain, and the two gates cannot be opened at the same time.
(2) The chain control of pyrolysis furnace gas valve and combustion chamber air valve opening with the temperature of the secondary combustion chamber. The temperature of the secondary combustion chamber is controlled at 850–1100 °C, and the temperature of the combustion furnace is maintained at the set temperature by adjusting the opening of the pyrolysis furnace gas valve and combustion chamber air valve.
(3) Secondary air volume and the secondary combustion chamber outlet flue gas oxygen concentration chain control. The secondary combustion chamber outlet flue gas oxygen concentration control in between 6–10%, the need to control the size of the secondary air volume, and the secondary combustion chamber outlet flue gas oxygen concentration to form a closed-loop control.
(4) The secondary combustion chamber temperature and burner chain control. When the heat of medical waste is low, the second combustion chamber temperature is difficult to maintain above 850 °C, the combustion support system must be started. Therefore, the light oil burner and the second combustion chamber temperature interlock to form a closed-loop control. When the temperature of

the secondary combustion chamber is lower than 850 °C, the controller automatically starts the light oil burner; when the temperature of the second combustion chamber is higher than 900 °C, the controller automatically shuts down the light oil burner. At the same time, a camera is set in the second combustion chamber to observe the combustion situation in the secondary combustion chamber on the monitor in the central control room.

(5) The chain control of negative pressure of gasifier and induced draft fan. The negative pressure of gasifier is automatically controlled at -100 Pa, and the speed of induced draft fan is controlled by frequency converter to maintain the constant negative pressure of combustion furnace.

(6) Emergency cooling tower outlet temperature and the amount of coolant sprayed into the chain control. The neutralization of the emergency cooling tower outlet temperature is controlled at 180–200 °C, and the emergency cooling tower outlet temperature and spray water emergency cooling tower coolant electric control valve chain closed-loop automatic control. That is, the emergency cooling tower outlet temperature fluctuations, the PLC PID adjustment module through the coolant supply pipeline electric control valve to achieve the control of the coolant supply, the normal work of the subsequent bag filter to provide a good temperature environment. When manual intervention is required (such as commissioning, maintenance, etc.), the amount of coolant can be adjusted directly through the central control room coolant manual controller.

(7) Chain-link control of HCl content and slaked lime addition in online monitoring system. By changing the dust concentration of slaked lime and activated carbon by frequency conversion, the amount of slaked lime and activated carbon addition is adjusted.

(8) Bag filter differential pressure and blowback solenoid valve chain control. With the accumulation of soot in the surface of the filter bag, bag filter net room and dust chamber pressure difference is increasing; resistance increases to a certain value (1500 Pa), which will inevitably lead to lower filtration efficiency, thereby affecting the total negative pressure of the system. Therefore, the interlock control of bag filter pressure difference and blowback solenoid valve is designed. When the differential pressure reaches the set upper limit, the PLC will automatically start the blowback control program in time for the bag filter blowback cleaning operation, in turn for each group of filter bags blowback cleaning. When the bag filter recovers the initial pressure, stop the back-blowing cleaning operation, then automatically start the spraying switch, and start the spraying operation on the bag filter.

(9) Emergency emission chimney and the linkage of the accident or emergency. Emergency emission chimney is set above the secondary combustion chamber, and the emergency linkage device is set so that it can only be opened in case of emergency or accident, such as power failure, induced draft fan failure, bag inlet flue gas over temperature for more than a certain period, the secondary combustion chamber positive pressure and other circumstances.

Computer monitoring system of all the measurement data, data processing results and facility operating status, should be able to display in the monitor.

4.1.5.7 Medical Waste Incineration Disposal Technology Optimization

4.1.5.8 Strictly and Effectively Carry Out Hazardous Waste Incineration Process Control

The high temperature decomposition stage is the main stage to control dioxin production. It should be ensured that a region with a particularly high temperature (above 850 °C) can inhibit the generation of dioxins under well- organized combustion conditions. On the other hand, to ensure adequate heat and mass transfer, so that the dioxin organic precursors in this region for the full oxidation of combustion, thus further eliminating the re-synthesis of dioxins. Many incinerators have subsequent combustion measures (two combustion chamber), the subsequent combustion temperature is generally controlled above 950 °C, to ensure the complete combustion of organic compounds. During incineration process, attention should also be paid to the oxygen content and the formation of low-temperature zone. Only with enough oxygen can organic pollutants be fully oxidized to eliminate its toxicity. The low temperature zone is caused by the uneven combustion conditions in the combustion chamber. The existence of the low temperature zone (less than 850 °C) causes the insufficient combustion of organic pollutants, and ultimately leads to the emission of many dioxins. Therefore, strict control of the operating parameters of the incineration process is an effective way to ensure the reduction of dioxins. Post-synthesis control, that is, minimizing the probability of dioxin synthesis and inhibition of the re-synthesis of incineration flue gas in the purification process, generally uses the method of controlling the flue gas temperature. Usually, when the incineration flue gas is discharged from the incinerator at a certain temperature (at this time, the temperature should not be lower than 500 °C), the flue gas is rapidly cooled to less than 200 °C (usually about 100 °C) within 1 s, thus leaping over the dioxin generation temperature zone. At the same time, the flue gas cleaning process must take certain measures to ensure the existence of a dioxin-free environment (Yan 2012; WHO 2005). The following is an analysis of the key factors involved in the process.

(1) Incineration temperature control

The incineration temperature is too low to destroy the waste and sufficiently produce dioxins; if the temperature is too high, it will waste fuel, while promoting the volatilization of heavy metals and the synthesis of nitrogen oxides. The role of the first combustion chamber is to burn or pyrolysis while the role of the second combustion chamber is to achieve complete combustion. Due to their different functions, the incineration temperature should also be specified differently. Incineration temperature should fully consider such factor as dioxins, waste ignition point, chlorine content, nitrogen oxides, destruction removal rate (DRE) and combustion efficiency (CE) and other factors. There are three dioxin generation mechanisms, namely;

the waste itself containing dioxins, waste incineration process to generate dioxins (precursors synthesized from scratch) and related but non-toxic small molecules (such as HCl, Cl_2) re-synthesis.

According to the current study, the theoretical destruction temperature is 500 °C for dioxins contained in the waste itself. When the actual operating temperature is greater than 850 °C and the residence time is more than 2 s, the dioxin destruction rate is greater than 99.99%. Experiments have shown that when the incineration temperature is 500–800 °C, it will promote the production of dioxins. When the temperature is greater than 900 °C, it will destroy the production of PCDD. Without PCDD production, the content of dioxins will drop drastically. When the temperature is around 1070 °C, almost no dioxin exists. The precursors synthesized from scratch are produced at 400–750 °C.

When the temperature exceeds 1200 °C, a large amount of NO_x will be produced, which will corrode the equipment and increase the operation cost. When the temperature of the first combustion chamber exceeds 870 °C, metal pollutants will be released into the second combustion chamber, and the volatile metal pollutants (such as Cu compounds) are catalysts for dioxins. When the temperature of the second combustion chamber is above 870 °C, the combustion efficiency is the highest. When the temperature reaches 870 °C, all wastes can be fully burned.

The United States, the European Union and other countries or organizations have specified the burning temperature of the first and second combustion chambers. It is recommended that the temperature of a combustion chamber should not be lower than 850 °C, and the best temperature range is between 850–870 °C. The second combustion chamber temperature is higher than 1100 °C, and the best range is between 1000–1200 °C; when the halide content exceeds 1% (mass), the temperature should not be lower than 1100 °C.

(2) Length of stay

The residence time determines the incineration effect and the size of the furnace volume. If the residence time is too short, the waste and flue gas combustion are not fully combusted, and DRE and emission flue gas does not meet the standard. Also, if the residence time is too long, the size of the furnace is too large.

The flue gas residence time and incineration temperature are inversely related. At present, the resident time of flue gas stipulated by the United States, the European Union and other countries or organizations is greater than 2 s, and its incinerator operation is good, indicating that this residence time can guarantee the effect of hazardous waste incineration. However, in the start-up phase of the incineration system, the premise that the incineration is not enough, can be considered to temporarily extend the flue gas residence time of the second combustion chamber.

In summary, the flue gas residence time in the second combustion chamber should be greater than 2 s. Also, considering the start-up phase of the incineration system, the residence time must be extended according to the actual situation.

4.1.5.9 Strengthen Exhaust Gas Purification Measures

Flue gases from incineration facilities include fly ash (particulate matter) containing heavy metals, dioxins, heat-resistant organic compounds, and gases such as nitrogen oxides, hydrocarbons, and hydrogen halides, with flue gas concentrations of approximately 2,000 ngTEQ/Nm generated by the uncontrolled mode (no flue gas cleaning)[3]. Therefore, flue gases are cleaned and the following flue gas cleaning measures must be combined with appropriate methods. The following flue gas cleaning measures must therefore be used in combination with appropriate methods to ensure the application of the best available technology:

(1) Separation of dust and non-volatile heavy metals

The separation of dust is often carried out using fiber filter membranes, electrostatic precipitators and fine wet scrubbers. Pre-cleaning of the flue gas can be done using cyclone dust removal, which is more effective for the separation of large particle sizes.

(2) Removal of HCl, HF, SO_2 and Hg

The removal of acidic components and mercury can be done by dry or wet adsorption (including activated coke or lime), or by scrubbing, generally in 1–2 scrubbing steps.

(3) Removal of NO_x

Primary measures include the use of low NO_x incinerators, staged combustion and reuse of flue gases. Secondary measures can be applied with SNCR and SCR technologies. In fact, a combination of primary measures (such as limiting their full process synthesis and optimizing combustion) and secondary measures, such as activated carbon filtration membranes, activated carbon and slaked lime injection, and catalytic oxidation, can be applied to reduce emissions of dioxins and other organic compounds.

4.1.5.10 Avoid Sensitive Temperature Intervals

A proper flue gas purification system is the guarantee for the flue gas to meet the emission standards. The main purpose of this system is to control the dioxins, acids and particulate matter produced. 300–500 °C is the best temperature range for dioxin synthesis from scratch, with 400 °C being the fastest rate. To be on the safe side, 180–550 °C is the sensitive range for dioxin production. Research shows that 250–350 °C is the best temperature range for dioxin re-synthesis, and the most amount is produced at 300 °C.

Medical waste contains PVC and some halogenated acids that produce HCl. In addition, the baghouse is sensitive to acids, so an acid removal device is necessary. Incineration will produce a large amount of particulate emissions. At the same time, the particulate matter is an effective carrier for dioxin resynthesis, so it is necessary to use a high-efficiency dust removal device. The emergency cooling unit, acid removal unit and dust removal unit are indispensable components of the flue gas cleaning system.

To make the flue gas cool down sharply in a short period of time, the key index of the emergency cooling equipment is the flue gas residence time, thus the temperature can be regulated in the temperature range of 180–550 °C, especially 200–350 °C.

For the water emergency cooling system, from the perspective of thermal transformation, the cooling rate is controlled at 250–500 K s^{-1}, and the recommended cooling rate in the temperature range of 250–400 °C is 350 K s^{-1}. Therefore, the residence time required to reduce the flue gas temperature from 550 to 180 °C in the emergency cooling equipment is 0.74–1.48 s. The longer the residence time, the more dioxin is produced, but less water sprayed; If the residence time is short, the production of dioxin is relatively small but the amount of water spray is large.

The key to the acid removal plant is the selection of the process, of which the available processes are dry, semi-dry and wet. From the comparison (Table 4.4), the removal efficiency of the dry process is not as good as the latter two. Although the removal efficiency of the wet process is better than the semi-dry process, but the energy and water consumption is large, but also produces wastewater, investment and operating costs are correspondingly expensive. The semi-dry process has good disposal effect and low cost, so it is the best choice for acid removal process.

There are two keys to the dust removal equipment: one is the equipment selection; the other is the working condition setting. Dust removal equipment including text scrubbers, electrostatic precipitators (ESP) and cloth bag filter (FF), etc. Text scrubber, electrostatic precipitator and bag filter performance comparison is shown in Tables 4.4 and 4.5.

The biggest disadvantage of the ESP is that it produces waste water, which requires secondary treatment, while the FF is highly efficient and does not produce secondary pollution.

The key operating conditions of the baghouse include the operating temperature and the flow rate of the activated carbon absorber. Shin and Chang et al. showed that even with the baghouse, high dioxin production occurs at temperatures above 230 °C. Brewster et al. showed that when only the baghouse is used, controlling the temperature at around 200 °C, dioxins can be controlled at 1.0 ngTEQ/Nm^3; if the control temperature is below 200 °C, dioxins can be controlled at 0.1 ngTEQ/Nm; if the activated carbon absorber is also added, dioxins can be maintained at 0.1 ngTEQ/Nm. If the control temperature is lower than 200 °C, the dioxin can be controlled at

Table 4.4 Comparison of dry, semi-dry and wet processes for acid removal units

Craftsmanship Category	Removal efficiency		Pharmaceuticals Consumption	Power consumption	Water consumption	Construction Costs	Running costs
	Individually	Matching Bag filter					
Dry method	50%	95%	120%	80%	100%	90%	80%
Semi-dry method	90%	98%	100%	100%	100%	100%	100%
Wet method	99%	–	100%	150%	150%	150%	150%

Table 4.5 Performance comparison of text scrubber, electrostatic precipitator and bag filter

Species	Efficiency (%)	Effective removal Particle diameter (μm)	Water requirement per unit of gas Volume (L/m)^3	Volume	Gas change Flow Impact		Operating temperature/°C
					Pressure	Efficiency	
Wentype scrubber	90–98	0.5	0.9–1.3	Small	Yes	Yes	70–90
Electrostatic precipitator	90–98	0.25	0	Big	No	Yes	–
Bag filter	95–99	0.4[a]/ 0.25[b]	0	Big	Yes	No	100–250

0.1 ngTEQ/Nm^3; if the activated carbon absorber is also added, the dioxin can be maintained at 0.1 ngTEQ/Nm^3 or less. Of course, the working temperature of the bag filter has a bottom line, which must be guaranteed to be 20–30 °C above the flue gas dew point temperature. The relationship between dust removal device and dioxin concentration is shown in Fig. 4.4.

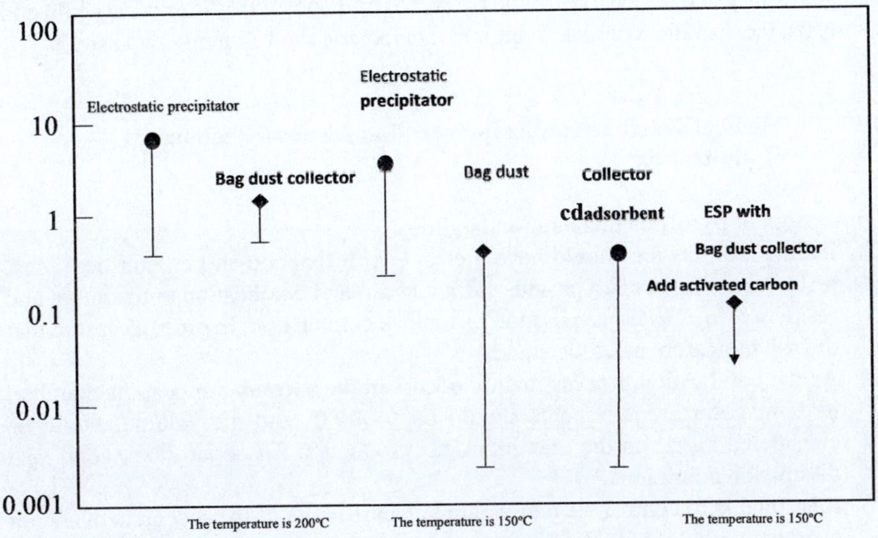

Fig. 4.4 Relationship between dust removal device and dioxin concentration

4.1.6 Medical Waste Incineration Disposal Technology Optimization Measures

4.1.6.1 General Requirements for the Optimization of Medical Waste Incineration Disposal Technology

When incineration is used to dispose medical waste, it is suggested to take the following combined measures to achieve dioxin emission reduction, taking corresponding measures for existing sources and new sources respectively, but the ultimate goal is the same, that is, to finally build a set of reasonably designed disposal facilities to meet the requirements of pollutant emissions in the incineration disposal process (Tsakona et al. 2007). Proposed incineration disposal technology optimization measures are shown in Table 4.6.

It can be seen from Table 4.6, that for incineration and disposal facilities, whether new facilities or existing facilities are constructed. The most fundamental goal is to achieve the completeness and progressiveness of the system from the technical point of view, to achieve the standard emission of dioxin and other pollutants. The final implementation of the construction and upgrading of disposal facilities will also consider local planning, construction costs and operating costs, and, of course, the vital feasibility and necessity. For the two types of disposal facilities, rotary kiln and pyrolysis, the specific technical requirements include the following two aspects.

4.1.6.2 Medical Waste Pyrolysis Incineration Disposal Technology Optimization

(1) Design of pyrolysis incineration facilities

(1) Incineration device should have a clear pyrolysis zone and combustion zone, residue burning zone to provide sufficient air, and combustion temperature and residence time, as far as possible to achieve combustion, in order to ensure that the ash indicators meet the standards.

(2) As the combustion process mainly occurs in the second combustion chamber, its temperature requirements should be 1400 °C, and the volume should be considered based on the maximum gas production for about 2.6 s, good heat dissipation conditions.

(3) It is suitable to control the temperature change range of the second combustion chamber within ±100 °C, which can be automatically adjusted, and the system sets the balance point of key parameters.

(4) The material of the equipment has a certain degree of adaptability, and can adapt to temperature changes, acid and alkali changes, etc.

(5) To configure a good automatic control system.

(6) Some technical approaches and requirements in rotary kilns are also suitable for pyrolysis incinerators.

(2) Flue gas cleaning of pyrolysis incineration facilities

Table 4.6 Medical waste incineration disposal technology optimization measures

Serial number	Technical requirements for new facilities	Existing facility renovation requirements pathway	Purpose
1	Configuration of the second combustion chamber	Adding a second combustion chamber	Guaranteed >2 s dwell time
2	Equipped with exhaust gas emission-furnace temperature-air conditioning feedback control system	Adding exhaust gas emission-furnace temperature-air conditioning feedback control system	Ensure enough oxygen content and adequate incinerator temperature
3	Equipped with automatic feeding system	Add automatic refilling system	Ensure stable feed
4	Equipped with flue gas reuse piping	Adding flue gas reuse pipes	Vent gas is well mixed to reduce excess air
5	Configuration of de-acidification facilities	Adding de-acidification facilities according to actual demand	Increase the exhaust gas purification performance of disposal facilities
6	Configure online monitoring and feedback control system	Add online monitoring and feedback control system	Timely grasp of exhaust gas emissions and timely adjustment of incineration conditions
7	Configuration of bag filter	Increase the configuration of the bag-type dust collector	Removal of particulate matter and particles adsorbed with dioxins
8	Equipped with activated carbon injection or activated carbon bed adsorption unit	Increase the configuration of activated carbon injection or activated carbon bed adsorption unit	Further adsorption of particulate matter, heavy metals and dioxins
9	Configuration of membrane fly ash washing system	Addition of membrane fly ash washing system	Cleaning of fly ash and proper collection
10	Equipped with advanced exhaust gas cleaning facilities such as SNCR	According to the actual, increase the advanced exhaust gas cleaning facilities	Increase the tail gas cleaning performance of disposal facilities
11	Configuration of low-temperature plasma facilities	Addition of low-temperature plasma system	Control of dioxin emissions through macromolecular bond breaking

(1) The tail gas treatment process gives priority to the dry or semi-dry process.
(2) Stable combustion, i.e., material, gas production, temperature, etc., are maintained at an equilibrium point as much as possible, so as to avoid large fluctuations, and making the actual operating condition be in line with the design condition as possible. When possible, continuous pyrolysis should be realized as much as possible.

(3) The gas production should be stable and uniform, otherwise the system spray emergency cooling water volume is too large and too small.
(4) Pyrolysis and incineration separation to avoid mutual interference. Strictly control the combustion air volume, pyrolysis temperature, and waste pyrolysis residence time in the pyrolysis zone.
(5) Ensure stable feeding and material balance during operation, without causing significant fluctuations, and the calorific value of materials should not be too high to avoid excessive thermal load on the secondary combustion chamber.

In addition, pyrolysis incineration technology is divided into vertical continuous pyrolysis, intermittent pyrolysis, and horizontal continuous pyrolysis. These three types of pyrolysis technologies have significant differences in process, and each type of technology has its own technical parameters and requirements, which need to be strengthened in research, classification, and comparison.

4.1.6.3 Optimization of Medical Waste Rotary Kiln Incinerator Disposal Technology

(1) Design of rotary kiln incineration facilities
(1) The incinerator should be designed to ensure its service life is not less than 10 years, the heat capacity intensity should be controlled at (60–90) × 104 kJ/m^3 h, the main material should be no less than Q235-A, the thinnest wall thickness should be no less than 16 mm (20–30 t/day furnace), the transmission device should be preferentially selected with large gear transmission and three-axis gearbox, and the speed regulation should adopt frequency conversion mode. The rotary kiln speed should be 0.2–2 r/min in the main power area of the driving motor.
(2) The technical performance of refractories used in incinerator should meet the requirements of combustion atmosphere of incinerator, the quality should meet the corresponding technical standards, and can bear the alternating thermal stress of incinerator working state. High alumina refractories should be given priority. The main components of refractories in rotary kiln should include Al_2O_3, SiO_2 and Cr_2O_3, of which Al_2O_3 content should not be less than 70%. Insulation bricks are not recommended in rotary kiln, and the service life of refractories should not be less than 8000h; The refractory material of the secondary combustion chamber can consider multi-layer structure, and should be equipped with insulation layer to improve thermal efficiency. The refractory material can appropriately reduce the content of Al2O3, but it should not be less than 40%, the service life should not be less than 36000h, and the metal wall thickness of the secondary combustion chamber should not be less than 10mm.
(3) There should be proper redundancy capacity, the redundancy capacity should be controlled within 115%, the waste feeding amount should be adjustable, and the feeding frequency should be controlled at about 40 times/h.

(4) The incinerator shall be equipped with explosion-proof doors or other explosion-proof facilities; An emergency discharge chimney should be set behind the combustion chamber, and a linkage device should be set so that it can only be started in an accident or emergency, such as power failure, induced draft fan failure, ultra-high and ultra-low boiler water level, over-temperature of bag dust removal flue gas inlet exceeding a certain time limit, positive pressure of secondary combustion chamber exceeding allowable value and other key equipment failure.

(5) It must be equipped with automatic control and monitoring system, which can display operating conditions (such as negative pressure, temperature, liquid level, flow rate, speed, equipment status, etc.) and tail gas emission parameters (smoke, nitrogen oxides, sulfur oxides, O_2, CO, HCl, CO_2) on line, and can automatically feedback and automatically adjust relevant main process parameters (exit temperature of the second combustion chamber, oxygen content, exit temperature of quench tower, etc.).

(6) Ensure that the oxygen content in the flue gas at the outlet of the incinerator reaches 6–10% (dry flue gas); The secondary combustion chamber should be set, and the residence time of flue gas above 1100 °C in the secondary combustion chamber should be more than 2s, and the flue gas quantity should be calculated under the maximum load condition.

(7) Slag ignition reduction rate should be less than 5%; The capacity of air combustion facilities should be able to meet the air distribution requirements for complete combustion of combustion materials in the furnace; Air heating device can be used; The number of fans shall be determined according to the setting requirements of incinerator; The maximum air volume of the fan should be 110–120% of the maximum calculated air volume; Air volume regulation should adopt continuous mode. The ability to start ignition and auxiliary combustion facilities should meet the requirements of ignition start-up and shutdown, and can support combustion when the calorific value of hazardous wastes is low. Auxiliary fuel burners should have good combustion efficiency, and should be equipped with automatic temperature control, temperature limit devices, flame monitoring, fire extinguishing protection and other safety devices. The auxiliary fuel should be determined according to local fuel sources.

(2) Flue gas purification of rotary kiln incineration facilities

(1) The semi-dry purification process includes semi-dry gas washing tower, activated carbon injection, bag filter and other treatment units, which should meet two requirements, namely, the residence time of flue gas in the reactor should meet the requirements of full reaction between flue gas and neutralizer; The temperature of flue gas at the outlet of the reactor should be above 130 °C to ensure that the flue gas in subsequent pipelines and equipment does not condense.

(2) The wet purification process includes quench scrubber and absorption tower (packed tower and sieve plate tower), which should meet two requirements, that is, wastewater treatment facilities must be equipped to remove harmful substances such as heavy metals and organic matter; Measures should be taken

to reduce the water content of flue gas before discharging through chimney to prevent the fan from carrying water.

(3) The flue gas purification device should also meet the following requirements: reliable anti-corrosion, anti-wear and anti-fly ash blocking measures; Suitable alkaline substances should be used as neutralizers for acidic pollutants such as hydrogen chloride, hydrogen fluoride and sulfur oxides, and neutralization reactions should be carried out in reactors; The temperature in the dust collector should be maintained above 30 °C above the dew point temperature of flue gas; Bag filter should pay attention to the selection of filter bag and bag cage materials, and the filter bag with film should be preferred, but the filter bag base material should be reasonably selected according to the corrosive gas components in flue gas, and the bag cage should be preferentially recommended to be made of stainless steel; The bottom of the dust collector should be equipped with a heating device, and the outside should be equipped with heat preservation, which should have the off-line automatic ash cleaning function; Long bag dust collector with low pressure pulse dust cleaning in separation line chamber should be preferentially used, and the gas-cloth ratio based on total gas is not more than 0.8 $m^2/m^3/s$. Tetrafluoroethylene-coated filter material should be used for bag filter, and the filter cloth has good hydrolysis resistance, oxidation resistance, high temperature resistance, fracture resistance and acid and alkali resistance.

(4) The high temperature flue gas produced by incineration of medical waste should be treated by rapid cooling, so that the flue gas temperature can be reduced to below 200 °C within 1.0 s minutes, and the residence time of flue gas in the temperature range of 200–600 °C can be reduced; The material of quench equipment is recommended as 316L, with a wall thickness of not less than 8 mm, and the scaling removal device and equipment insulation in the equipment should be considered; Two-fluid nozzle with 316L material is preferred for quenching atomization nozzle. The average Shaw diameter (SMD) of atomized liquid particles should be less than 140 μm, and the maximum particle diameter should be less than 200 μ m.

4.1.6.4 Medical Waste Incineration Disposal Technology Optimization

China's medical waste management problem must be solved in the face of China's national conditions. How to solve the technical problems of existing facilities and the technical selection of construction projects within the plan must have a clear understanding of the problem (Wang et al. 2019). According to the "Medical Waste Treatment and Disposal Pollution Prevention and Control Best Available Techniques Guide (Trial)" requirements, medical waste incineration treatment and disposal pollution prevention and control feasible technology combination as shown in Fig. 4.5.

In terms of feasible process parameters, using pyrolysis incineration technology, a combustion chamber temperature control is used in the reduction of heat absorption

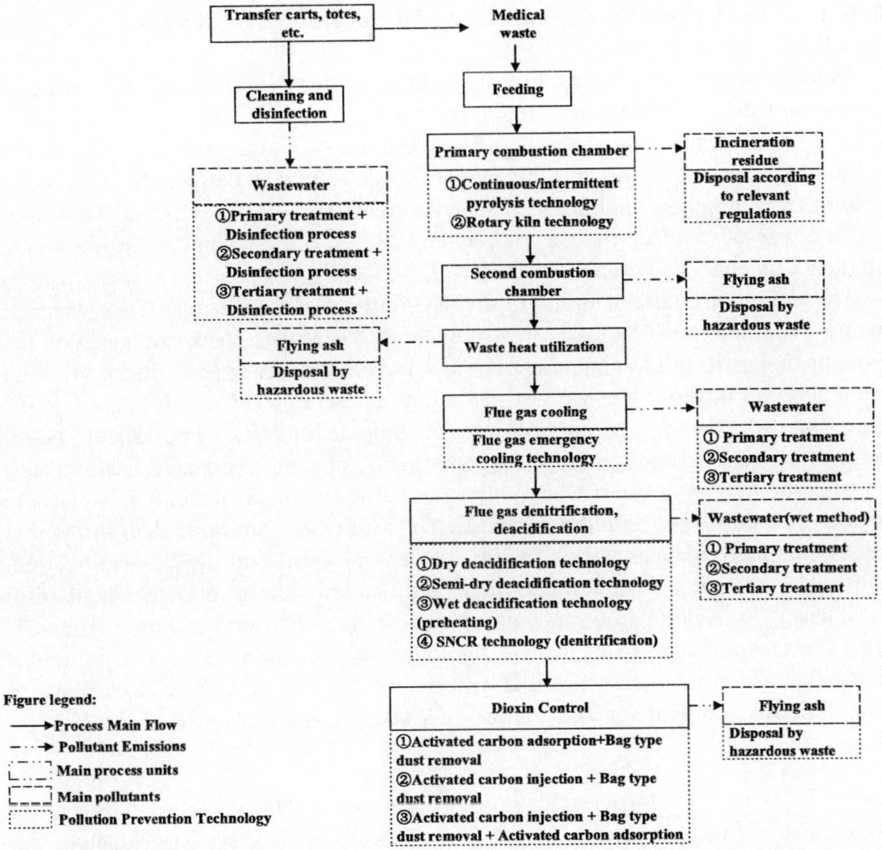

Fig. 4.5 Medical waste incineration treatment and disposal of pollution prevention and control of feasible technology combinations

stage at 35–350 °C. The oxidation exothermic stage of the furnace temperature should not be higher than 800 °C. When using rotary kiln incineration technology, a combustion chamber temperature should be controlled at 600–900 °C.

The temperature of the secondary combustion chamber should not be lower than 850 °C (chemical and pharmaceutical waste, not less than 1100 °C), and the flue gas residence time should not be less than 2 s. The pressure difference during the combustion at the beginning of the secondary combustion chamber is controlled at -10 mmH_2O, and the pressure differential during self-ignition period is controlled at -12 mmH_2O.

The high temperature hot flue gas enters the waste heat recovery unit and the temperature of the flue gas drops to about 600 °C after recovering most of the energy. The recovered waste heat can be used for baghouse heating, domestic heating, etc. The high temperature flue gas from the waste heat recovery device should be disposed of by emergency cooling, so that the flue gas temperature can be reduced to below

200 °C within 1 s, and the retention time of the flue gas in the temperature zone of 200–500 °C can be reduced.

In terms of the economic applicability of the medical waste incineration disposal technology, the rotary kiln incineration technology should be selected, and the daily disposal scale of the single unit should not be more than 10t. Pyrolysis incineration technology is selected, and the daily disposal scale should be between 5–10 T. Incineration technology is suitable for large-scale centralized disposal of medical waste, and the type of medical waste is more adaptable. Specific technical and economic applicability analysis is shown in Table 4.7.

The pollutant emission limits in the "Pollution Control Standards for Medical Waste Treatment and Disposal" just released in December 2020, are based on the existing domestic and foreign medical waste incineration flue gas purification, using technology to improve the level of particulate matter control which is used as the basis for strengthening the control of sulfur dioxide and NO_x. The existing large-scale incineration facilities, adopt the combination of semi-dry deacidification, activated carbon and bag filter. Under strict operation and management, it is possible to achieve compliance with the standards. The available monitoring data shows that large-scale medical incineration facilities equipped with semi-dry de-acidification, activated carbon injection and bag filter facilities are able to meet the standard in most cases. Activated carbon injection plus bag dust collection technology, SNCR and other technologies are the main technologies suitable for the control of pollutants in the flue gas of medical waste incineration and disposal in China. Its technology comprehensive comparison and applicability analysis is shown in Table 4.8.

Table 4.7 Economic applicability analysis of medical waste incineration disposal technology

Technology Type	Disposal Costs		Technical features and applicability
	Running costs/ (yuan/t)	Investment costs (equipment and installation)/ (RMB/t)	
Pyrolysis incineration Technology	1500–2500	100–150	Low flue gas volume, high heat utilization rate, metal is not easily oxidized to promote the formation of dioxin metal ion catalyst during the reaction under reducing conditions; suitable for the disposal of all medical waste with a single unit size of 5–10t/d
Rotary kiln burning Burning technology	2500–3500	150–200	Strong adaptability to medical waste, large treatment capacity, high heat utilization rate, complete combustion, mature technology, stable control; applicable to the disposal of all medical waste with a single disposal scale of 10 t/ d or more

Table 4.8 Comprehensive comparison and applicability analysis of flue gas pollutant control technologies

Project Name	De-acidification technology	Activated carbon injection with cloth bag Dust collection technology	SNCR Technology
Waste type	Applicable to all categories	Applicable to all categories	Applicable to all categories
Factory size	Suitable for any size	Suitable for any size	Suitable for any size
New construction/ remodeling	All facilities are to use	All facilities are to use	Any system can be added flexibly
Usage	Can be used alone or in combination with other technologies (semi-dry/ dry process generally used in combination with bag filter)	Can be used alone or in combination with other technologies	Can not be used alone, combined with activated carbon injection and bag dust collection technology
Flue gas pollutant removal efficiency	Acid removal efficiency: 80–95% Wet method can remove highly volatile heavy metals	Heavy metals, particulate matter: 99% Removable dioxins	NO_x: 30–80%

In the current incineration standards implemented in the course of 20 years, medical waste incineration disposal technology and pollution control method have been developed, especially for particulate matter, NO_x, SO_2 and other pollutants emission control, has formed a more complete and effective technical system. For the existing medical waste incineration facilities, the general flue gas purification process is a combination of semi-dry deacidification + activated carbon + cloth bag dust collector.

4.1.6.5 Medical Waste Incineration Disposal Technology Management Practice Cases

4.1.7 Medical Waste Incineration Disposal Technology

4.1.7.1 Case Introduction

The technical case is selected from a company in Shandong Province. The design and construction scale is 16,500 t/year (50 t per day of medical waste incineration production line). The annual operation days are not less than 330 days, with 40–120% load continuous operation capacity. The case of rotary kiln incineration disposal technology is shown in Fig. 4.6.

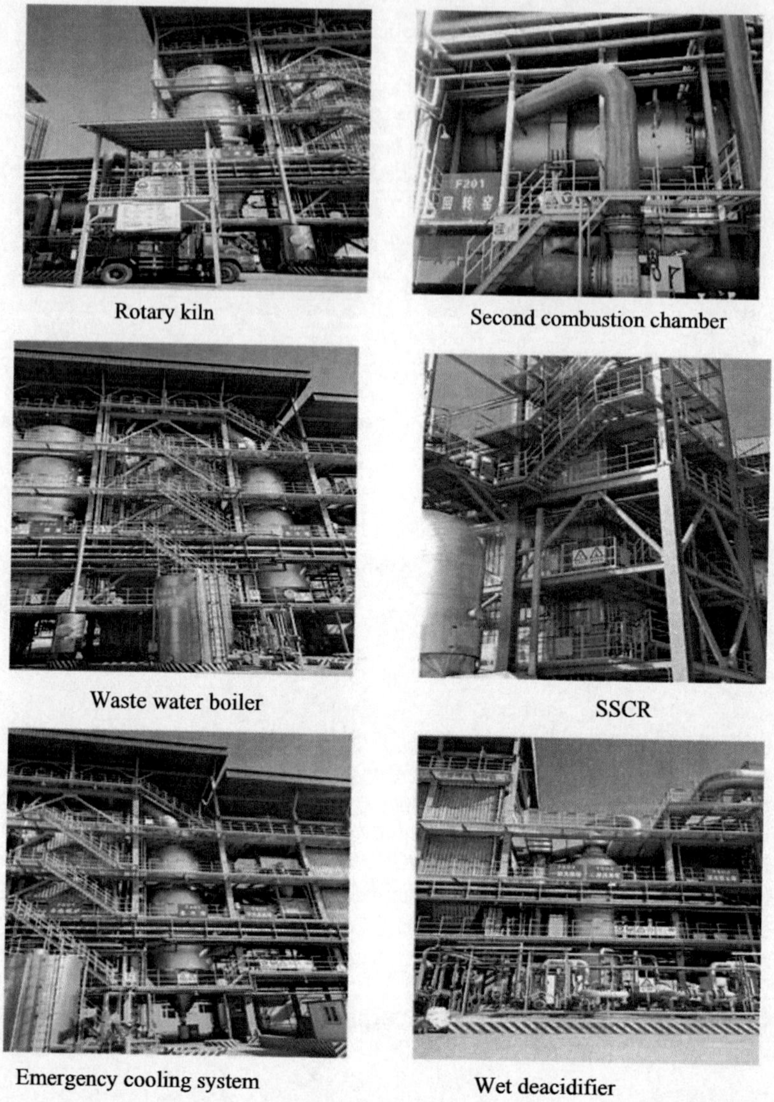

Fig. 4.6 Rotary kiln incineration disposal technology case application site equipment situation

4.1.7.2 Treatment Scale

At present, the company has built an incineration line with a daily treatment capacity of 50t.

50t/d medical waste incineration line as a backup line.

4.1.7.3 Processing Equipment

The main equipment and secondary pollution control equipment consists of rotary kiln, second combustion chamber, waste heat boiler, SCR, emergency cooling tower, bag filter, wet de-acidification equipment, etc. Equipment specifications and main technical parameters are shown in Tables 4.9 and 4.10.

4.1.7.4 Secondary Pollution Control

(1) Exhaust gas treatment unit

After a flue gas from the medical waste incinerator is treated by a set "emergency cooling + dry de-acidification + activated carbon adsorption + bag filter + two-stage wet de-acidification + wet electric dust removal + flue gas reheating + SCR denitrification" device, a 60 m high flue gas is discharged at the chimney.

(2) Wastewater/waste liquid treatment unit

Domestic wastewater and domestic sewage are collected separately, and the wastewater treatment process is disinfection + membrane bioreactor (MBR). Incineration flue gas scrubbing wastewater is mainly treated by evaporation of triple-effect evaporation device. Medical waste workshop rinsing wastewater is treated by one set of integrated sewage treatment device and sent to the park sewage treatment plant after reaching qualified discharge standards. The domestic sewage is discharged into the municipal sewage network and treated in the sewage treatment plant of the park.

(3) Solid waste treatment and disposal unit

The incineration fly ash is sent to the landfill of hazardous waste of the existing project; the sludge of the sewage station is sent to the incinerator of medical waste for incineration; the waste denitrification catalyst has not yet been generated and will be sent to qualified units for recycling and disposal; the incineration slag is general industrial solid waste and is sent to the landfill of hazardous waste of the existing project; the domestic garbage is removed by the sanitation department (Yang et al. 2004).

4.1.7.5 Treatment and Disposal Effects

The exhaust emissions meet the standards of Table 3 of the Hazardous Waste Incineration Pollution Control Standards. Table 1 of the Regional Comprehensive Emission Standards for Air Pollutants (DB 37/2376-2019) with the limits of the key control area, and Table 2 of the Environmental Health Standards for Medical Waste Incineration (GB/T 18773-2008). The waste water discharge meets the requirements of Table 1B of the Water Quality Standards for Sewage Discharge into Urban Sewers (GB/T 31962-2015). Noise meets the Class 3 of the Environmental Noise Emission Standard for Industrial Enterprise Factory Boundaries.

Table 4.9 Technical parameters of the main equipment

Serial number	Equipment name	Item	Content
1	Rotary kiln	Specification size (outside diameter x length)	–
2		Rotational Speed	0.1–1.0 r/min
3		Power	37 kW, IP55,F class, frequency control
4		Processing Capacity	≥50 t/d
5		Daily running time	24 h
6		Annual running time	≥8000 h
7		Continuous operation working time	≥2900 h/year
8		Main design life	>20 years
9		Refractory material service life	≥2 years
10		Combustion operation temperature	≥850 °C
11		Maximum operating temperature	>1300°C
12		Kiln head negative pressure	−50 Pa
13		Material residence time	30–120 min
14		Slag thermal scorch rate	≤5%
15		Material feeding method	Feeding method of screw conveyor into the furnace
16		Slag delivery method	Scraper + water seal. The ash is dropped into the water-filled scraper chute for water cooling and then sent out via the scraper conveyor
17		Combustion air conditioning method	Primary fan of rotary kiln feeds combustion air to combined burner and kiln head hood
18		External temperature	The outer surface temperature of rotary kiln shell: 180–230 °C, 55 °C for kiln head shell

(continued)

Table 4.9 (continued)

Serial number	Equipment name	Item	Content
19	Second combustion chamber	Daily running time	24 h
20		Annual running time	≥330d
21		Main design life	>20 years
22		Refractory material service life	>2 years
23		Combustion operation temperature	≥850 °C
24		Maximum operating temperature	≤1300 °C
25		Pressure in the furnace	−100 Pa
26		Slag thermal scorch rate	≤5%
27		Flue gas residence time	≥2 s
28		Surface temperature	≤55 °C
29		Natural gas burners	2 sets
30		Outlet smoke temperature	≥850 °C
31		Pressure in the furnace	−100 Pa
32		Rated evaporation capacity	8.2t/h
33		Rated pressure	1.6 MPa
34		Saturated steam temperature	201 °C
35		Daily running time	24 h
36		Annual running time	330d
37		Design life	>20 years
38		Inlet flue gas temperature	≥850 °C

(continued)

Table 4.9 (continued)

Serial number	Equipment name	Item	Content
39		Outlet flue gas temperature	≥500 °C
40		Maximum operating temperature	≤1300 °C
41		Discharge rate	2%
42	Emergency cooling system	Daily running time	24 h
43		Annual running time	330d
44		Shell design life	>20 years
45		Nozzle life	≥1 year
46	Emergency cooling system	Inlet flue gas temperature	500 °C
47		Outlet flue gas temperature	<200 °C
48		Rapid cooling time	<1 s
49		Surface temperature	≤60 °C

4.1.8 *Medical Waste Pyrolysis Incineration Disposal Technology*

4.1.8.1 Case Introduction

The technology case is selected in a company in Jiangxi, which designed and constructed a medical waste pyrolysis and combustion production facility with a daily capacity of 15t. The case of pyrolysis incineration disposal technology is shown in Fig. 4.7.

4.1.8.2 Treatment Scale

At present, the company has built a pyrolysis incineration line with a daily processing capacity of 15t.

Table 4.10 Technical parameters of secondary pollution control equipment

Serial number	Equipment name	Requirements Content
1	Bag filter	Fabric bag area 1728 m^2
2		Filtration speed 0.5 m/min
3		Daily running time 24 h, annual running time ≥330d
4		Fabric bag specification ϕ160 × 6000 mm
5		Design life >20 years Normal service life of filter bag ≥22,500 h
6		Design resistance ≤1500 Pa
7		The steel structure of the bag filter is designed for a temperature of 200 °C
8		Dust removal efficiency ≥99.99%
9	Wet deacidification equipment	Specification ϕ3000 × 12200 mm
10		Material 316L
11		Number of spraying layers 3 layers
12		Number of nozzles 5
13		Nozzle specifications 1.5 inch (1 inch ≈ 33.33 mm)
14		Acid removal efficiency ≥95%
15	Electrostatic precipitator	Outlet soot concentration <10 mg/Nm^3
16		Electric field flow velocity ≤1.0 m/s, residence time ≥6 s
17		Dust removal efficiency ≥85%
18		Wet type electric dust collector body pressure drop ≤500 Pa
19		Wet type electric dust collector body leakage rate ≤2.0%
20	SCR	GGH
21		Quantity 1pc
22		Type: rotary gas–air heat exchanger
23		Cold end flue gas inlet/outlet temperature: 160 °C/200 °C
24		Pressure loss 1000 Pa
25		Burners
26		Quantity 1pc
27		Burner arrangement: Horizontal combustion
28		Main fuel: Natural gas
29		Burner normal operation fuel consumption ≤55 Nm^3/h
30		SCR reactor
31		Quantity 1pc
32		Minimum allowable operating temperature 230 °C
33		Maximum allowable operating temperature 350 °C
34		Flue gas resistance 1000 Pa
35		Catalyst

(continued)

Table 4.10 (continued)

Serial number	Equipment name	Requirements Content
36		Type: Honeycomb
37		Layer 2 + 1
38		Active temperature range 230–350 °C
39		Specific surface area 670 m^2/m^3
40		Base material TiO_2
41		Active substance V O_{25}/WO_3
42		4 modules per catalyst layer
43		Flue gas flow rate in the catalyst 5.12 m/s
44		Soot Blower
45		Type: Sonic type
46		Quantity 2pcs
47		Equipment and piping insulation and anti-freezing shall be covered with stainless steel material

4.1.8.3 Processing Equipment

(1) Main body equipment and secondary pollution control equipment

The main equipment and secondary pollution control equipment consists of pyrolysis furnace, second combustion chamber, waste heat boiler, emergency cooling tower, acid removal tower, bag dust collector, etc. The main technical parameters of the disposal process are shown in Table 4.11.

(2) Incineration system

(1) Automatic loading machine. Medical waste is transported to the entrance of the incinerator charging bin through the metering and automatic lifting loading system, which can realize the action of automatic opening, turning, dumping and returning of the crates, and the empty crates are flushed and disinfected for temporary storage and reuse, so that the operators do not meet medical waste and are far away from pollution sources.

(2) Hopper. The hopper is installed on the entrance of the pyrolysis furnace, which is a temporary storage bin for medical waste and plays the role of sealing. The hopper inlet has a cover plate to ensure the hopper is sealed when no material is fed. A TV monitor is installed above the hopper, so that the operator can clearly understand the material level in the hopper in the central control room, and so that the material can be added in time to maintain the necessary material level height in the hopper. The bottom of the hopper is equipped with an automatic fire extinguishing device, which automatically controls the start of the solenoid valve by detecting the temperature of the lower part of the hopper, to enable the fire extinguishing medium (liquid carbon dioxide) to be sprayed

Incinerator

Second combustion chamber

Flash cooling tower and acid removal

Fig. 4.7 Pyrolysis incineration disposal case site equipment situation

Table 4.11 Main technical parameters of the disposal process

Serial number	Item	Parameter
1	Negative pressure of furnace chamber	20–50 Pa
2	Second combustion chamber temperature	1100–1200 °C
3	Boiler outlet carbon monoxide	<30 mg/m^3
4	Boiler outlet oxygen content	8–9%
5	Bag filter inlet temperature	150–190 °C
6	Online oxygen content	10–11%

into the hopper for emergency fire extinguishing. The incinerator is equipped with a special automatic feeding system, according to different medical waste packaging boxes using different fixtures. The material is automatically sent to the incinerator charging bin, and the material is continuously fed into the incinerator through the charging device for harmless incineration treatment, feeding, and charging system without leakage.

(3) Double-roller feeder. Double-roller feeder slowly rotates the medical waste in the bin for initial crushing and continues evenly into the furnace, while with the rotation of the furnace body, so that the medical waste in the furnace is evenly distributed to ensure the stability of the furnace incineration conditions. The hopper and the double-roller feeder cooperate with each other to ensure that the unburnt medical waste does not spill harmful gases, while keeping a certain volume of medical waste into the furnace in the hopper channel to prevent the smoke from the furnace from spilling out of the hopper.

(4) Secondary combustion chamber. Incinerator generates mixed combustible flue gas into the second combustion chamber for aerobic combustion. Second combustion chamber is a gas combustion chamber, and its main function is to ensure that the incinerator generated by high-temperature combustible gas combustion, the degree of flue gas and oxygen mixing, is the most important of the design of the secondary combustion chamber. The second combustion chamber operates at a temperature of about 1100 °C, with the flue gas residence time of more than 2 s, and can minimize the emission of harmful substances, especially dioxin emissions. The main body of the secondary combustion chamber consists of: flue gas inlet, secondary air distribution port, burner flame inlet, flue gas outlet, ash sink chamber, cleaning door, etc. The supplemental air inlet (secondary blast) provides the required oxygen in proportion to the burner, which is equipped with a diesel (or natural gas) burner to provide heat for combustion during the start-up phase and in the event of insufficient pyrolysis gas.

(5) Combustion air system. The combustion air includes the primary combustion air of the incinerator (primary air) and the secondary combustion air (secondary air). The combustion air is fed into the flue side of the second combustion chamber, and the air required for auxiliary fuel. Equipment includes air supply fan (primary fan, secondary fan), the corresponding air volume adjustment system (frequency converter, control system) and various pipelines, valves, etc.

(6) Auxiliary fuel system. Auxiliary fuel system consists of oil nozzle, storage tank, oil pump, ignition device, the corresponding automatic control system and connection piping, etc., and has two functions of auxiliary combustion and starting combustion.

(7) Explosion-proof system. During the operation of the system, there will be special conditions (deflagration, power failure, equipment failure, etc.) when the system is in operation, the incinerator is being pyrolyzed, the secondary combustion chamber is burning, and the flue gas in the system easily escapes. In this case, an effective explosion-proof measure is required. It is installed on the flue connecting the incinerator to the secondary combustion chamber and has both

explosion-proof and emergency release functions. The emergency explosion-proof device is used for emergency flue gas venting in case of general accidents and power outages, and has both emergency and safety venting functions.

(3) Flue gas treatment system

(1) Waste heat boiler. Waste heat boiler is a special boiler supplied as a package with incineration process equipment. The flue gas outlet temperature of the waste heat boiler is 600 °C, avoiding the temperature range of 200–500 °C for dioxin synthesis. The flue gas exit from the waste heat boiler and enters the emergency cooling tower. The original one-piece semi-film wall semi-convection tube bundle boiler is changed to independent vertical film wall boiler + independent convection tube bundle water tube boiler. The reason of equipment adjustment is that the original integrated semi-film wall semi-convective tube bundle boiler is horizontal due to the space and space problem, and the ash accumulation is serious, especially in the convective tube bundle part of the ash accumulation which cannot be handled by online ash cleaning, and needs to be cleaned manually, and this result to frequent start and stop of the whole production line, which affects the production efficiency. The new boiler is vertical and the ash accumulation and cleaning effect is obvious. In addition, due to the adoption of the split process, the convection tube bundle, which is more prone to corrosion, can be regularly repaired and replaced to reduce equipment costs.

(2) Emergency cooling tower. Hollow tower structure, consists of a tower body, inlet flue, outlet flue, ash drop bucket and nozzle, etc. The design inlet flue gas temperature is (600 ± 20) °C, and the design outlet flue gas temperature is (200 ± 5) °C (in 1 s time from 600 °C cooling to 200 °C). Avoid the generation of dioxin again.

(3) Acid removal spray tower. The NaOH solution supplied by the lye system is atomized into 200 μm droplets by the atomizing nozzle, and then by the upward hot flue gas, and a high-density area of alkaline droplet suspension is formed in the upper part of the tower to neutralize the acidic substances in the flue gas. The reacted flue gas in the tower enters the lower stage equipment with the dry powder dust of the reaction products.

(4) Activated carbon feeder. Generally, a screw conveyor is used to feed the activated carbon into the flue through the screw conveyor and send the activated carbon into the flue through the spiral piece in the pipe, which will disperse the activated carbon in the flue gas in the process of flue gas flow and make full contact with the flue gas to achieve the purpose of adsorption of dioxins and heavy metals.

(5) Cloth bag dust collector. Bag dust collector according to "pulse blowing class bag dust collector technical conditions" (JB/T 8532-1997) for manufacturing, inspection and acceptance. Bag filter is used to remove medical waste after incineration of dust, heavy metals and dioxins in the flue gas. It is the last line of defence for tail gas treatment. The bag filter is composed of silo body, bag, bag cage, blowback system and automatic control system. The concentration of flue gas emission after bag filtering and the service life of bag filter depend on the design parameters of the bag, and the filtering area cannot be large or small. Bag filter is not only for dust removal, but also for dioxin suppression. Adding a

screw conveyor in the ash discharge system to transport the fly ash directly to the hazardous waste storage room can reduce the environmental pollution caused by the transfer of fly ash in the field and reduce the danger to the operator.

(4) Wastewater treatment measures

Canteen wastewater enters the wastewater treatment system after oil separation, and office living wastewater enters the wastewater treatment system after septic tank.

The wastewater of flushing in the factory is mixed with canteen wastewater and domestic wastewater, and then stored in the reuse pool after sedimentation + parthenogenic + disinfection + carbon filtration, and all of them are reused in the production, and there is no external discharge.

4.1.8.4 Lessons Learned

(1) Waste heat boiler adjustment

The original one-piece semi-film wall semi-convection tube bundle boiler was changed to an independent vertical film wall boiler + independent convection tube bundle water tube boiler. The reason for adjusting the equipment is that the original one-piece semi-film wall semi-convection tube bundle boiler is horizontal due to the space and space problem, and the ash accumulation is massive, especially in the convection tube bundle part of the ash accumulation which cannot be handled by online ash cleaning. It requires manual cleaning, resulting in frequent starting and stopping of the whole production line, which affects the production efficiency. The new boiler is vertical and the ash accumulation and cleaning effect is obvious. In addition, due to the adoption of the split process, the convection tube bundle, which is more prone to corrosion, can be regularly repaired and replaced to reduce equipment costs.

(2) Bag filter equipment adjustment

The existing bag system is modified, and a screw conveyor is added to the ash discharge system to transport the fly ash directly to the hazardous waste storage room. This can reduce the environmental pollution caused by the transfer of fly ash in the field and reduce the hazard to the operators.

(3) Adding flue gas washing system

Flue gas washing system can more effectively remove particulate matter, sulfur dioxide, nitrogen oxides and other harmful substances dissolved in the washing liquid, and reduce the flue gas temperature, so as to control the flue gas pollutants and further reduce the pollutant emission value.

4.1.8.5 Medical Waste Incineration Disposal Flue Gas Pollutant Control New Technology

Due to the complexity and uncontrollability of medical waste components, as well as the variability of the incineration process, the type and content of pollutants in the flue gas generated by the incineration of medical waste have a wide range of variation, and its wastewater emissions can meet the requirements of the standard under the strengthening of operational management. Therefore, the pollutant control technology for medical waste incineration and disposal mainly focuses on flue gas pollutant control technology (Remme 1998).

(1) New Dioxin Control Technology

Dioxins are the general term for chlorinated dibenzo-p-dioxins and chlorinated dibenzofurans, a group of 210 isomers containing chlorinated tricyclic aromatic compounds. The main sources of dioxins in flue gas are dioxins contained in medical waste and other incoming materials; chlorine, chlorine phenols, polychlorophenols, which are contained in the medical waste itself or produced in the combustion process are converted into dioxins. During the incineration process, hydrocarbons are polymerized and cyclized to form polycyclic aromatic hydrocarbons, and then react with chlorine to produce dioxins. Flue gas dioxin control and removal measures are mainly as follows: First, to ensure that the flue gas in the combustion chamber operates at a temperature of >850 °C with residence time of more than two seconds, so that dioxins can be decomposed. Second, the decomposition of dioxins can be 247–597 °C when the dioxins re-synthesize, as well as hydrogen chloride and monomeric chlorine at 200–300 °C, due to the catalytic effect of metals to generate dioxins, by using rapid cooling. The method to minimize the time of the flue gas in the temperature zone of 200–600 °C. Third, the use of activated carbon adsorption removal. At present, the use of low-temperature plasma treatment technology dioxins to achieve more satisfactory results. After the flue gas re-heater is heated to 130 °C into the low-temperature plasma dioxin processor, integrated dioxin treatment integrated treatment technology is based on plasma super oxidation, bond breaking mechanism as the core, to achieve dioxin removal in flue gas.

(2) New technology for NOx control

NO_x includes NO, NO_2, N_2O and NO_{23}, and the NO_x in flue gas is mainly NO, which is mainly generated by the decomposition of nitrogen-containing compounds in medical waste and the high-temperature oxidation of nitrogen in the air for combustion. At present, the domestic general use of SCR and SNCR two NO_x treatment technology. Large-scale medical waste incineration and disposal enterprises can also use SCR as NO_x treatment technology.

References

Alvim Ferraz M C M, Cardoso J I B, Pontes S L R. Concentration of atmospheric pollutants in the gaseous emissions of medical waste incinerators [J]. Journal of the Air and Waste Management Association, 2000, 50(1): 131-136.

Coutinho M, Rodrigues R, Duwel U, et al. The DG European dioxin emission inventory-stage II: characterization of the emissions of 2 hospitals waste incinerators in Portugal [J]. Organohalogen Compounds, 2000(46): 287-290.

Guidelines for the management of biomedical waste in Canada [S]. Ottawa: Canadian Council of Ministers of the Environment, 1992.

Kilgroe J D. Control of dioxin, furan, and mercury emissions from municipal waste combustors [J]. Journal of Hazardous Materials, 1996(47): 163-194.

Liu Z. S. Process introduction and environmental analysis of medical waste pyrolysis gasification incinerator system [J]. Chemical Engineering and Equipment, 2008(8): 124-127.

Malkan S, Nelson J. Global trends in responsible healthcare waste management: a perspective from healthcare without harm [J]. Waste Management, 2005(25): 567-574.

Medical waste management and disposal [R]. Pollution Technology Review No. 200, Noyes Data Corporation, Washington, D.C., USA, 1986.

Yan Mi. Study on dioxin generation inhibition and environmental impact of incinerator during medical waste incineration [D]. Hangzhou: Zhejiang University, 2012: 36.

Niessen W R. Combustion and incineration processes [M]. New York: Marcel Dekker Inc., 2002: 301-308.

Pei Zhaotang, Wu Weixiang. The application of rotary kiln incinerator in medical waste treatment [J]. Environmental Protection Science, 2007, 25(2): 55-56.

Remme T V. Evaluation of the available air pollution control technologies for achievement of the MACT requirements in the newly implemented new source performance standards (NSPS) and emission guidelines (EG) for hospital and medical/infectious waste incinerators[J]. Waste Management, 1998(18): 393-402.

Stockholm convention on persistent organic pollutants (POPs) [R], 2001.

Sun ZP, et al. Experimental study of pyrolysis and gasification of typical components of medical waste [J]. Power Station System Engineering, 2005(9): 13–15, 18.

Science Times. Best disposal for SARS medical waste: Burn! [EB/OL]. [2003-05-14]. http://www.cas.cn/zt/jzt/kpzt/zkywfzfdzgx/zlff/200305/t20030514_2666762.shtml.

Tsakona M, Anagnostopoulou E, Gidarakos E. Hospital waste management and toxicity evaluation: a case study [J]. Waste Management, 2007(27): 912-920.

Tudor T L, Noonan C L, Jenkins L E T. Healthcare waste management: a case study from the national health service in Cornwall, United Kingdom [J]. Waste Management, 2005, 25(6): 606-615.

Uysal F, Tinmaz E. Medical waste management in Trachea region of Turkey: suggested remedial action [J]. Waste Management and Research, 2004(22): 403-407.

Vesilind P, Worrell W, Reinhart D. Solid waste engineering [M]. California, CA: Thompson Learning Inc., 2002.

Walker B L, Cooper C D. Air pollution emission factors for medical waste incinerators [J]. Journal of the Air and Waste Management Association, 1992(42): 784-791.

Wang YANGYANG, Zhang YUE, Zhang YUEHONG, et al. Practice and exploration of the innovation of medical waste total supervision system management in Guiyang city [J]. Environmental Protection Science and Technology, 2019, 25(1): 52–57.

WHO. Safe healthcare waste management policy paper by the World Health Organization [J]. Waste Management, 2005(25): 567–574.

WHO. Assessment of small-scale incinerators for healthcare waste [S], 2004.

Windfeld E S, Brooks M, Su L. Medical waste management: a review [J]. Journal of Environmental Management, 2015, 163(1): 98-108.

Xie R, Li W J, Li J, et al. Emissions investigation for a novel medical waste incinerator [J]. Journal of Hazardous Materials, 2009(166): 365-371.

Yang YN, Xiong YS, Yang J, et al. Solid waste treatment and disposal engineering and management [M]. Beijing: Science Press, 2004.
Chen Yang, Liu Liyuan, Feng Qinzhong, et al. Key issues study on the operation management of medical waste incineration disposal facilities [J]. Procedia Environmental Sciences, 2012(16): 208-213.
Zhang Y, Xiao G, Wang G X. Medical waste management in China: a case study of Nanjing [J]. Waste Management, 2009(29): 1376-1382.
Zhao L Y, Zhang F S, Wang K S, et al. Chemical properties of heavy metals in typical hospital waste incinerator ashes in China [J]. Waste Management, 2008(29): 1114-1121.
Zhu H M, Yan J H, Jiang X G. et al. Study on pyrolysis of typical medical waste materials by using TG-FTIR analysis [J]. Journal of Hazardous Materials, 2008(153): 670–676.

Chapter 5
Medical Waste Non-incineration Optimized Treatment Technology

In recent years, medical waste non-incineration treatment technology has been developed significantly. This chapter introduces the pollution control technology of high-temperature steam treatment of medical waste, chemical disinfection treatment pollution control technology, microwave disinfection treatment pollution control technology and high-temperature dry heat treatment technology. Furthermore, it introduces the newly emerged ethylene oxide treatment technology, hot melt disinfection curing molding treatment technology, as well as mobile treatment technology and small decentralized treatment technology. The pollution control measures of various non-incineration treatment technologies of medical waste are comprehensively analysed, and corresponding optimization measures are proposed. The typical non-incineration treatment technologies of medical waste and application practical cases are also introduced and analysed.

5.1 Medical Waste High-Temperature Steam Treatment Pollution Control Technology (Cheng et al. 2007)

5.1.1 Technology Overview

The main hazard of medical waste is infectious pathogenicity. Medical waste is exposed to a certain temperature of water vapor atmosphere and stay for a certain period, and using of high temperature steam to kill pathogenic microorganisms in medical waste, eliminates the potential infectious hazards of the treatment method. This technology is applicable to the treatment of infectious waste, injury waste and pathological waste such as human tissue and pathological wax blocks discarded after pathological sectioning in the "Medical Waste Classification Catalogue", and it is not applicable to the treatment of pharmaceutical waste and chemical waste.

Y. Chen et al., *Innovative Disposal Technology and Management Practice for Medical Waste*, https://doi.org/10.1007/978-981-99-6786-5_5

5.1.2 *Treatment Process*

The typical flow of high-temperature steam treatment process typical process is shown in Fig. 5.1. The process is as follows: steam treatment equipment preheating → loading medical waste → treatment equipment cavity vacuum → into the steam treatment → exhaust gas discharge and drying waste → unloading medical waste → mechanical treatment (crushing or compression).

The high temperature steam centralized treatment process for medical waste should be equipped with at least one process link to enhance the thermal penetration (Dengchao et al. 2021) and thermal homogeneity of the steam, including but not limited to: ① steam treatment before the disinfection bin vacuum operation; ② steam treatment before the medical waste crushed to a smaller particle size; ③ steam treatment process stirring medical waste. The main process units include; receiving and storage unit, steam supply unit, feeding unit, high temperature steam treatment unit, crushing unit, compression unit, disinfection treatment residue disposal, cleaning and disinfection unit, exhaust gas treatment unit, wastewater treatment unit, solid waste treatment and disposal, noise control, etc.

The use of high-temperature steam to kill pathogenic microorganisms in medical waste is the main part of the high-temperature steam treatment process of medical waste. For medical waste management considerations, to avoid the illegal use of medical waste and recycling, generally require the need to be disfigured, while the disfigured medical waste in a certain degree of sensory improvement. To reduce the cost of transporting waste after high-temperature steam treatment, it is usually supplemented by compression measures. The core equipment of the high-temperature steam treatment system is the high-temperature steam treatment equipment, which also includes the pre-vacuum treatment equipment, mechanical crushing treatment

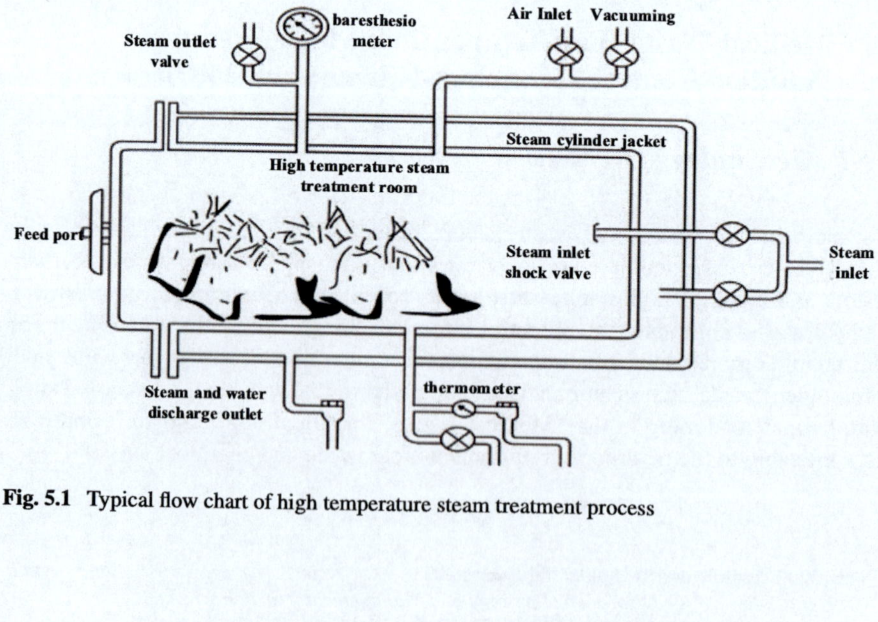

Fig. 5.1 Typical flow chart of high temperature steam treatment process

equipment. The main types of industrial processes include: ① steam treatment before the disinfection bin vacuum operation; ② steam treatment before the medical waste crushed to a smaller particle size; ③ steam treatment before the disinfection bin vacuum operation + steam treatment while mixing or crushing medical waste.

In addition, the treatment of waste gas and waste liquid mainly includes two aspects: First, the medical waste in the heating, humidification before part of the untreated extracted gas and leakage of liquid disinfection treatment. The second, is the possibility of heating and humidification process with the precipitation of volatile organic compounds and heavy metal substances for treatment.

5.1.3 Pollutant Emissions

High-temperature steam treatment of medical waste often adopts two process forms: steam treatment before crushing, steam treatment and crushing simultaneously. For the process of steam treatment before crushing, the processing unit includes feeding, pre-exhaust, steam supply, disinfection, exhaust pressure relief, drying, crushing and other process units, while for the steam treatment and crushing process at the same time, the processing unit includes feeding, steam supply, stirring crushing + disinfection, exhaust pressure relief, drying and other process units.

Medical waste high-temperature steam centralized treatment process generated by the exhaust gas mainly comes from the high-temperature steam treatment and the vacuum extractor, storage, feeding and unloading, crushing and other links before and after treatment. The main pollutants in the exhaust gas generated in the vacuum extraction process are VOCs, particulate matter, etc., which are organized emissions. The main pollutants are VOCs and odor gas, which are unorganized emissions, in storage, feeding, unloading, crushing, etc. The wastewater from centralized treatment process mainly comes from high temperature steam treatment, cleaning and disinfection of transportation vehicles and crates, cleaning and disinfection of production areas such as unloading area and storage area, cleaning and disinfection of high temperature steam treatment and crushing equipment, as well as initial rainwater and accident wastewater from production area and wastewater treatment area. The main pollutant indicators are pH, five-day biochemical oxygen demand (BOD_5), chemical oxygen demand (COD_{Cr}), suspended solids (SS). The solid waste generated from the centralized treatment process mainly includes disinfection treatment residue, failed fillers from the exhaust gas treatment unit, and sludge from wastewater treatment.

Typical high temperature steam treatment process flow and pollution production nodes are shown in Figs. 5.2 and 5.3.

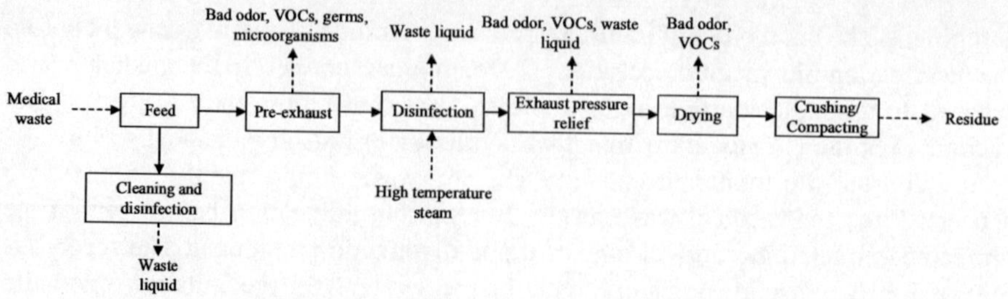

Fig. 5.2 Medical waste high-temperature steam technology first steam treatment and then crushing process flow and pollution production nodes

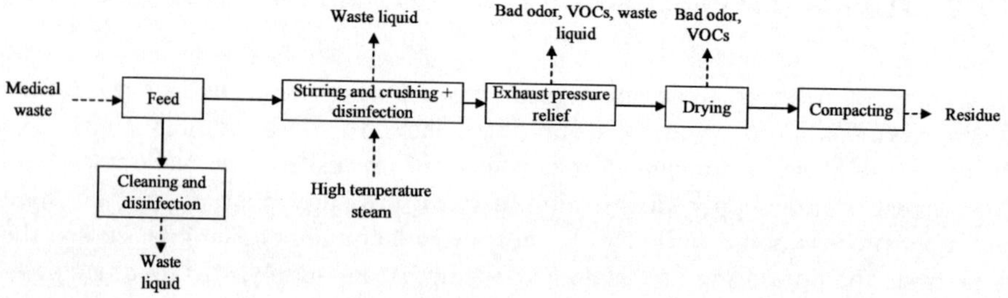

Fig. 5.3 Medical waste high-temperature steam technology steam treatment and crushing at the same time process flow and pollution nodes

5.1.4 High Temperature Steam Treatment Technology Analysis

To promote the development of this medical waste disposal work, to ensure the effectiveness of disposal, the main factors to be considered for the management of the link are as follows:

(1) Ensure that the process parameters of the treatment facilities meet the corresponding standard requirements. In terms of high-temperature steam treatment facilities, and the conventional process parameters for the temperature, pressure, processing time, etc., the general requirements of the steam treatment process should be in the disinfection temperature ≥ 134 °C, pressure ≥ 0.22 MPa (gauge pressure) conditions, and the corresponding disinfection time should be ≥45 min.
(2) Strengthen the source of medical waste classification management, to eliminate radioactive waste, chemical waste and pharmaceutical waste mixed into the disposal, to ensure that medical waste is safely disposed off.

(3) The disposal effect of medical waste should be tested regularly to ensure that the logarithm of bacteriophage killing meets the requirements of the corresponding standards. The logarithmic value of killing for the bacterium Bacillus thermophilus should not be less than 4.
(4) Other pollution control should be strengthened, such as odor, VOCs and particulate matter, etc. Necessary tail gas purification measures should be taken to reduce their pollution to the environment.

5.2 Pollution Control Technology for Chemical Treatment of Medical Waste

5.2.1 Technology Overview

Medical waste chemical disinfection (Chen 2015) treatment technology is the use of chemical disinfectants to kill pathogenic microorganisms in medical waste, so that it eliminates the potential for infectious hazards of treatment methods. This technology is applicable to the treatment of infectious waste, injurious waste and some pathological waste in the Medical Waste Classification Catalogue; it is not applicable to the treatment of pharmaceutical waste and chemical waste. Chemical disinfection treatment technology has a long history and a relatively wide application in disinfection and sterilization. For centralized medical waste chemical disinfection treatment facilities, dry chemical disinfectants and ethylene oxide are preferred as chemical disinfectants. Dry chemical disinfection technology is the use of calcium oxide, neutralizer and other compound dry chemical disinfectants to kill pathogenic microorganisms in medical waste, and its function is to eliminate the potential infectious hazards of treatment. Ethylene oxide disinfection technology is based on ethylene oxide that can react with biological proteins, so that DNA and RNA non-specific alkylation, at room temperature can kill pathogenic microorganisms in medical waste which enables it to eliminate the potential hazards. It is a treatment method to eliminate potential infectious hazards. Dry chemical disinfection technology has been commercially available in the United States and other countries since the mid-1980s. Ethylene oxide disinfection technology is a new technology that has emerged in China in recent years.

5.2.2 Treatment Process

The typical process flow of dry chemical disinfection treatment and ethylene oxide disinfection treatment are shown in Figs. 5.4 and 5.5, respectively. Dry chemical disinfection treatment uses the process of crushing and chemical disinfection at the same time while ethylene oxide disinfection treatment uses the process of disinfection first and then crushing.

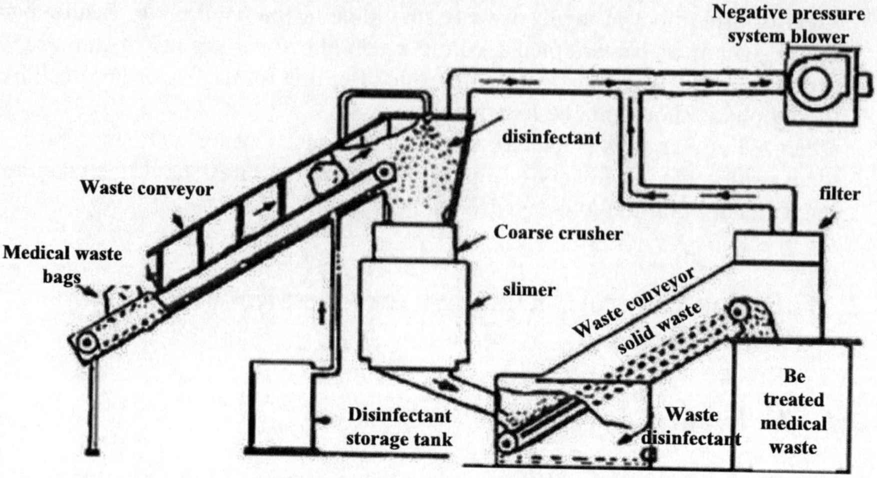

Fig. 5.4 Typical process flow diagram of dry chemical disinfection treatment

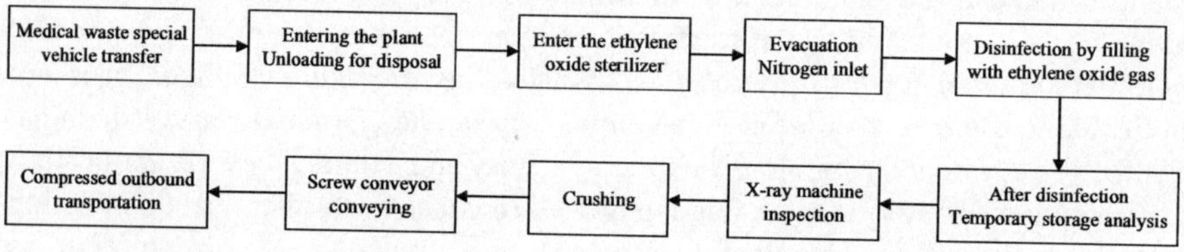

Fig. 5.5 Typical process flow diagram of ethylene oxide disinfection treatment

The efficiency of dry chemical disinfectants to inactivate microorganisms is related to contact time, temperature, chemical disinfectant concentration, pH (acidic and alkaline environment), and the number and type of microorganisms killed. Chemical disinfectant inactivation effect must ensure that the chemical disinfectant and medical waste disinfection surface has sufficient reaction contact time. When killing bacterial spores, the contact time and temperature of the reaction should be increased, which usually takes one or several hours. Typically, the effectiveness of disinfection increases as the temperature increases. Therefore, to ensure disinfection, adequate contact time and reaction temperature must be ensured.

The following are the advantages of dry chemical disinfection treatment technology. They include: simple process equipment and operation, with low one-time investment, low operating costs, and high capacity reduction rate of waste, site selection is convenient, it can be mobile processing, and environmental pollution is very small and so on. However, the crushing system requirements are high and the operation process (automation level) is very demanding.

For ethylene oxide disinfection treatment technology, disinfectant ethylene oxide is mainly used in the liquid form of tank equipment. After the pre-vacuum degree, system temperature and relative humidity of the ethylene oxide disinfection cabinet reach the specified requirements, the system is filled with a certain concentration of ethylene oxide gas, and the disinfection time is maintained for 4 h to achieve the purpose of medical waste disinfection, which is the first application of ethylene oxide in the field of medical waste treatment in China.

5.2.3 *Pollutant Emissions*

Medical waste chemical disinfection disposal system equipment generally includes receiving storage unit, feeding unit, crushing unit, disinfectant supply unit, chemical disinfection treatment unit, discharge unit, disinfection treatment residue disposal, cleaning and disinfection unit, exhaust gas treatment unit, wastewater treatment unit, solid waste disposal, noise control, etc.

The exhaust gas from the treatment process mainly comes from chemical disinfection treatment and vacuuming, storage, into the unloading, crushing and other links before and after the treatment. The pollutants are mainly malodors, VOCs, particulate matter, etc. The disinfection process using ethylene oxide as disinfectant also releases ethylene oxide gas. Wastewater from the treatment process mainly comes from chemical disinfection treatment, cleaning and disinfection of transport vehicles and crates, cleaning and disinfection of operating areas such as unloading sites and storage sites, chemical disinfection treatment and cleaning and disinfection of crushing equipment, as well as initial rainwater and accident wastewater from production areas and wastewater treatment areas. The main pollutant indicators are pH, five-day biochemical oxygen demand (BOD_5), chemical oxygen demand (COD_{Cr}) and suspended solids (SS). The solid wastes generated from the treatment process are mainly disinfection treatment residues, failed fillers from the exhaust gas treatment unit, and sludge from wastewater treatment. The dry chemical disinfection treatment and ethylene oxide disinfection treatment process flow and pollution production nodes are shown in Figs. 5.6 and 5.7, respectively.

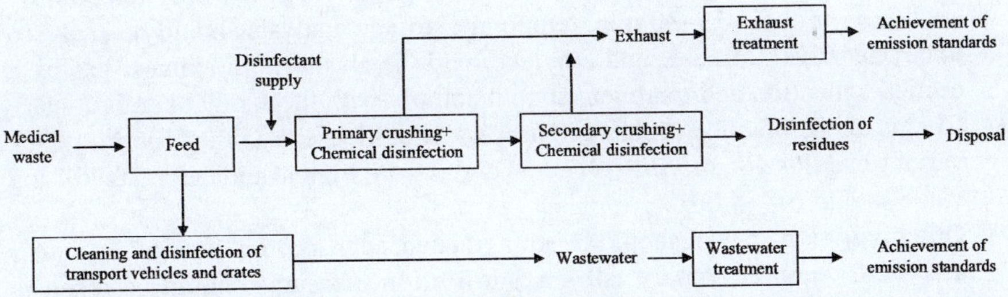

Fig. 5.6 Dry chemical disinfection process flow and pollution production nodes

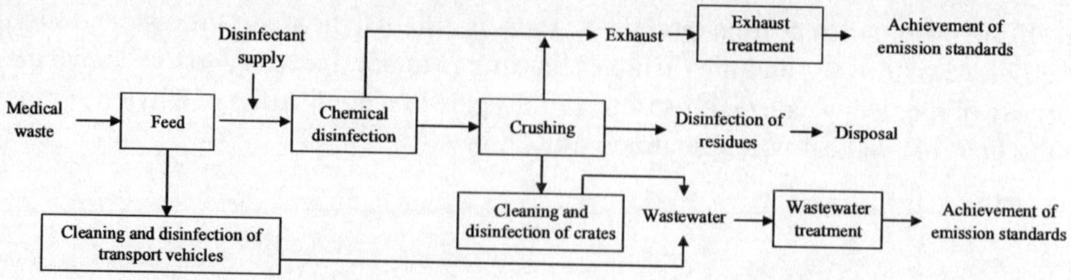

Fig. 5.7 Ethylene oxide disinfection treatment process and pollution production nodes

5.2.4 *Chemical Disinfection Treatment Technology Analysis*

(1) For dry chemical disinfection treatment technology, the key is to use highly effective chemical disinfectants, and should ensure that the concentration of the selected agent and the content of bacteria, viruses and fungi tested remain at normal levels to achieve the required disinfection effect. The content of calcium oxide in the dry chemical disinfectant used should be more than 90%, and the particle size of calcium oxide should not be greater than 200 mesh. The dry chemical disinfectant dosage should be greater than 0.075–0.12 kg dry chemical disinfectant/kg medical waste, and the water spray ratio should be 0.006–0.013 kg/kg medical waste to ensure that the disinfection temperature is ≥90 °C and the pH of the strongly alkaline environment controlled by the reaction, is in the range of 11.0–12.5. 11.0 ~ 12.5. Furthermore, the total contact reaction time between dry chemical disinfectant and crushed medical waste should be >120 min.

(2) For dry chemical disinfection treatment technology, the purity of ethylene oxide should be greater than 99.9%, the concentration of ethylene oxide should be ≥900 mg/L, the disinfection temperature should be controlled in the range of (54 ± 2) °C, the disinfection time should be ≥4 h, the relative humidity should be controlled in the range of 60~80%, and the initial pressure of −80 kPa vacuum environment. After disinfection of medical waste has been temporarily analysed for about 15–30 min, temporary storage analysis should be analysed under negative pressure, and ethylene oxide analysis room exhaust gas by a unified collection and treatment should comply with the standard emissions.

(3) Chemical disinfection treatment process should use Bacillus *subtilis* black variant bacillus (*B. subtilis ATCC 9372*) as a biological indicator. Its kill log value should be ≥4.00.

(4) Other pollution control should be strengthened, such as odor, particulate matter and VOCs, etc. Necessary tail gas purification measures should be taken to reduce their pollution to the environment.

5.3 Medical Waste Microwave Disinfection Treatment Pollution Control Technology (Ling et al. 2007; Luo et al. 2020; Feng'e, 2019)

5.3.1 *Technology Overview*

Medical waste microwave disinfection technology is the use of microwave action to kill pathogenic microorganisms in medical waste. Its function is to eliminate the potential for infectious hazards of treatment methods. The microwave treatment system is wetted under controlled conditions and the waste is crushed and placed in a tank, and the waste is disinfected by microwaves, reducing the volume of waste by 60–90%, and the treated waste is indistinguishable from other waste. Bacillus microbial spore test showed that this method can achieve harmless treatment of medical waste. This technology is suitable for the treatment of infectious waste, injurious waste and unidentifiable pathological waste such as discarded human tissues and pathological wax blocks after pathological sectioning in the "Medical Waste Classification Catalogue", but not for the treatment of pharmaceutical waste and chemical waste. The application of this technology may be due to microwave leakage and other potential occupational risks, in terms of operating and investment costs may be higher than high-temperature steam and chemical treatment, but lower than the incineration method.

5.3.2 *Microwave Treatment Process and Disinfection Mechanism*

The process of microwave disinfection of medical waste centralized treatment project can decide to choose microwave disinfection treatment process or microwave and high-temperature steam combination disinfection treatment process. The typical process flow of microwave disinfection treatment of medical waste is shown in Fig. 5.8.

5.3.3 *Pollutant Emissions*

Medical waste microwave disinfection treatment technology generally consists of receiving storage system, feeding unit, crushing unit, disinfection treatment unit, discharge unit, disinfection treatment residue disposal, cleaning and disinfection unit, exhaust gas treatment unit, wastewater treatment unit, solid waste treatment and disposal, noise control and other components.

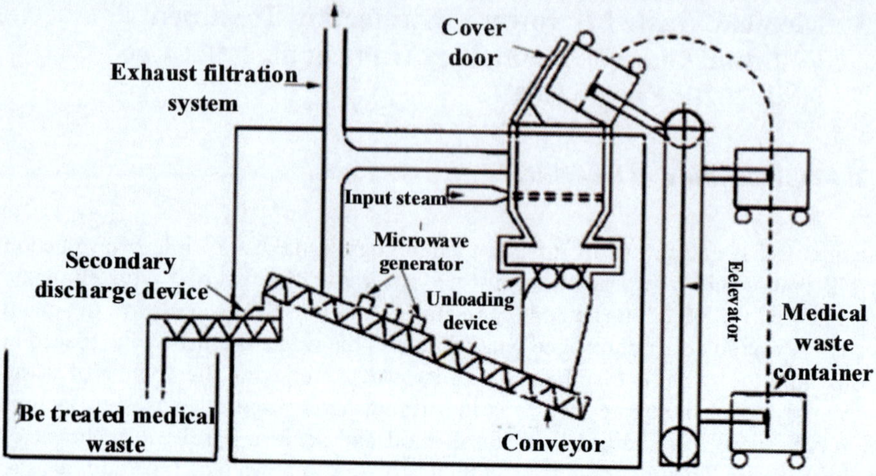

Fig. 5.8 Typical process flow diagram of microwave disinfection treatment

Medical waste microwave disinfection process generated by the exhaust gas are mainly from the microwave disinfection process, and the pollutants are mainly particulate matter, VOCs, etc. The wastewater from the centralized treatment process mainly comes from the cleaning and disinfection of medical waste transportation vehicles and crates, cleaning and disinfection of production areas such as unloading sites and storage sites, microwave disinfection treatment, and shredding equipment cleaning and disinfection, as well as the initial rainwater and accident wastewater from production areas and wastewater treatment areas. The main pollutant indicators are pH, five-day biochemical oxygen demand (BOD_5), chemical oxygen demand (COD_{Cr}) and suspended solids (SS). The solid wastes generated in the centralized treatment process are mainly disinfection treatment residues and solid wastes generated in the wastewater and waste gas treatment processes.

Typical microwave disinfection treatment process and pollution production nodes are shown in Figs. 5.9 and 5.10.

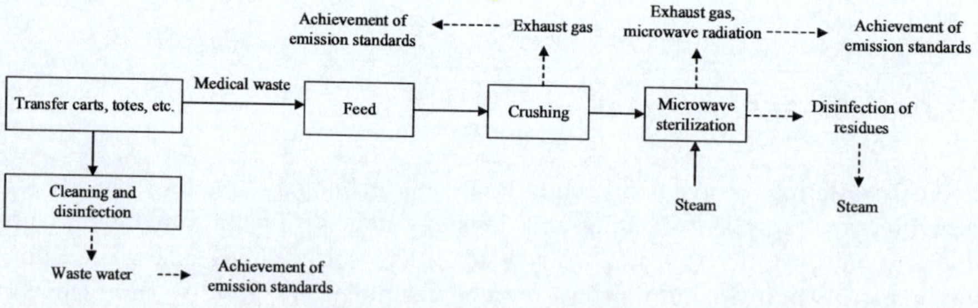

Fig. 5.9 Typical process flow and pollution production nodes for microwave disinfection treatment

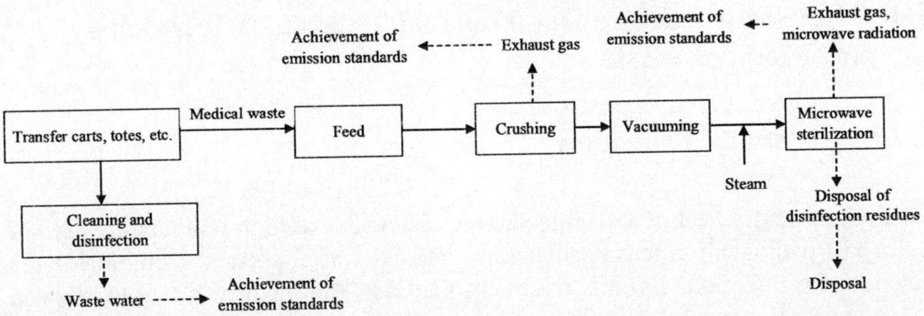

Fig. 5.10 Typical process flow and pollution production nodes for combined microwave and high-temperature steam disinfection treatment

5.3.4 Analysis of Microwave Disinfection Treatment Technology

(1) During the use of microwave disinfection process, microwave frequency of about (915 ± 25) MHz or (2450 ± 50) MHz should be used, and the disinfection temperature should be ≥ 95 °C with disinfection time ≥ 45 min.
(2) During the use of microwave and high-temperature steam combined disinfection process, microwave frequency of about (2450 ± 50) MHz, pressure ≥ 0.33 MPa, temperature ≥ 135 °C, and disinfection time ≥ 5 min should be applied.
(3) Microwave disinfection process can apply *Bacillus subtilis* black variant (*B. subtilis ATCC 9372*) bacterium as the indicator strain. Microwave and high temperature steam combined disinfection process can apply *Bacillus* thermophilus (*Bacillus ATCC 7953*) bacterium and *Bacillus subtilis* black variant (*B. subtilis ATCC 9372*) bacterium as the indicator strain. The log kill value should be ≥4.00.
(4) Other pollution control should be strengthened, such as odor, VOCs and particulate matter, etc. Necessary tail gas purification measures should be taken to reduce their pollution to the environment.

Due to the application of microwave technology, the equipment uses a special sterilization chamber design, which can effectively prevent microwave leakage and discharge phenomena caused by metal fragments in medical waste. Feeding, crushing, conveying and other key processing aspects are negative pressure safety process, and will not bring about secondary pollution to the environment and operation. However, the key concern is that this technology cannot treat chemical waste, pharmaceutical waste and radioactive waste. Therefore, it would be more scientific to have specific identification and alarm devices for the wastes that cannot be handled by this technology. This work has gained practical application with the progress of science and technology.

5.4 High-Temperature Dry Thermal Treatment Technology for Medical Waste

5.4.1 Technical Principles

Medical waste dry heat treatment technology is the use of high-temperature dry hot air to kill pathogenic microorganisms in medical waste, so that it eliminates the potential for infectious hazards treatment method. The medical waste is exposed to negative pressure and high temperature for a certain period of time after high intensity grinding, and the heat is efficiently transferred to the medical waste to be treated using a precise conduction procedure, which causes protein denaturation and coagulation of the pathogenic microorganisms carried in the medical waste, thus leading to the death of the pathogenic microorganisms in the medical waste and rendering the medical waste harmless and achieving safe disposal. This technology is applicable to the treatment of infectious waste, injury waste and unidentifiable pathological waste such as human tissue and pathological wax blocks discarded after pathological sectioning in the "Medical Waste Classification Catalogue", and is not applicable to the treatment of pharmaceutical waste and chemical waste.

5.4.2 Treatment Process

The high temperature dry heat process is divided into medical waste treatment system, extraction + gas purification system, heating system and automatic control system. The medical waste treatment process includes the process of loading, grinding, disinfection and discharging. The typical process flow of high-temperature dry heat treatment of medical waste is shown in Fig. 5.11.

5.4.3 Pollutant Emissions

Medical waste high-temperature dry heat treatment technology generally includes feeding, vacuum, grinding, dry heat sterilization, electrostatic decontamination and other process units. Medical waste high-temperature dry heat process in the vacuum will produce bad odor, VOCs, germs microorganisms, noise, etc. The specific discharge nodes are shown in Fig. 5.12.

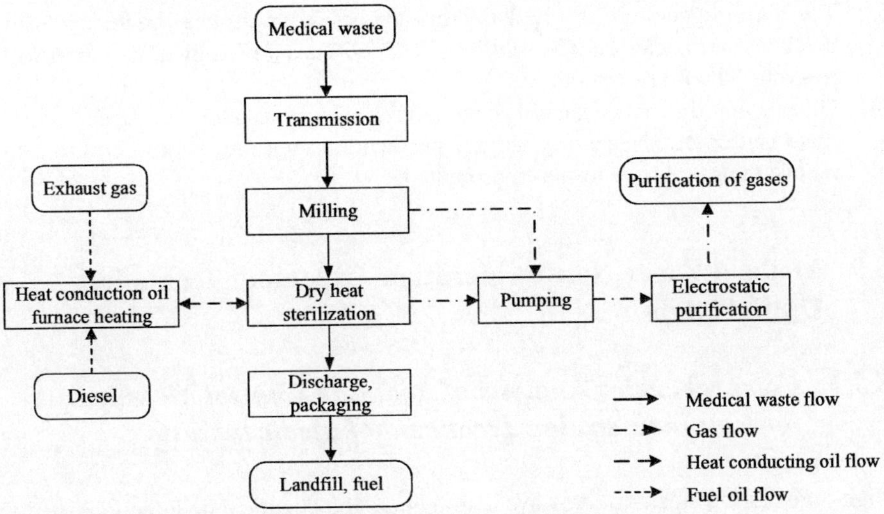

Fig. 5.11 Typical process flow diagram of high-temperature dry heat treatment of medical waste

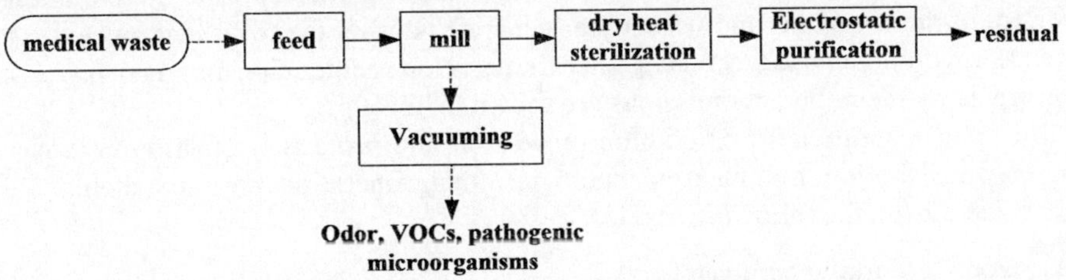

Fig. 5.12 Medical waste high-temperature dry heat treatment process and discharge node diagram

5.4.4 High-Temperature Dry Heat Treatment Technology Analysis

High-temperature dry heat treatment requires medical waste to be exposed to a certain temperature for enough time to achieve a certain microbial inactivation efficiency. To improve the efficiency of microbial inactivation, it is necessary to improve the efficiency of heat transfer to the interior of the material, in order to ensure efficient and safe sterilization, the operation of high-temperature dry heat facilities should meet the following requirements:

(1) The pressure inside the sterilizer should be 300 Pa, close to vacuum.
(2) The temperature inside the sterilizer should be 180~200°C, and the processing time should not be less than 20 min, and the mechanical stirring device should be stirred at a speed of not less than 30 r/min.

(3) The high temperature dry heat treatment process should use *Bacillus subtilis* black variant bacterium (*B. subtilis ATCC 9372*) as a bioindicator, and its kill log value should be ≥4.00.
(4) Other pollution control should be strengthened, such as odor, VOCs and particulate matter, etc. Necessary tail gas purification measures should be taken to reduce their pollution to the environment.

5.5 Medical Waste Non-incineration Treatment Technology Optimization

5.5.1 *Comprehensive Analysis of Pollution Control Measures for Non-incineration Treatment of Medical Waste*

High-temperature steam, microwave disinfection and chemical disinfection are the most widely used treatment technologies around the world, and the pollution control of the treatment process should consider the following issues: the scope of application of the treatment object, the treatment effect of medical waste, the pollution control of the waste gas and wastewater generated during the treatment process, as well as environmental safety management issues. Non-incineration treatment process generated by the pollutant emissions are shown in Table 5.1.

In order to promote the pollution prevention and protection of different types of non-incineration treatment technologies, their operational safety should be considered from the following aspects.

(1) Control working parameters

Different disposal technologies have certain operating parameters, including overall temperature, pressure, and disinfection time. In addition, different processes also include different control requirements.

(2) Pollutant emission control

In terms of non-incineration treatment technology, the core elements of pollution control are the following three aspects:

(1) Infectiousness of disinfection treatment residues, i.e., the problem of disinfection effect detection. In terms of disinfection effect detection, there are no standards and methods specifically for the detection of treatment effect of non-incineration technology for medical waste disposal. However, the amount of microorganisms contained in the treated residue is the fundamental channel for disinfection detection. Therefore, appropriate indicator strains should be used to test the microbial content of the residue in order to determine the final disinfection effect. The indicator bacteria must be selected to ensure the safe disposal of medical waste, i.e., the treatment effect should be such that the most difficult microbial strains are killed. Based on the resistance of Bacillus subtilis

Table 5.1 Non-incineration treatment technology pollutant emissions

Preface no	Contaminants risk type	High temperature steam	Chemical disinfection	Microwave sterilization	High temperature dry heat	Control measures
1	Atmosphere	Will produce odors and VOCs and other air pollutants	Will produce odor, VOCs and dust and other air pollutants	Will produce odor, VOCs and dust and other air pollutants	Will produce odor, VOCs and dust and other air pollutants	Set up with sterilization, disinfection, and can effectively remove microorganisms, VOCs and other pollutants in the exhaust gas purification facilities
2	Wastewater	Auxiliary liquid materials, residual liquid in medical waste, holding containers cleaning and disinfection, etc	Auxiliary liquid materials, residual liquid in medical waste, cleaning and disinfection of holding containers and wet chemical disinfectants	Auxiliary liquid materials, residual liquid in medical waste, holding containers cleaning and disinfection, etc	Residual liquid in medical waste, holding containers cleaning and disinfection, etc	A variety of practical disposal technologies should be used to achieve the corresponding type of pollution emission requirements
3	Disinfection service residue management	It may cause the treatment effect to exceed the standard	It may cause the treatment effect to exceed the standard	It may cause the treatment effect to exceed the standard	It may cause the treatment effect to exceed the standard	The microbial content of the residue should be tested using the appropriate indicator strain
4	Noise	Crushing equipment, fans, etc	Crushing equipment, fans, etc	Crushing equipment, fans, etc	Crushing equipment, fans, etc	Noise reduction measures should be taken accordingly
5	Security	High-pressure steam burns, medical waste infections	Potential for chemical disinfectant leaks, medical waste infections	Microwave radiation problems, medical waste infections	Dry heat equipment burns, medical waste infections	Should take the appropriate safety protection facilities

black variant and Bacillus thermophilus to chemicals and heat, respectively. These two strains are usually chosen as indicator strains internationally, and they are also used as indicators of disinfection effect in China. Among them, chemical disinfection treatment is mainly suitable for chemical resistance. It is recommended to use the strongest chemical resistance of Bacillus subtilis black variant bacillus as the indicator bacteria, heat-based high-temperature steam treatment effect detection using the strongest resistance to thermophilic fatty bacillus bacillus as the indicator bacteria. The microwave treatment technology can use Bacillus subtilis black variant bacillus or Bacillus thermophilus bacillus as the indicator bacteria. The high temperature dry heat treatment technique can use Bacillus subtilis black variant bacillus.

Regarding the kill criteria for non-incineration treatment of medical waste, the microbial kill levels were defined in the first STAATT meeting as follows:

Tier 1: Sterilization rate of 6lg or higher for propagating bacteria, fungi and lipophilic viruses.

Tier 2: Sterilization rates of 6lg or greater for propagating bacteria, fungi, lipophilic/hydrophilic viruses, parasites and mycobacteria.

Level 3: Sterilization rate of 6lg or more against propagating bacteria, fungi, lipophilic/hydrophilic viruses, parasites and mycobacteria. The sterilization rate of Bacillus thermophilus or Bacillus subtilis black variant bacillus is 4lg or more.

Level 4: Kills propagating bacteria, fungi, lipophilic/hydrophilic viruses, parasites, Mycobacterium and thermophilic lipophilic bacilli at a rate of 6lg or more.

Tier 3 is the minimum standard recommended by STAATT.

6lg kill rate is the 10^6-kill rate, which is equivalent to the possible survival rate of microorganisms in one million, or 99.9999% of microorganisms after treatment, i.e. the kill log value of 6.

The 4lg kill rate is the 10^4-kill rate, which corresponds to the possible survival rate of one in 10,000 microorganisms, or 99.99% of microorganisms killed after treatment, i.e. the log kill value of 4.

Although China has not yet issued mandatory national standards, the adoption of tertiary standards is also the basic starting point for the application of non-incineration treatment technologies. Considering the current testing capability and level in China, the sterilization rate requirements for propagating bacteria, fungi, lipophilic/hydrophilic viruses, parasites and mycobacteria can be disregarded, and only the killing rate of Bacillus thermophilus bacillus or Bacillus subtilis black variant bacillus of 4lg or more can be considered.

(2) Air pollutant control in the disposal process. Exhaust gas is mainly generated in the non-incineration process produces odor, VOCs, particulate matter and other pollutants. In terms of air pollution control, all non-incineration technology for the disposal of medical waste processes, should set up waste gas purification device, the device should be able to effectively remove VOCs, dust and other pollutants in the waste gas, and according to the actual needs of the deodorization

device. Purification technology can choose activated carbon adsorption, biological filtration, UV photo-oxidation catalyst, low-temperature plasma and other technologies, and according to the characteristics of the exhaust gas and emission requirements alone or a combination of settings. It is necessary to regularly check the operating status of the exhaust gas purification facilities, and timely adjustment of operating conditions. The inspection includes inlet and outlet valve opening and closing status, pressure meter display and fluctuation status, exhaust gas flow, flow rate, temperature, pressure, etc.

Using low-temperature plasma treatment technology, the voltage, current, frequency and other working parameters should be adjusted in a timely manner, and do a good job of reactor maintenance, maintenance and repair; using activated carbon adsorption technology should be strictly controlled by the flue gas temperature and dust content, regularly check whether the adsorbent saturated, and timely replacement of adsorption materials. When using granular activated carbon, the wind speed of the adsorption layer should be 0.20~0.60 m/s; when using activated carbon fiber felt, the wind speed of the adsorption layer should be 0.20~0.60 m/s; using biological filtration technology, the actual nature of the gas should be based on the screening, domestication of microorganisms, real-time monitoring of various information on microbial metabolic activities; using UV photo-oxidation catalytic technology, the light source, catalyst, temperature, humidity and residence time should be adjusted in a timely manner. Using UV photo-oxidation technology, the working parameters such as light source, catalyst, temperature, humidity, and residence time should be adjusted in time, and the reactor should be well maintained, serviced and repaired; using spraying technology, the corresponding drenching solution should be accurately prepared and added, and the working parameters such as working pressure, pressure-holding time, spraying time and spraying volume should be adjusted in time.

(3) The safety protection of the treatment process. Several different non-incineration treatment technologies in the safety protection reflects different characteristics, including high-temperature steam treatment, is mainly the possible impact of pressure steam on the operator. Microwave treatment mainly helps to prevent the impact of microwave radiation on the operator during the disposal of medical waste, chemical disinfection treatment is mainly to prevent the impact of chemical disinfectants on the operator, and high-temperature dry heat is mainly to prevent thermal contact and the operator. The main purpose of high temperature dry heat is to prevent the risk of burns to the operator due to heat contact. The relevant protective measures include occupational disease protective equipment, protective equipment and supplemented by corresponding management measures.

5.5.2 *Optimization of Non-incineration Treatment Technologies for Medical Waste*

5.5.2.1 General Requirements for Optimization of Non-incineration Treatment Technologies

Medical waste non-incineration technology is suitable for treatment plants with a daily treatment scale of 10t or less, which can treat infectious, injurious and partially pathological medical waste, with VOCs and odor as the main pollutants, and no dioxin-like pollutants. Non-incineration treatment facilities with low investment costs, suitable for medical waste collection, better classification, less economically developed regions. Specific technical and economic applicability analysis is shown in Table 5.2.

Table 5.2 Economic applicability analysis of non-incineration treatment technologies for medical waste

Disposal technology	Disposal costs		Technical and economic applicability
	Running costs/ (yuan/t)	Investment costs (equipment and Installation)/ (million yuan/ t)	
High temperature steam Vapor treatment	1800~2300	60~80	Effective disinfection, and no acid gas, heavy metals, dioxins and other toxic and harmful substances produced, and low cost, simple operation and maintenance; suitable for disposal scale of 10 t/d or less infectious and injurious medical waste disposal
Chemical treatment	1500~2000	45~55	Low investment, low operating costs, simple operation, small environmental pollution; suitable for the scale of 10 t/d below the disposal of infectious and injurious medical waste
Microwave processing	1200~1500	50~60	Germicidal spectrum, no residue, good deodorization effect, clean and sanitary; suitable for the disposal of infectious and injurious medical waste with a scale of 10 t/d or less
High temperature dry heat	1000~1200	60~70	Low investment, low operating costs, simple operation, small environmental pollution; suitable for the scale of 10 t/d below the disposal of infectious and injurious medical waste

The core aspects of the optimization of non-incineration treatment technology should include the control of working parameters and pollutant emission control. Therefore, the optimization problem should also be strictly focused on these two aspects. The following three typical non-incineration treatment technologies are combined to discuss their disposal technology optimization issues.

5.5.2.2 High-Temperature Steam Treatment Technology

High-temperature steam treatment requires medical waste to be exposed to a certain temperature for enough time to achieve a certain microbial inactivation efficiency. The best feasible process flow for pollution prevention and control of high-temperature steam treatment of medical waste is shown in Fig. 5.13.

In order to ensure the safety of the high temperature steam treatment, the following process parameters should be ensured:

(1) The steam treatment process should be in the disinfection temperature ≥134 °C, pressure ≥0.22 MPa (gauge pressure) conditions, and the corresponding disinfection time should be ≥45 min.
(2) The steam pressure entering the disinfection bin should be in the range of 0.3~0.6 MPa.
(3) High-temperature steam processing equipment should have a drying function, and the water content of the material after drying should not be greater than 20% of the total weight.

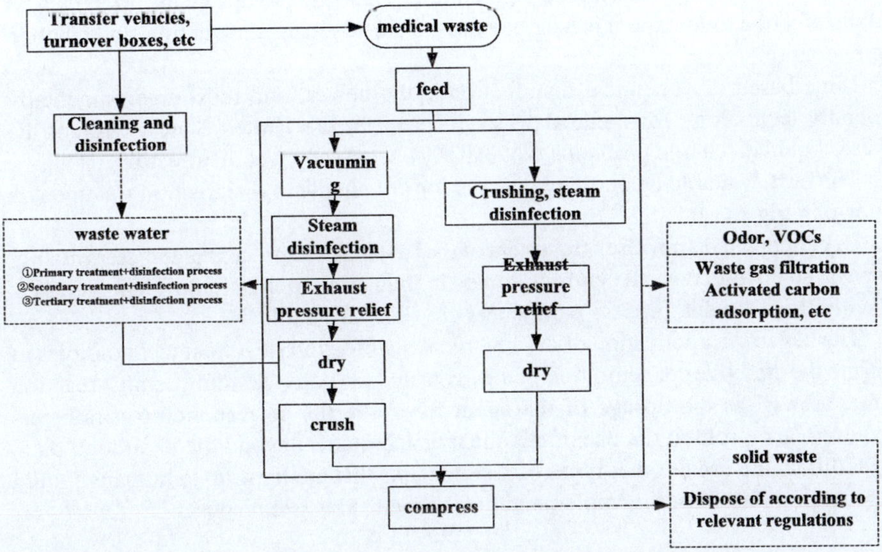

Fig. 5.13 Medical waste high temperature steam treatment pollution prevention best feasible process flow

(4) The steam should be saturated steam, and the non-condensable gas it contains should not exceed 5% (volume fraction).
(5) The exhaust gas purification treatment technology can choose activated carbon adsorption, exhaust gas biological purification device and other technologies, and according to the characteristics of the exhaust gas and emission requirements alone or a combination of settings.
(6) Crushing equipment should be able to crush hard materials and soft materials at the same time, material crushing particle size should generally not be greater than 5 cm. if the first crushing cannot meet the requirements, and should be set up secondary crushing.

5.5.2.3 Optimization of Dry Chemical Disinfection Treatment Technology

Dry chemical treatment of medical waste is usually used in combination with mechanical shredding. The efficiency of chemical disinfectant inactivation of microorganisms is related to contact time, temperature, chemical disinfectant concentration, pH (acidic and alkaline environment), and the number and type of microorganisms killed. Chemical disinfectant inactivation effectiveness must ensure sufficient reaction contact time between the chemical disinfectant and the medical waste disinfection surface. The USEPA test procedure for killing propagating bacteria, fungi and mycobacteria should require a contact time of 10 min at a reaction temperature of 20 °C. When killing bacterial spores, the contact time and temperature of the reaction should be increased, usually by one or several hours, and the effectiveness of disinfection usually increases with the temperature. Therefore, to ensure the effectiveness of disinfection treatment, it is necessary to ensure adequate contact time and reaction temperature.

Lime based dry chemical disinfection is the newest and most environmentally friendly technology for medical waste disposal in the United States, without the waste liquid discharge problems that usually accompany wet disinfection.

The best available technology process for dry chemical disinfection treatment is shown in Fig. 5.14.

Dry chemical disinfection treatment of the best feasible process parameters mainly lies in the disinfectant, the current domestic industrial application of disinfectant is mainly lime powder, process parameters are shown in Table 5.3.

The key to the application of dry chemical disinfection treatment technology is to ensure the effective concentration of the agent and the corresponding contact reaction time, as well as the dosage of the agent to ensure the corresponding parameters required, and prohibit the use of chemical disinfectants beyond the expiration date. In addition, the use of other types of disinfectants that are harmful to humans should also be prohibited for the implementation of medical waste disposal.

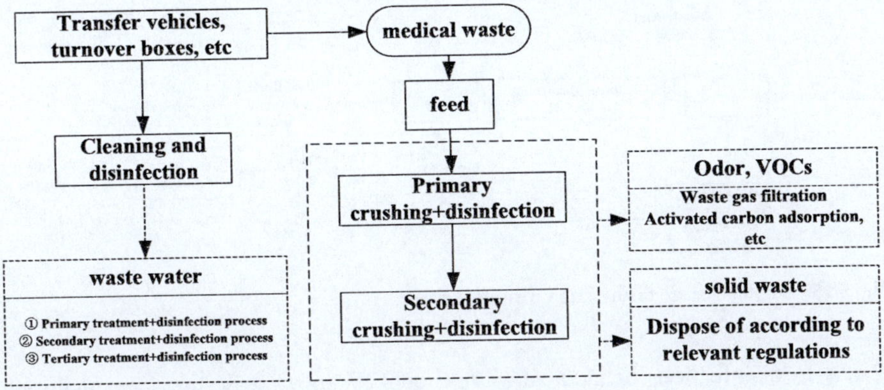

Fig. 5.14 Best practicable technology process for dry chemical disinfection of medical waste

Table 5.3 Dry chemical treatment process disinfectant parameters

Disinfectants	Concentration/purity	Contact time	Dosing amount of pharmaceuticals/kg medical waste	pH
Lime powder	>90%	>120 min	>0.075 kg	11.0~12.5

5.5.2.4 Optimization of Ethylene Oxide Disinfection Treatment Technology

Medical waste disinfection process is as follows: medical waste is packaged in its original form and pushed into the ethylene oxide disinfection cabinet, which is injected with an effective concentration of 893 mg/L of ethylene oxide in a vacuum environment, with a pre-vacuum of −80 kPa, system temperature (54 ± 2) °C, relative humidity 50 + 10%, and disinfection time of about 4 h. medical waste is temporarily stored and analyzed after disinfection, and then box by box put into the conveyor belt, through the X-ray machine, reciprocating hoist, automatic conveying system, automatic feeding system, secondary crusher, shaftless screw conveyor, and finally into the compression car compression system. After filling the compressor truck, the disinfected and crushed medical waste will be sent to the domestic waste disposal plant for final disposal.

The process flow of ethylene oxide disinfection treatment is shown in Fig. 5.15.

In order to ensure the safety of EO disinfection treatment, the following process parameters should be ensured:

(1) The purity of ethylene oxide should be greater than 99.9%, the concentration of ethylene oxide should be ≥900 mg/L, the disinfection temperature should be controlled within the range of 54 ± 2 °C, the disinfection time should be ≥4 h, the relative humidity should be controlled within the range of 60~80%, and the initial pressure is −80 kPa vacuum environment.

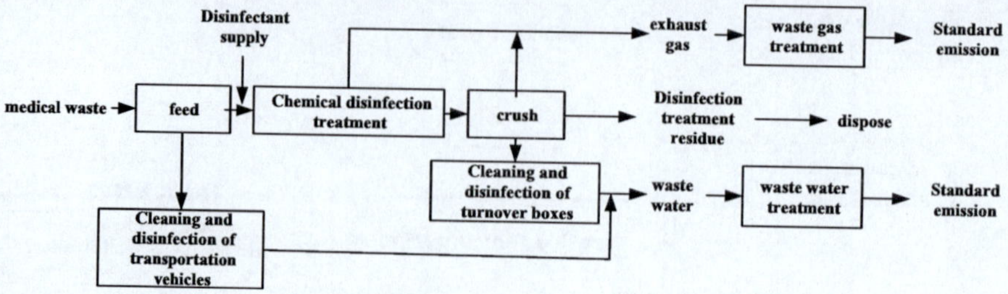

Fig. 5.15 Ethylene oxide disinfection treatment process

(2) After disinfection, the medical waste should be temporarily stored and analysed for 15~30 min, and the temporary analysis should be operated under negative pressure, and the exhaust gas from the ethylene oxide analysis room should be collected and treated in a unified manner to meet the standard.
(3) Ethylene oxide supply unit, disinfection unit, crushing unit and ethylene oxide storage site should be equipped with ethylene oxide gas concentration alarm device.
(4) The disinfectant adding nozzle should be evenly set at the top of the disinfection bin and configured with internal circulation and heat preservation devices to ensure the concentration and temperature balance of ethylene oxide in the bin.
(5) Disinfection bin and crushing space should be ventilated with nitrogen to replace the oxygen in it to prevent the occurrence of deflagration.
(6) Disinfection bin should be set up with explosion-proof door or pressure relief port, in order to reduce the impact of deflagration accident range.

5.5.2.5 Microwave Disinfection Processing Technology Optimization

1) Best available technology process for microwave disinfection of medical waste

The best feasible process flow for microwave disinfection of medical waste is shown in Fig. 5.16.

2) The best feasible process parameters for microwave disinfection treatment
(1) When using microwave disinfection process, microwave frequency of (915 ± 25) MHz or (2450 ± 50) MHz should be maintained. Disinfection temperature should be ≥95 °C, disinfection time ≥45 min.
(2) While using microwave and high-temperature steam combined disinfection process, microwave frequency of (2450 ± 50) MHz, pressure ≥0.33 MPa, temperature ≥135 °C, disinfection time ≥5 min should be maintained.

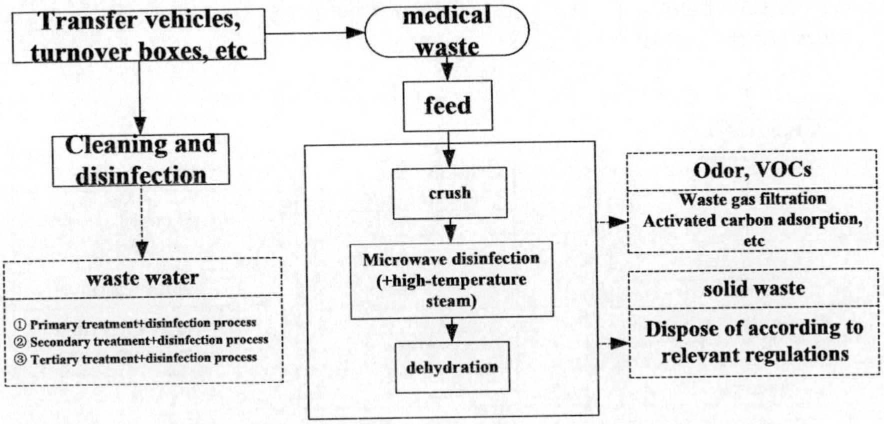

Fig. 5.16 Medical waste microwave disinfection treatment best feasible process flow

5.5.3 Optimization of High-Temperature Dry Heat Treatment Technology

1) Medical waste high-temperature dry heat treatment technology process

(1) The disposable carton or bag containing medical waste is placed on the lift. The medical waste is lifted to the top and automatically opens the equipment inlet bin door, after which the bin door is automatically closed.

(2) The medical waste is fed into the mill for grinding, after grinding, medical waste is reduced by 80%, to achieve the purpose of destruction, and grinding 300 kg takes about 7~10 min.

(3) The pumping device to pump the disinfector. The pressure inside the disinfector is 300 Pa, close to vacuum. The pumping + gas purification process includes the pumping and gas purification process. The pumping equipment has three pumps: two liquid surround vacuum pumps and one electric pump. The vacuum set has a cooling function, which is mainly needed to ensure the normal operation of the pumping unit, with a rated power of 20 kW.

(4) The heat from the heat-conducting oil can raise the temperature inside the sterilizer to 180~200 °C. After heating to make medical waste completely dehydrated. Sterilization through a certain time (some countries in Europe and the United States sterilization is for 20 min), especially the crushing of medical waste, so that more penetration of the waste for sterilization is achieved, in order to ensure the sterilization effect.

After disinfection, the disinfector pumping valve is automatically closed, the discharge bin door is opened and the material is discharged to the conveyor belt, which can be collected and sent to the landfill.

2) Optimal process parameters for high temperature dry heat treatment

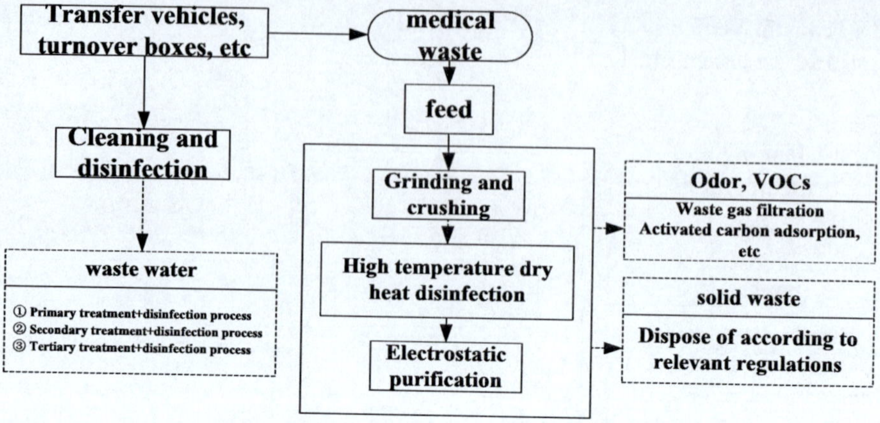

Fig. 5.17 Best available technology flow for high temperature dry thermal treatment of medical waste

Medical waste dry heat treatment process requires that the processing temperature in the sterilization room is not less than 180 °C, the pressure is not higher than 1000 Pa (gauge pressure) conditions, and the corresponding processing time should not be less than 20 min.

3) Best available technology process for high temperature dry heat treatment of medical waste

The best feasible technology process for high-temperature dry thermal treatment of medical waste is shown in Fig. 5.17.

5.5.4 Medical Waste Non-incineration Treatment Technology Optimization Measures

For medical waste, high-temperature steam treatment technology, dry chemical disinfection treatment technology, ethylene oxide disinfection treatment technology, microwave disinfection treatment technology, high-temperature dry heat and other technologies can be used. But the non-incineration treatment technology can only deal with the "medical waste classification catalog" and "National Hazardous Waste List" in the infectious waste, injury waste, as well as the pathological section after the waste of human tissue, pathological wax blocks and other unidentifiable pathological waste, not for the treatment of pharmaceutical waste, chemical waste. For the final treatment of medical waste to meet the corresponding standard requirements, and requires a large capacity of landfill to accommodate the treated waste. Combined with the medical waste management practices of foreign developed countries, and combined with China's national conditions, the following measures are proposed for the application of non-incineration treatment technology for medical waste.

(1) Further improvement for the application of non-incineration medical waste treatment technology management system construction

At present, although the promulgation of medical waste chemical disinfection, microwave and high-temperature steam treatment technology engineering construction specifications, but there is lack of a corresponding support system to promote the widespread of non-incineration treatment technology application. Therefore, we should establish and improve the technical verification and evaluation system, treatment effect monitoring system, supervision and management system, and technical training system for non-incineration treatment technology of medical waste according to the need to promote the establishment of the whole process management system in this field.

(2) Further strengthen the research and development of non-incineration treatment technology and promote the localization of non-incineration treatment technology

Strengthen international exchanges and cooperation, understanding and learning from Europe and the United States and other developed countries and regions of non-incineration treatment technology management model, treatment facilities construction status and advanced management and facilities construction experience, for non-incineration treatment technology in China's R & D and application to provide reference. Based on fully absorbing and learning from foreign developed countries in non-incineration treatment technology, equipment and advanced operation and management experience, and further combining China's resources and technology status, promotes the development and promotion of non-incineration treatment technology and equipment, and realize the localization of non-incineration treatment equipment.

(3) Actively promote the effective interface between the separate collection of medical waste and the non-incineration treatment process of medical waste

The biggest problem with the application of non-incineration treatment technology for medical waste is its scope of application. Therefore, the application of non-incineration treatment technology must be based on an effective waste classification, and the current medical waste classification and management system of medical institutions from the classification of medical waste, to the collection of medical waste, handover, in-hospital transfer and temporary storage, and the handover of medical waste centralized treatment centre and other links It is a complex system project, which requires a strict management system and dedicated personnel for the work. However, due to the lack of personnel or funding, existing medical institutions are not equipped with dedicated personnel to collect, transfer and temporary storage of medical waste, making loopholes in the management of medical waste, especially for non-incineration treatment technology, which is more strict in the classification of waste, the only way to eliminate the environmental risks of medical waste, which is a problem that must be solved in the application of non-incineration treatment technology.

(4) Increase efforts to study and promote medical waste treatment effectiveness testing and monitoring and supervision of major pollutants.

In terms of medical waste treatment effect testing methods, there are currently no standards and methods specifically for the application of non-incineration medical waste treatment technology related to testing methods, and should continue to study and develop corresponding standards and methods to promote the development of the work. On the other hand, the process of non-incineration of medical waste will also produce secondary pollution such as odor and VOCs, and for these pollutants, the country also lacks the corresponding standards and practical detection methods. Therefore, cannot ensure effective supervision and management of these pollutants, becoming another source of human health and environmental hazards. Supervision and management are a prerequisite and basis for the application of non-incineration treatment technologies, and should further strengthen the capacity of supervision and management to ensure the application of such technologies from the perspective of environmental safety, and to improve the overall level of law enforcement to ensure the standardized operation and management of non-incineration treatment facilities.

(5) Actively promote the co-disposal of medical waste, and promote the complementary advantages of non-incineration and incineration technologies.

Both incineration and non-incineration technologies have a certain scope of application, but also have their own irreplaceable advantages. In terms of incineration technology, the most important feature of its scope of application, for all types of medical waste have applicability, but the cost of construction of treatment facilities and treatment costs are high, and will produce dioxins and heavy metals and other pollutants. In terms of non-incineration treatment technology, the cost of construction and operating costs of its treatment facilities are much lower, and does not produce dioxins and heavy metals and other pollutants, cleaner, and no international conventions require. Therefore, two sets of disposal facilities can be built at the same disposal site, an incineration facility and a non-incineration treatment facility, to achieve the advantages of both complement each other. This is particularly relevant for the technical transformation of incineration disposal facilities with insufficient disposal capacity. In addition, it can also promote the technical advantages of complementary management between cities, and promote the coordinated development of medical waste management and disposal between cities, the establishment of inter-municipal collaboration, promote the role of the provincial hazardous waste treatment center, and thus promote the collaborative disposal of medical waste.

(6) Further strengthen the construction of personnel training system and improve the management level of relevant managers.

In terms of medical waste non-incineration treatment technology, because of the many types involved, the process varies greatly, and should be a comprehensive promotion of medical waste non-incineration treatment technology in the field of training, from the training system, training materials and training requirements, for facility operation managers, operators and environmental supervision and management personnel to carry out relevant management and technical training, in order to improve the management of non-incineration treatment technology and to provide technical support to improve the management of non-incineration treatment technology and the operation level of facilities.

(7) Increase efforts to promote the application of non-incineration medical waste treatment technology in remote areas to solve the problem of medical waste management in remote areas of China

According to the National Plan for the Construction of Hazardous and Medical Waste Disposal Facilities, its planning scope basically does not include remote areas such as suburban counties. Therefore, it is possible to promote the use of non-segregated incineration disposal technology in suburban counties to promote the management and disposal of medical waste in remote areas of China. This also has significance for the management and disposal of medical waste in central and western China.

5.6 Medical Waste Non-incineration Treatment Technology Management Practice Cases

5.6.1 High-Temperature Steam Treatment Technology for Medical Waste

5.6.1.1 Case Unit Overview

A company in Hebei, relying on foreign advanced "high-temperature steam sterilization process" and independent research and development of "Internet of things control system", with the help of two innovative management model, and the establishment of urban and rural areas covering the entire area of medical waste collection and transportation, disposal management system, the entire process, the whole process of supervision, leaving no dead ends, can be traced back to the source.

At present, the company has two high-temperature steam sterilization production lines with a daily processing capacity of 10t, 10 medical waste transfer vehicles, with a professional medical waste collection, transfer and disposal team which is responsible for the collection and transportation of medical waste and environmental disposal of medical institutions throughout Langfang, with a daily processing capacity of 9 to 12 t.

The company employs 45 people, including 8 managers, 3 technicians, 34 operators, 3 environmental engineering professionals and related professional intermediate titles, and has more than 3 years of experience in solid waste pollution management.

The high temperature steam treatment technology case study site is shown in Fig. 5.18.

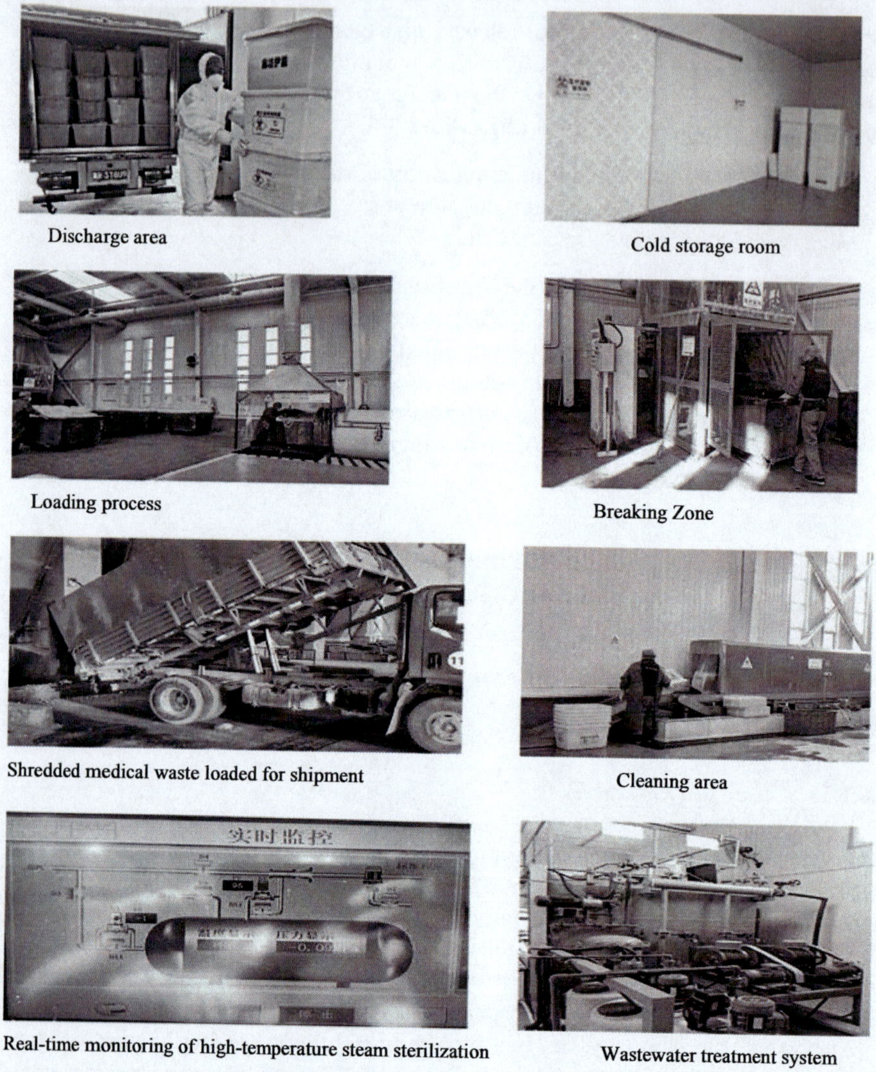

Fig. 5.18 High-temperature steam treatment technology case study site equipment

5.6.1.2 Processing Equipment

(1) Main Body Equipment

The main equipment consists of high-temperature steam processing pot, gas steam boiler, crusher, automatic cleaning line for crates, sterilization trolley, circulating pool cooling tower, condenser, etc. Equipment specifications and main technical parameters are shown in Table 5.4.

Table 5.4 Main equipment and technical parameters of high-temperature steam treatment technology

Name	Specification model	Design capabilities/(t/year)	Number quantity	Technical parameters
High temperature steam treatment pot	MWC-1000 × 6	3650	1	Working temperature 120~140 °C;Working pressure 0~220 kPa; Steam temperature 160~170 °C; Steam pressure 0.4~0.8 MPa; Disinfection time 30~45 min; Microbial inactivation rate 99.9999%
Gas steam boiler	WNS2-1.25-Y(Q)	–	1	Steam volume 2 t/h; rated working pressure 1.25 MPa; rated steam temperature 193 °C
Crusher	GS-330X900	8760	1	Host power 30 kW
Automatic cleaning line for crates	GX1000-8.5B	–	1	
Sterilization trolley	–	–	18	Volume 1 m^3
Circulation pool cooling tower	DBNL-50T	–	1	
Condenser	LNQ670	–	1	

(2) Secondry pollution control equipment

The secondary pollution control equipment mainly includes MBR sewage treatment device, centrifugal fan, activated carbon adsorber, etc. The main technical parameters are shown in Table 5.5.

Table 5.5 Secondary pollution control equipment and main parameters of high-temperature steam treatment technology

Name	Specification model	Design capabilities/(t/year)	Number quantity	Technical parameters
MBR wastewater treatment plant	HJ/M-0.5	3300	1	Disposal capacity 0.5 t/h; single membrane area 0.8 m^2
Centrifugal fans	4–72 6 °C	–	1	Flow rate 16187 m^3/h; speed 2240 r/min
Activated carbon adsorber	QL-HXT	–	1	

5.6.1.3 Operation Management

Steam treatment process of medical waste stirring, stirring strength to achieve effective breakage of the outer bag of medical waste; steam treatment process should be carried out under the conditions of disinfection temperature ≥134 °C, pressure ≥0.22 MPa (gauge pressure), the corresponding disinfection time ≥45 min; high-temperature steam treatment according to the process status of the material pressure relief, cooling treatment, effectively reducing the temperature of the discharge material. Under the control of such process parameters, the disinfection effect of medical waste can be guaranteed.

For the disinfection effect, *Bacillus* thermophilus *ATCC 7953* was used as a bioindicator to ensure its kill log value ≥4.00, and the overall must meet the requirements of the Evaluation Methods and Standards for Disinfection and Sterilization Effect (GB 15,981–1995) and the Hospital Hygiene Disinfection Standard (GB 15,982). The disinfection effect is verified by B-D test, testing reagents and test cards, and the frequency is one batch per day, and the third-party self-monitoring is once a year.

Pollution control facilities configuration and treatment management, timely detection, supervision of TVOC, odor, particulate matter and other exhaust pollutants treatment effect, detection, supervision of wastewater treatment and noise emission requirements, and timely resolution of problems. The monitoring plan of the facilities is annual for external inspection and every 3 years for internal inspection; once a year for safety valves; once every six months for pressure gauges, which must be tested by local special equipment testing institutions. When the sterilization pot fails, the electric and steam switches should be closed immediately, the safety valve and exhaust valve should be opened, and the relevant person in charge should be reported in time. After the sterilization pot is repaired normally, the vacuum and sterilization degree must be tested before it can be restarted, and the medical waste that has not been sterilized should be re-autoclaved.

High-temperature steam treatment facilities operating process management requirements are: ① steam pressure into the disinfection warehouse should be 0.3~0.6 MPa range; ② steam should be saturated steam, which contains non-condensable gases should not exceed 5% (volume fraction); ③ steam supply should be able to meet the needs of full-load operation of the processing project; ④ annual steam supply days should not be less than 350 d, and continuous interruption of supply time should not more than 48 h; ⑤ steam provided by self-contained boilers, boiler design, production, installation, commissioning, use and inspection should meet the relevant standard requirements. Steam supply system set up pressure regulating devices to reduce steam pressure disturbances on the impact of high-temperature steam processing equipment.

5.6.1.4 Lessons Learned from Management Practices

High-temperature steam treatment technology is simple to operate, but the treatment cycle is relatively long, intermittent work, investment and operating costs are low.

The project medical waste disposal process mainly includes collection and transportation, unloading, temporary storage, vehicle, crate cleaning and disinfection, high-temperature steam sterilization, crushing, inspection and other processes.

1) Collection and transportation

The project has a total of 10 medical waste transfer vehicles, the transfer vehicle with a fully enclosed refrigerated micro-negative pressure design. The box is equipped with ultraviolet disinfection and automatic spraying device, and can be regularly disinfected on the transfer of medical waste.

In order to ensure that the whole process of medical waste transfer supervision, 10 transfer vehicles are equipped with Beidou positioning and video monitoring system, real-time data directly to the Hebei Provincial Department of Transport management platform. Through the vehicle dynamic monitoring system can always grasp the vehicle running state to ensure that the driver does not exceed the speed limit and fatigue driving. Each vehicle is equipped with a driver and escort, the driver and escort must hold a dangerous goods transport qualification certificate before they can start work.

Through the electronic scale, the car can go to a fixed medical point for collection and transportation. When collecting and transporting the medical waste, the medical institutions has classified and packaged the medical waste into a special crate. The hand over parties confirm the type and weight of the medical waste, reserve the quantity of the special crate, and fill in the five- sheet and two sheets respectively according to the regulations.

2) Discharge medical waste is transported to the unloading area of the plant by the transport truck, and the medical waste is unloaded into the sterilization truck, and the medical waste not yet processed is temporarily stored in the cold storage in crates, and after the previous batch of medical waste is processed, it is transported from the cold storage to the sterilization cart and unloaded into the sterilization truck for the disposal of the next batch of medical waste.
3) Vehicle crate cleaning and disinfection.

Medical waste transfer vehicles and crates after completion of unloading, are cleaned and disinfected before leaving the factory to use.

4) Loading

Medical waste is poured into the sterilization car by the staff, high-temperature steam processing pot warehouse mouth to the warehouse equipped with sterilization car driving track, medical waste poured into the sterilization car by the warehouse mouth is pushed into the warehouse, close the processing pot warehouse mouth, waiting for sterilization processing.

5) High temperature steam sterilization (Zhaotang et al. 2009).

(1) Pre-vacuum. After pushing the sterilization car into the warehouse and closing the door, the pressure in the high-temperature steam treatment pot is pumped to −0.09 MPa through the steam ejector, and the extracted air is mixed with the high-temperature steam from the boiler and discharged into the condenser for rapid condensation. The condensed exhaust gas is discharged after filtering with activated carbon, and the condensed wastewater formed after steam condensation is discharged into the plant's wastewater treatment station.
(2) High temperature steam sterilization. After pre-vacuum, the air inlet valve is opened and the steam enters the sterilization bin to heat the medical waste, so that the temperature in the sterilization bin rises to 134 °C and the pressure rises to 0.22 MPa, and when the temperature in the sterilization bin reaches the set value, it turns into the sterilization stage, in which the air inlet valve in the sterilization bin is controlled by the pressure and temperature in the bin together to ensure that the temperature in the bin is maintained at 134 °C and the pressure is maintained at 0.22 MPa, and the medical waste in the sterilization bin is sterilized by high-temperature steam, and the sterilization time last for 45 min.
(3) Post-vacuum. After the high-temperature steam treatment process is completed, the post-vacuum process is carried out in the bin to extract the water vapor from the sterilization bin, dry the medical waste and reduce the temperature of the medical waste to reduce the odor of the medical waste. The condensed exhaust gas is discharged after high efficiency filter + activated carbon filtration, and the condensate formed after steam condensation is discharged to the plant sewage treatment station for treatment.

6) Discharge and testing.

Open the door of the processing bin, the sterilization car along the track to push out the sterilization bin, the sterilization of medical waste sterilization effect testing, testing standards into the crushing process, unqualified materials back to the high-temperature steam processing pot again for sterilization treatment.

7) Breakage

The crushing unit is composed of elevator, crusher and screw conveyor. After sterilization of medical waste, the crusher will crush and cut cotton, gauze, plastic or glass bottles, needles, scalpels, etc. in the waste to form particles less than 50 mm × 50 mm. The crushed material enters the screw conveyor with the discharge port and is sent to the conveyor.

8) Transit

As the project is close to the landfill, the compressor unit is not considered. The crushed medical waste is directly transported to the crushed waste for incineration and power generation by CLP Power Generation Co.

9) Wastewater treatment

The wastewater from this project is medical waste high temperature steam treatment, including process wastewater and domestic sewage. After the disinfection of high temperature water, the process wastewater, domestic wastewater and part of the rainwater are removed most of the floating materials and large particles by the grille screen, and then enter the comprehensive wastewater conditioning tank by self-flow. The COD and BOD in the wastewater are relatively low and easy for biochemical treatment. The wastewater after homogenization and homogenization is lifted by the submersible pump into the A-agent tank, and the wastewater after homogenization is lifted by the submersible pump into the A-agent and O-agent biological contact oxidation tank for microbiological treatment.

The wastewater biochemical treatment mainly consists of hydrolysis acidification tank, biochemical tank, MBR tank, sludge tank and so on. In the hydrolysis acidification tank, micro-aeration agitation is set up to control DO at 0.3~0.5 mg/L under anaerobic and partly aerobic conditions, and some of the large molecule organic matter in the water is decomposed into small molecule organic matter. After this treatment, the mixed water body enters the contact oxidation tank. Contact oxidation tank is a kind of biological treatment device mainly by biofilm method and also by activated sludge method. The oxygen source is provided by the aeration blower, and the DO is controlled at 2.5~3 mg/L. The organic matter in this device is adsorbed and degraded by the microorganisms, so that the water quality is purified. The aerobic tank adopts polyethylene composite filler. The packing has a large specific surface area, which is not easy to make the biofilm agglomerate, and itself has the characteristics of fine bubbles, high oxygen utilization rate, and uniform air distribution.

After oxidation, the mixture flows into the MBR tank, which uses activated sludge to further degrade the organic molecules in the water body, so that the pollutants in the water body can be decomposed again under the action of activated sludge, and then the water body can be finally purified through the MBR membrane.

The water from the MBR tank enters the disinfection tank, where the disinfection and sterilization solution is added to further remove the germs in the water body, and the water quality reaches the urban miscellaneous water quality standard.

Hazardous waste generation: The wastewater biochemical treatment mainly consists of hydrolysis and acidification tank, biochemical tank, MBR tank and sludge tank. In the process of biochemical treatment, residual sludge is generated.

5.6.2 *Medical Waste Microwave Treatment Technology*

5.6.2.1 Case Unit Overview

The technical case was selected in a company in Jiangxi, mainly treating infectious and injurious medical waste, with a treatment scale of 5t/d. The company has 16 employees, including 2 managers, 5 technicians and 6 general staff. Among the technical staff, there are 3 professionals in environmental science, 1 in chemical engineering, and 1 in engineering management.

The microwave processing technology case study site is shown in Fig. 5.19.

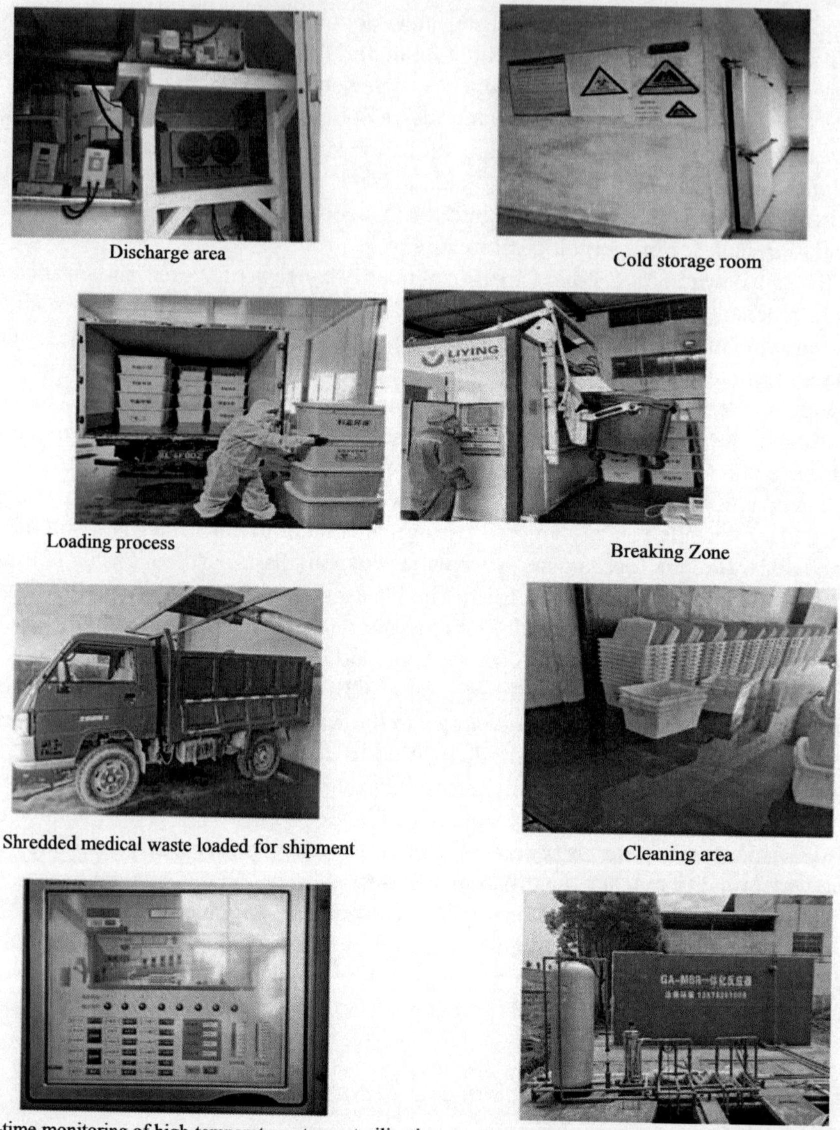

Discharge area; Cold storage room; Loading process; Breaking Zone; Shredded medical waste loaded for shipment; Cleaning area; Real-time monitoring of high-temperature steam sterilization; Wastewater treatment system

Fig. 5.19 Microwave processing technology case study site equipment

5.6.2.2 Processing Equipment

1) Main Body Equipment

The main equipment is medical waste microwave sterilization equipment, which can realize automatic operation, including feeding system, crushing system, microwave sterilization system, discharging system, steam supply system, exhaust gas treatment system and automatic control system.

The main equipment parameters are as follows: Specification: MDU-5B; Design capacity: 5 t/d; Quantity: 1 unit.

The technical parameters of microwave disinfection equipment are as follows:

Size: 7615 mm × 2840 mm × 3551 mm; Medical waste processing volume/capacity: 5t/d; Power: 115 kW; Sterilization temperature: 95~100 °C; Total microwave power: 12 kW; Sterilization logarithmic value (ATCC9372 biological indicator): ≥4.

2) Secondary pollution control equipment

(1) Sewage treatment equipment.

Equipment name: GA-MBR integrated reactor.

Specification Model: Customized.

Design capacity: 1 t/h of wastewater treatment.

Quantity: 1 set.

(2) Exhaust gas treatment equipment.

① Equipment name: Square cyclone tower pretreatment equipment.

Specification Model: AH-XL-10000.

Design capacity: 10,000 m^3/h of treated air volume.

Technical parameters: Size: 1350 mm × 1100 mm × 3300 mm; power 3.7 kW; circulating water: 0.5t; absorbent solution: 3‰ baking soda solution.

② Name of equipment: UV photo-oxidation catalytic purification equipment.

Specification Model: AH-UV-10000.

Design capacity: 10000 m^3/h.

Technical parameters: Size: 1400 mm × 1100 mm × 1250 mm; Power: 3.6 kW; UV band: DUV.

5.6.2.3 Operation Management

During the process of operation, in the microwave disinfection treatment facilities, timely check the medical waste reception, check the medical waste unloading, check the medical waste storage, check the medical waste cleaning and disinfection to ensure that the feeding unit, crushing unit, discharge unit, disposal process, waste gas treatment unit, wastewater/waste liquid treatment unit, solid waste treatment and disposal unit, process control unit are in stable operation.

Microwave disinfection process operating process parameters: During microwave disinfection process, microwave frequency of (915 ± 25) MHz or (2450 ± 50) MHz, disinfection temperature ≥95 °C, disinfection time ≥45 min should be maintained; using a combination of microwave and high-temperature steam disinfection process, microwave frequency should be used at (2450 ± 50) MHz, and pressure ≥0.33 MPa with a temperature ≥135 °C, and disinfection time ≥5 min. Microwave processing equipment around the set shield blocks the spread of microwaves, and should be set up with automatic alarm function of the immediate monitoring device to prevent microwave leakage to the operator caused by personal injury.

For the disinfection effect, Bacillus subtilis black variant bacterium was used as a bioindicator to ensure its kill log value ≥ 4.00, and the overall requirements must meet the requirements of the Evaluation and Standards of Disinfection and Sterilization Effect (GB 15,981–1995) and the Hospital Health Disinfection Standards (GB 15,982–2012). The monitoring frequency is 2 times/year.

The main components of the microwave disinfection system are odor gases (odor concentration, ammonia, hydrogen sulfide) and volatile organic gases (VOCs). There is a closed dust collection hood at the feed inlet to make the crushing in a closed environment, and at the same time, the microwave disinfection system forms a micro-negative pressure inside to prevent the escape of malodorous gases. When the equipment is fed, the fan is turned on, and the odorous exhaust gas from crushing and microwave disinfection is extracted by the fan, and the extracted exhaust gas is purified by the secondary filtration membrane and activated carbon adsorption and then led to a 25 m high exhaust pipe. The wastewater is treated by integrated biofilm reactor (MBR system) + disinfection and then partially reused for production, while the rest (about 2.95 m^3/d) is discharged.

5.6.2.4 Lessons Learned from Management Practices

Medical waste microwave disinfection treatment technology equipment with minimal investment, high automation, less labour-intensive, simple operation, short processing cycle, high efficiency, microwave action is more energy efficient.

(1) Management of contaminated areas

The medical waste temporary storage area and loading area are contaminated areas. In order to reduce the risk of occupational infection, the unit has developed a comprehensive management program that combines practical organizational management and protective management measures. Every year, regular medical check-ups and vaccinations are conducted for the staff to improve their immunity and resistance to diseases.

The staff is required to use chlorine disinfectant to flush with high-pressure water gun daily to ensure the appropriate temperature and humidity. Activate the negative pressure device in the workshop to ensure the ventilation of the working area. Ensure that new air passes through the workplace first. When using chlorine-containing disinfectants to soak equipment, make sure the lid is covered to prevent the evaporation of irritating gases from the disinfectant.

(2) Management of electromagnetic radiation protection

Microwave disinfection workshop is equipped with microwave leakage online detection system, a set value will exceed the shutdown alarm, microwave disinfection equipment made of sandwich color steel bins for radiation control area, the equipment cannot open the barn door during operation, or into the interior of the equipment bins. Regularly do electromagnetic radiation detection, to ensure that the power density of electromagnetic radiation to meet the requirements of national regulations on electromagnetic radiation protection.

Require staff to disinfect the workstations with 75% alcohol wipe daily. Enable the negative pressure device in the workshop to ensure the ventilation of the working area. Enable the air conditioning system to maintain the temperature and humidity of the workshop.

5.6.3 *Medical Waste Dry Chemical Disinfection Treatment Technology*

5.6.3.1 Case Unit Overview

Shandong a medical waste disposal centre, treatment scale: 17320 t/year, design capacity of 20 t/d, the number of installations 2 sets. Medical waste from Weifang City, infectious waste, injury waste and part of the pathological waste. The company has 43 employees, including 5 professional and technical personnel.

The dry chemical disinfection treatment technology case study site is shown in Fig. 5.20.

5.6.3.2 Processing Equipment

Stationary disposal equipment has medical waste storage system, feeding system, crushing and disinfection system, discharge system, residue collection system, cleaning system, sewage treatment system, etc. The medical waste is treated harmlessly, and the process index fully conforms to the requirements of the Technical Specification for Chemical Disinfection of Medical Waste Centralized Treatment Project, and no waste water and exhaust gas are generated during the whole treatment process.

1) Key Parameters
(1) Model: UEGDC-1000(fixed); UEYDC-600(mobile).
(2) Processing capacity: 800~1000 kg/h; 400~600 kg/h.

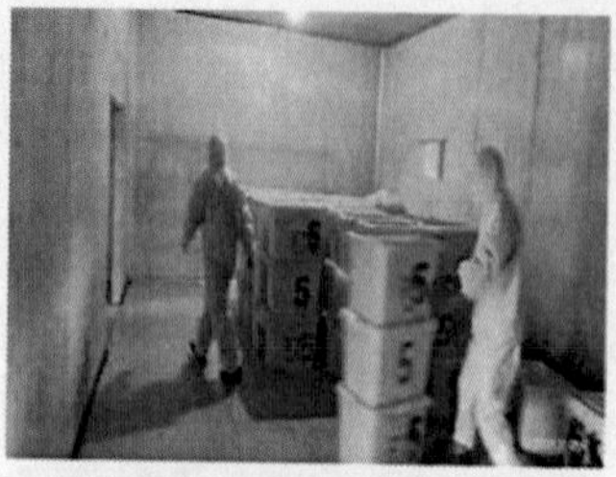

Discharge area

Temporary storage area

Crushing systems

Multi-stage automatic transfer device for treated residues

Shredded medical waste loaded for shipment

Cleaning area

Process control system for exhaust gas treatment

Wastewater treatment system

Fig. 5.20 Dry chemical disinfection treatment technology case study site equipment

(3) Power supply: Power system AC380V/50 Hz, control system AC220V/50 Hz.
(4) Installed power: 150 kW; 120 kW.
(5) Dimension: 15 m × 8 m × 4 m; 8 m × 2.3 m × 2.4 m.
(6) Weight: 22 t; 16.5 t.

2) Exhaust gas control measures

The exhaust gas treatment adopts air collection hood collection + high efficiency air filter (with activated carbon and secondary filtration membrane) + 18 m high exhaust pipe process to treat alkaline chemical disinfection exhaust gas.

The exhaust gas is purified by high efficiency air filter (HEPA system) and primary filtration membrane filtration (filtration size $\leq 0.2\ \mu m$, temperature resistance not less than 140 °C, filtration efficiency above 99.999%), 99.999% of the bacteria-containing particles are adsorbed by the two-stage filtration device, and activated carbon is applied on the filtration membrane to adsorb malodorous substances through activated carbon, and finally discharged through an 18 m high exhaust pipe. The exhaust gas emission can meet the requirements of the Technical Specification for Chemical Disinfection of Medical Waste Centralized Treatment Project (HJ/T 228-2006), and the exhaust gas generated in the process of crushing and disinfection of medical waste is purified by the first and second stage filtration membranes. It can meet the requirements of relevant pollutant emission standards.

The technical parameters of the filter membrane are as follows:

(1) One-stage filter membrane: 590 mm × 284 mm × 20 mm, $> 5\ \mu m$, the filtration level of primary plate filter is G4, the cardboard type is made of high-strength waterproof paper frame, the filter cotton is made of high-quality activated carbon filter cotton, the plate filter is lined with galvanized metal mesh on one side, and the deviation of filter shape size is ≤ 2 mm.
(2) High efficiency filter. 12 in. × 24 in. × 12 in. (1 in. = 25.4 mm), $>0.3\ \mu m$, high efficiency filter grade 99.999, filter material using ultra-fine water-resistant glass fibre filter paper, high efficiency filter frame using galvanized steel material, aluminium partition, filter sealant using polyurethane sealant, filter shape size deviation ≤ 1 mm and the deviation of filter shape size is ≤ 1 mm.
3) Wastewater treatment measures

The wastewater treatment of the wastewater treatment station adopts the treatment process of "grating + conditioning + hydrolysis acidification + contact oxidation + ultra-filtration system".

After the biochemical treatment, most of the bacteria such as E. coli and Streptococcus faecalis still exist in the sewage, except for some bacteria which are precipitated with the sludge, and must be disinfected. Further disinfection is carried out by chlorine dioxide method. The disinfection tank adopts a stratospheric contact reaction device to improve the contact time and achieve a better disinfection effect.

5.6.3.3 Operation Management

Medical waste transport vehicles and crates/barrels of cleaning and disinfection are used to spray disinfectants. For example, the use of effective chlorine concentration of 1000 mg/L chlorine disinfectant solution, and time at above 30 min. medical waste transfer vehicles, crates in every operation should be disinfected, and cleaned. Unloading facilities, operating sites, storage room floors and 2 m high walls are regularly disinfected, also by using a concentration of 1000 mg/L of chlorine dioxide disinfection solution.

The dry chemical disinfection process is as follows: The first-stage crushing unit is equipped with a low-speed, high-torque crushing device with a speed of 20 revolutions per minute and a micro-negative pressure response environment. The crushing performance is good for soft materials (such as infusion tubes, plastic bags, cotton swabs and gauze, etc.) and hard materials (such as scalpels, needles, etc.). The size of the crushed material is below 8 cm, and then it enters the secondary crushing system. The secondary crushing system is a high speed and low torque crushing, with a speed of about 400 rpm, and the reaction environment is a micro-negative pressure environment. The residue discharged after treatment is usually 3~5 cm in length, and the final volume of treated medical waste is reduced by about 75% and is unrecognizable. The pH is continuously monitored throughout the treatment process to ensure that the pH of the treated medical waste is between 11.0 and 12.5 when it leaves the outlet. The pH monitoring head is connected to the discharge conveyor terminal and is connected to the built-in computer.

In pollution prevention and control facilities configuration and treatment management, the use of Bacillus subtilis black variant bacterium as a biological indicator to ensure that its kill log value ≥4.00, in order to achieve the disinfection effect requirements; timely detection, supervision of TVOC, odor, particulate matter and other waste gas pollutants treatment effect, detection, supervision of wastewater treatment and noise emission requirements, and timely resolution of problems.

Dry chemical disinfection process parameters required are as follows: ① dry chemical disinfectant dosage should be greater than 0.075~0.12 kg dry chemical disinfectant/kg medical waste, spray water ratio of 0.006~0.013 kg/kg medical waste, to ensure that the disinfection temperature is ≥90 °C, and the reaction control of strong alkaline environment pH in the range of 11.0~12.5; ② dry chemical disinfectant and the total contact reaction time between the dry chemical disinfectant and the crushed medical waste is >120 min; ③ the disinfection of prion-contaminated medical waste should be appropriately increased by the disinfectant dosage, and the disinfection time should be appropriately extended.

Dry chemical disinfection process management, dry chemical disinfectant must guarantee the disinfection effect requirements, to ensure the disinfection process to achieve the killing or inactivation of infectious bacteria; dry chemical disinfectant supply must ensure its effective concentration and dosage, not to use more than the expiration date of the chemical disinfectant; dry chemical disinfectant content of calcium oxide should be more than 90%, and calcium oxide particle size should be 200 mesh.

5.6.3.4 Lessons Learned from Management Practices

The dry chemical disinfection treatment technology has short operation time, energy saving, low operation cost, high automation of the main equipment and low labor force.

For the chemical disinfection method, it is crucial to use highly effective chemical disinfectants and to ensure that the concentration of the selected agents and the levels of bacteria, viruses and fungi tested are maintained at normal levels in order to achieve the required disinfection effect. Chemical disinfection method is applicable to the treatment of "medical waste classification list" (Ministry of Health and State Environmental Protection Administration issued [2003] No. 287) in the infectious waste, injury waste and pathological waste, but not for the treatment of pharmaceutical and chemical medical waste, not suitable for the treatment of pathological waste in human organs and infectious animal carcasses, etc. It is a must to ensure that the treated medical waste does not have the characteristics of hazardous waste, that is, the pH value should be less than 12.5, and otherwise it will become hazardous waste. Therefore, the process should be strictly controlled and the treatment effect should be strictly tested to ensure that the residue generated after disposal is disposed off together with domestic waste in a safe and risk-free manner. For this technology, in addition to the above-mentioned problems, the most important issue is how to ensure the actual effect of the chemical disinfection of medical waste. In addition, the application of chemical disinfection technology can also generate air pollutants such as odor, TVOC and dust.

5.6.4 *Medical Waste Ethylene Oxide Disinfection Treatment Technology*

5.6.4.1 Case Unit Overview

A company in Hangzhou, has a scale of treatment of 40,000 t/year. Medical waste comes from infectious waste, injury waste and some pathological waste in the region. The company is equipped with 100 collectors, 60 disposal personnel, 20 logisticians, 20 management and customer service personnel. Among them, there are 20 professional and technical personnel.

An example of ethylene oxide treatment technology is shown in Fig. 5.21.

5.6.4.2 Processing Equipment

(1) Main Body Equipment

The main equipment consists of ethylene oxide disinfection cabinet, reciprocating elevator, automatic feeding system, duplex crusher, shaftless screw conveyor, double-channel automatic cleaning machine, nitrogen production unit and other parts. Equipment specifications and main technical parameters are shown in Table 5.6.

(2) Secondary pollution control equipment

Discharge area

Temporary storage area

After disinfection, the conveyor belt is sent to the crushing process

Breaking Zone

Shredded medical waste loaded for shipment

Cleaning area

Exhaust gas treatment facilities

Wastewater treatment system

Fig. 5.21 Ethylene oxide disinfection treatment technology case equipment situation

Table 5.6 Main equipment and technical parameters of ethylene oxide disinfection treatment technology

Number	Equipment name	Specification model	Design capabilities	Quantity	Technical parameters
1	Ethylene oxide Disinfection cabinet	42 m^3	20 t/d	8	Pressure: –80~+30 kPa Temperature: 20~80 °C Humidity: 20~100%RH EO concentration: 300~1100 mg/L
2	Reciprocating lifter		1200 cases/h	4	Power: 4 kW Lifting rate: 1200 cases/h
3	Automatic feeding system		1200 cases/h	4	
4	Duplex crusher	DDWK-1	5~20 t/h	4	Power: 75 kW × 2 System pressure: 16~20 MPa Rotation speed: 16~25 r/min Knife set: 25 sets + 30 sets
5	Shaftless spiral conveyor	650 mm	10~20 t/h	8	Power: 4 kW Conveying capacity: 10~20 t/h
6	Dual-channel automatic cleaning machine		1200 cases/h	2	Power: 37 kW Function: Wash, rinse, disinfect, blow dry
7	Nitrogen production unit		500 m^3/h	1	Power: 200 kW Nitrogen output: 500 m^3/h Nitrogen purity: 99%

The secondary pollution control equipment mainly includes the exhaust gas pickling spray treatment system and the sewage MBR membrane biochemical disinfection treatment system, and the main technical parameters are shown in Table 5.7.

Table 5.7 Secondary pollution control equipment and main parameters of ethylene oxide sterilization treatment technology

Number	Name of equipment	Design capabilities	Quantity	Technical parameters
1	Exhaust gas pickling spraying treatment system	8000 m^3/h	1	Exhaust height: 8 m System power: 10 kW
2	Wastewater MBR membrane biochemical disinfection treatment system	50 t/d	1	Treatment process: coagulation and sedimentation + biochemical + MBR membrane + chlorine dioxide disinfection

5.6.4.3 Operation Management

In order to verify the effect of ethylene oxide disinfection on medical waste treatment, the company commissioned an authoritative institution to conduct an environmental technology verification evaluation. According to the technical characteristics and evaluation objectives, the test parameters were divided into environmental effect parameters, operation process parameters and maintenance management parameters. The specific test parameters are shown in Table 5.8.

The number of samples collected and the number of valid data measured in this validation test were counted in the following table, with 462 samples collected and 782 valid data obtained. The samples were collected and measured according to the methods specified in the relevant national standards (GB) and environmental protection standards (HJ). The statistics of the number of validation test samples and the number of valid data are shown in Table 5.9.

5.6.4.4 Validation Evaluation Results

This technology is a chemical disinfection treatment technology for infectious, pathological and injurious wastes in medical waste. The medical waste is packaged in its original form into the ethylene oxide sealed disinfection cabinet under the following conditions: effective concentration of ethylene oxide 893 mg/L, pre-vacuum −80 kPa, system temperature (54 ± 2) °C, relative humidity 50 ± 10%, disinfection time of about 4 h. The medical waste is disinfected by ethylene oxide and shredded in secondary form and transported to a designated domestic waste incineration plant or sanitary landfill in a harmless and non-reusable form. Waste incineration plant or sanitary landfill for disposal. The technology was evaluated to achieve the following results:

The logarithmic value of killing Bacillus subtilis black variant is stable up to ≥4.0.

After the exhaust gas generated by this technology is treated, the concentration of ethylene oxide emission from the exhaust gas outlet is lower than the relevant requirements of the Occupational Exposure Limits for Hazardous Factors in the Workplace

Table 5.8 Test parameters of ethylene oxide sterilization treatment technology

	Test subjects		Specific parameters
Environmental effect parameters	Disinfection effect (bioassay)		Bacillus subtilis black variant bacteriophage killing log values
	Pollutant emissions	Atmospheric pollutants	Ethylene oxide, VOCs, Hg, particulate matter, malodor; Total airborne bacteria in the work area
		Water pollutants	pH, COD_{Cr}, BOD_5, SS, Hg, total residual chlorine, ammonia nitrogen, volatile phenols, fecal coliform count
		Solid waste	Solid waste emissions after medical waste treatment
	Noise		Continuous Equivalent A Sound Level
Process operating parameters	Processing systems		Ethylene oxide concentration
			Pre-vacuum
			System temperature
			Relative Humidity
			Disinfection time
	Processing scale		Processing Capacity
Maintenance management parameters	Energy consumption		Water quantity, electricity
	Raw material consumption		Amount of ethylene oxide, amount of nitrogen, amount of chlorine dioxide, amount of liquid enzyme

Table 5.9 Statistics on the number of validation test samples and the number of valid data in ethylene oxide sterilization treatment technology

Test subjects		Number of samples/pc	Number of valid data/pc
Disinfection effect (bioassay)		408	408
Atmospheric pollutants	Exhaust emission port	21	189
	Unorganized emissions	14	140
	Total number of airborne bacteria	12	12
Water Pollutants		7	21
Noise		–	12
Total		462	782

(Chemical Factors) (GBZ 2.1-2007); the concentration of VOCs is lower than the requirements of the relevant standard limit values in the Medical Waste Pollution Control Standards, and the concentration limit values of VOCs in the unorganized emission are lower than the requirements of the standard limit values by reference to the implementation of the Comprehensive Emission Standards for Atmospheric Pollutants (GB 16,297-1996) in the VOCs standard limit value, lower than the standard limit value requirements; Hg, particulate matter is lower than the "Comprehensive Emission Standards for Atmospheric Pollutants" of the relevant standard limit value requirements; odor gas is lower than the "Comprehensive Emission Standards for Odor Pollutants" (GB 14,554-1993) of the relevant standard limit value. 1993); the total number of bacteria in the working area is 2.1~19.3 CFU/dish.

The process parameters of the technical system are stable and reach: the effective concentration of ethylene oxide is 893 mg/L, the pre-vacuum is -80 kPa, system temperature is (54 ± 2) °C, relative humidity is $50 \pm 10\%$, and disinfection time in the disinfection cabinet is about 4 h.

This technology treats 1t of medical waste, discharges 0.18t of wastewater, and produces 0.974 t of treated harmless medical waste.

The technology treats unit weight medical waste with water consumption of 0.18 t/t, electricity consumption of 40.24(kW–h)/t, ethylene oxide consumption of 8.82 kg/t, nitrogen consumption of 47.77 m^3/t, liquid enzyme consumption of 0.5 L/t, chlorine dioxide consumption of 0.13 kg/t. The cost of treatment was calculated to be 312.31 RMB for 1t of medical waste.

The whole process of the verification and evaluation work was carried out in strict accordance with the "General Rules for Verification and Evaluation of Environmental Protection Technology", "Specification for Verification and Testing of Environmental Protection Technology" and "Verification and Evaluation Program", and the corresponding documents were filed in each step.

5.6.4.5 Lessons Learned from Management Practices

1) Production environment control 5S system standard

To create a good working environment for employees, Hangzhou Dadiwei Kang Medical Environmental Protection Co., Ltd. strengthens the management of the production work environment, improves the efficiency of production, and prevents safety accidents, and sets the 5S standard for production sites.

(1) The specific duties of each type of personnel are as follows:

① Crushing management personnel. Supervise operators in accordance with 5S classification and positioning standards, medical waste placed in the designated area, and stacked neatly and firmly; supervise operators in accordance with 5S cleaning and institutionalized standards regularly clean the main production site and auxiliary production site; supervise operators in accordance with 5S

cleaning and institutionalized standards for key production sites (crushing plant, crushing line) to do timely cleaning and sweeping cleaning work; implementation of 5S Standardization standards, regular inspection and correction of the implementation of the operator.
② Disinfection control personnel. Supervise the collection department unloading personnel in accordance with 5S classification and positioning standards, medical waste placed in the designated area, and stacked neatly and firmly; supervise operators in accordance with 5S cleaning and institutionalization standards regularly clean the unloading pile area, ethylene oxide disinfection oven, disinfection staging room; implement 5S cleaning and institutionalization standards regularly clean the disinfection control room; implement 5S standardization standards, regular implementation of operators Conduct inspection and correction for operators.
③ Cleaning management personnel. Supervise the operators according to 5S classification and positioning standards, the different categories of packaging containers neatly stacked in the designated place; supervise the operators in accordance with 5S cleaning and institutionalization standards, the key production site (cleaning line, cleaning workshop, etc.) to do timely cleaning and cleaning work; implementation of 5S standardization standards, regular inspection and correction of the implementation of the operators.
④ Operators. Implement 5S classification and positioning standards, place items (medical waste or packaging containers) in designated areas and stack them neatly and securely; implement 5S cleaning and institutionalization standards to regularly clean the main production site and auxiliary production site; implement 5S cleaning and institutionalization standards to achieve timely cleaning and clean-up work on key production sites.

(2) The 5S standard for production site environment includes five aspects: classification, positioning, cleaning, institutionalization, and standardization.

① Classification. The production site is divided into key production site, main production site and auxiliary production site.
② Positioning. The key production site, the main production site, and the auxiliary production site are clearly positioned.
③ Cleaning. The key production site, the main production site, and the auxiliary production site set clear cleaning work.
④ Institutionalization.

a. Key production sites. Implement both timely cleaning and sweeping cleaning.
b. Main production site. Regular cleaning is carried out. The warehouse is cleaned once a day, and the miscellaneous storage room is cleaned once a shift. The aisle and compressed vehicle parking site are cleaned twice a shift.
c. Auxiliary production site. Implement a regular cleaning method. Clean once a day.

⑤ Standardization.

a. Key production sites. The standardization of key production sites is verified in three levels: the first level is self-inspection by the responsible managers of each department (crushing managers, disinfection control personnel, cleaning managers). Not less than 2 times per shift, and must be checked after completing the production task of the shift. The second level is regularly inspected by the person in charge of the production department, once a day. The third level is randomly inspected by the production manager from time to time.
b. Main production site and auxiliary production site. The standardization of the main production site and auxiliary production site is verified at two levels. The first level is self-inspection and self-correction by the responsible managers of each department (crushing managers, disinfection control personnel, cleaning managers). Check once after completing the production task of the shift; the second level is randomly checked by the manager of the production department from time to time.

2) Medical waste treatment and disposal technology management practice recommendations.

Centralized medical waste treatment and disposal enterprises regardless of which disposal method for the collection and disposal of medical waste, the following elements need to focus on:

(1) The most significant characteristic of medical waste is its infectious nature, and infectious waste accounts for a significant proportion of medical waste. Therefore, it is crucial to reduce the infection in the secondary link in each link. The main links are as follows:

① The packaging of medical waste must be standardized. Good packaging can effectively reduce the likelihood of secondary infection for operators. This was particularly evident during the outbreak.
② Reduce the intermediate dumping of medical waste. This is an effective measure to reduce the secondary infection of medical waste, both from the internal collection system of hospitals and the disposal process of centralized medical waste disposal enterprises.
③ Reduce the direct contact of operators in the collection and disposal process. The use of existing automation technology can be considered for adaptation to reduce manual contact, which can improve the safety of the collection and disposal process.

(2) The classification of medical waste can be further refined. Because the non-incineration disposal technologies for medical waste have different types of adaptation, sometimes a category of medical waste is not necessarily all adapted. And medical waste centralized treatment and disposal enterprises generally do not enter the analysis of the requirements for the prevention of infectious risk of medical waste, but also limit the analysis of medical waste that has not been

disposed of. Therefore, further refinement of the classification of medical waste, can reduce the disposal process because of the category of non-adaptive risk, but also to facilitate compliance with the implementation of medical institutions.

(3) Use the existing internet of things information technology to carry out information management on the whole process of medical waste collection and disposal. This can improve the management level of medical waste generation, collection and disposal enterprises, but also to facilitate the supervision of the competent authorities, the overall improvement of medical waste management.

5.6.5 *High-Temperature Dry Thermal Treatment of Medical Waste Technology*

5.6.5.1 Case Unit Overview

The technical case was selected in a medical waste treatment centre in Liaoning, with a capacity of 5 t/d. The medical waste treatment centre has 16 employees.

The case study site of high temperature dry heat treatment technology is shown in Fig. 5.22.

5.6.5.2 Processing Equipment

The main equipment is described below.

Design capacity: 4~5 t/d. Technical parameters: Disinfection temperature 170~200°C, disinfection time 20 min, internal pressure of disinfection tank stable at 4200~4600 Pa, capacity reduction rate of 80%, reduction rate of 30%. The overall system consists of three parts: pumping system, gas purification system and heating system.

The function of the pumping equipment is to pump out the waste gas generated in the process of medical waste treatment to the exhaust gas purification system. There are three pumps in the pumping equipment: two liquid surround vacuum pumps and one electric water pump. This vacuum set has a cooling function, mainly to ensure that the pumping unit can work properly, with a rated power of 20 kW.

The gas purification system consists of three parts:

(1) Sterilization. The gas extracted from the pumping equipment first passes through a filtration device with disinfection solution, in which the gas is initially sterilized, after which the gas enters the electrostatic purifier, where it is further sterilized, and finally the gas is filtered through activated carbon fiber to achieve complete sterilization.
(2) Adsorption of particles. Due to the continuous release of high voltage electrostatic in the electrostatic purifier, the dust and particles are positively charged and then adsorbed by the negative electrode plate. The dust collected by the

Fig. 5.22 High-temperature dry heat treatment technology case study site equipment

electrostatic purifier is landfilled as domestic waste together with the treated medical waste.

(3) Adsorption of chemical gases. The top of the equipment is equipped with high efficiency filter, which can instantly adsorb chemical odor and different kinds of harmful gases.

The heating system is fueled by diesel oil and heat-conducting oil as medium. Power: 60000 kcal/h. Equipped with 100 L expansion tank (with heat-conducting oil inside). The boiler exhaust gas is discharged through the exhaust pipe.

The main system and major units are shown in Fig. 5.23.

5.6.5.3 Case Unit Operation Management

In order to verify the performance of the high temperature and high heat treatment facility, the company commissioned an authority to conduct an environmental technology verification evaluation. According to the requirements of "Environmental Protection Technology Verification Test Specification", the test parameters are divided into three categories: environmental effect parameters, operation process parameters, and maintenance management parameters. In this evaluation, suitable parameters were selected according to the characteristics of high-temperature dry heat treatment technology and evaluation objectives, and the test parameters are shown in Table 5.10.

During the validation test, the samples collected were atmospheric samples and disinfection effect test samples, totalling 158.

The above samples were collected and measured in accordance with the methods specified in the relevant national standards (GB), environmental protection standards (HJ) and technical specifications for disinfection.

Empirically, the following conclusions were drawn:

(1) Bacillus subtilis black variant bacteriophage killing log value >5, to achieve the disinfection effect of Bacillus subtilis black variant bacteriophage killing log value ≥4.
(2) The exhaust gas emission meets the emission limit requirements in the "Comprehensive Emission Standards for Air Pollutants" and "Comprehensive Emission Standards for Odor Pollutants". The test results of atmospheric pollutants are shown in Table 5.11.
(3) The object of treatment is infectious medical waste, injury medical waste, part of the pathological medical waste (unidentifiable).
(4) The operating parameters of the facility were normal, the disinfection temperature was stable at 170~210°C, the disinfection time was 20 min, the agitation speed was 21 r/min, and the internal pressure of the disinfection tank was 4200~4600 Pa.

Fig. 5.23 Main system and main unit of high temperature dry heat treatment

Table 5.10 Test parameters of high-temperature dry thermal treatment technology

Parameter category	Object	Specific parameters
Environmental effect parameters	Disinfection effect	Bacillus subtilis black variant bacteriophage killing log values
	Atmospheric pollutants	Odor, VOCs, particulate matter, mercury and its compounds (in Hg), hydrogen chloride, chlorine, detection of the total number of bacteria in the air near the discharge outlet of the exhaust gas purification system and at the discharge port of the treatment equipment at sensitive locations in the workshop
Process operating parameters	High temperature dry thermal treatment system for medical waste	Disinfection time
		Disinfection temperature
		Stirrer speed
	Processing scale	Unit time processing capacity
Maintenance management parameters	Comprehensive energy consumption (in standard coal) for processing unit weight of medical waste	Electricity consumption
		Diesel consumption

Table 5.11 Atmospheric pollutant test results

Test items	Test results	Emission limit value	Unit	Compliance rate/%
VOCs	2.2~6.1	20	mg/m^3	100
Stench	<10 (sampling in the exhaust pipe)	10 (Standard value of plant boundary level)	Dimensionless	100
Particulate matter	18~22	120	mg/m^3	100
Mercury and its compounds (in Hg)	Not detected	0.012	mg/m^3	100
Hydrogen chloride	<0.9	100	mg/m^3	100
Chlorine gas	<0.2	65	mg/m^3	100

Note According to the sampling analysis of fixed source organized emissions, the detection limit of chlorine gas is 0.2 mg/m^3 and the detection limit of hydrogen chloride is 0.9 mg/m^3

(5) The treatment capacity of the system is 0.3 t/h. If it is calculated by working 18 h per day, the treatment capacity is 5.4 t/d; if it is calculated by working 24 h per day, the treatment capacity is 7.4 t/d.
(6) The power consumption of the system is 27.31 kWh/t(medical waste), the diesel consumption is 16.89 kg/t(medical waste), the comprehensive energy consumption is 27.94 kg(standard coal)/t(medical waste); there is no water consumption in the disinfection process (excluding water for cleaning medical waste transfer box).

The final operating process parameters of high temperature dry heat treatment process are determined as follows: (1) the pressure inside the disinfector is 300 Pa, close to vacuum; (2) the temperature inside the disinfector is 180~200 °C, the processing time should not be less than 20 min, the mechanical stirring device is stirred at a speed of not less than 30 r/min; (3) the operation of dry heat treatment equipment should be prevented from human interference, to make sure that medical waste disinfection treatment is not completed before artificially stopping the operation.

In the operation of high-temperature dry heat treatment facilities, timely inspection of medical waste reception, inspection of medical waste storage, inspection of medical waste cleaning and disinfection to ensure that the feeding unit, crushing unit, discharge unit, disposal process, waste gas treatment unit, wastewater/waste liquid treatment unit, solid waste treatment and disposal unit, and process control unit are in stable operation.

5.6.5.4 Lessons Learned from Management Practices

High temperature dry heat treatment equipment technology consists of feeding crushing system, sterilization system, exhaust gas purification system, heating system and automatic control system, which is an all-in-one fully enclosed processing equipment system. The system works under 200 Pa pre-vacuum, the disinfection temperature is stabilized at 170~200°C, the disinfection time is 20 min, the internal pressure of the disinfection tank is stabilized at 4200~4600 Pa, the average logarithmic value of the carrier bacteria is >6.00. The disposal process does not require water, the amount of odor and VOCs generated is very small, the capacity reduction rate reaches 80%, the reduction rate reaches 30%, the sterilization temperature is lower than 220 °C, and no harmful substances such as dioxins and other POPs are produced. The sterilization of medical waste is done in the process of turning, crushing, sterilizing and discharging in the same system, with no secondary pollution sources, no dioxins, small odor concentration, no waste water, no acid gas and other waste gases, fully automatic control, simple operation and stable effect.

5.7 Medical Waste Non-incineration Treatment Technology

5.7.1 Medical Waste Hot Melt Disinfection Curing Molding Treatment Technology

5.7.1.1 Technology Overview

Medical waste hot melt disinfection curing molding treatment technology is a composite medical waste treatment method. Medical waste through chemical disinfection, thermal disinfection, physical curing three-stage treatment process, chemical disinfectants and mechanical grinding of their own heat to achieve high-temperature disinfection after extrusion molding, molding solids can achieve harmless treatment.

The main process flow of the technology is shown in Fig. 5.24.

(1) Chemical disinfection

After the medical waste is crushed by grinding, the chemical disinfection process is carried out first, in which lime powder is mixed with the water in the medical waste to disinfect and inhibit bacteria in an alkaline environment. The crushed medical waste and lime powder are in full contact in this process, and the final product also maintains alkaline characteristics to avoid regeneration of bacteria and viruses.

(2) Thermal disinfection

The internal heat generated by the mechanical grinding of medical waste can reach 160~180 °C, the liquid inside the cell evaporates, the cell wall ruptures and the protein denatures, thus achieving the purpose of sterilization. Since the heat comes from the mechanical grinding of the waste, the heat is uniform and stable.

(3) Physical Curing

When the medical waste reaches 160~180 °C, it will be in a semi-molten state, which can be used for secondary use after high die-casting molding. After molding, the solid surface is relatively dense and smooth, with small volume and high density.

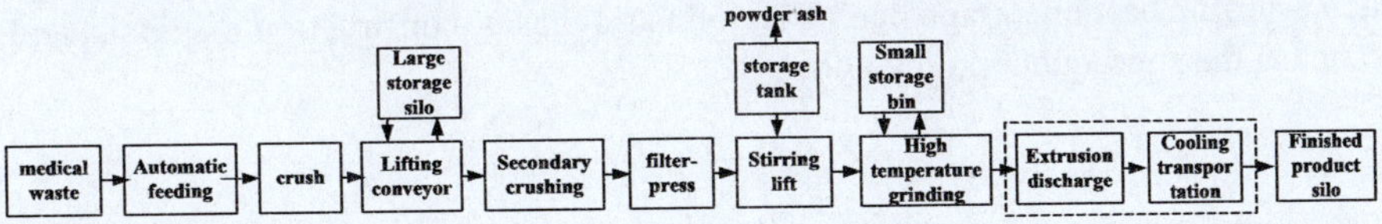

Fig. 5.24 Process flow diagram of the main body of the technology

5.7.2 Technology Innovation Analysis

(1) Innovation 1: The first international composite medical waste treatment technology.

This technology combines chemical disinfection technology, thermal disinfection technology, and physical curing technology, and is a composite medical waste treatment technology. Each single technology is applied to several stages of the treatment process, and after the superposition of technologies, the disinfection effect of medical waste is improved and the defects of single treatment technology are compensated, and the scale of treatment, treatment time, and treatment effect are greatly improved in many aspects.

(2) Innovation 2: Internal source heat instead of external source heat supply.

This technology replaces the external heat supply with the internal heat generated by the self-friction of medical waste, so that the medical waste under treatment can be guaranteed to reach a uniform high temperature in the disinfection container and solve the problem of heat decay in the heat transfer that exists in the external heat supply. At the same time, due to the relatively uniform temperature in the disinfection container, the size of the disinfection container can be increased accordingly to improve the unit disposal volume of medical waste. Due to the internal heat generated by friction, the overall energy consumption of this technology is significantly less than that of high-temperature steam disinfection treatment using external heat.

(3) Innovation 3: The treated solids have anticorrosive properties and high stability.

Usually, medical waste is disinfected by other non-incineration disposal methods, as the treated medical waste is prone to secondary bacterial growth, so it cannot be stored for a long time and should be immediately sent to a domestic waste disposal plant for landfill or incineration. The medical waste treated with this technology completely changes its original form, and the solid material formed after treatment avoids the secondary growth of bacteria and viruses due to the combined effect of alkaline environment and high density of lime powder, and the leaching rate of pollutants is extremely low, which makes long-term storage easy. The treated solid material has a high calorific value and can be used as an alternative fuel for domestic waste incineration plants and cement plants, and can be developed for various applications such as traffic barriers, grape pan piles, industrial bases, construction materials, and so on, as the subsequent policy allows.

5.7.3 Technology Validation Evaluation Analysis

(1) Test site

In 2016, the company upgraded the original process technology and added the medical waste hot melt disinfection and solidification molding treatment technology as a backup auxiliary production line. At present, the daily treatment capacity is 100 t, the annual operation accumulates 60 d, with an annual treatment capacity of 6000 t.

(2) Test conditions

The test was completed before the commissioning of the relevant processes at the test site, the validation evaluation agency and the validation test agency jointly visited and confirmed that the technical process line and supporting environmental protection facilities are operating normally, medical waste and the required raw and auxiliary materials are prepared, lime dust removal equipment has been emptied, water and electricity and other energy measuring instruments have been recorded, the laboratory has been equipped with Bacillus subtilis black variant bacillus, Bacillus thermophilus, VOCs Hg, particulate matter, odor, pH, COD_{Cr}, BOD_5, SS, total residual chlorine, ammonia nitrogen, volatile phenols, and fecal coliform count. The tests were conducted as scheduled in accordance with the General Rules for Environmental Protection Technical Validation Tests and the Validation Evaluation Program.

(3) Test parameters

According to the technical characteristics and evaluation objectives, the test parameters are divided into environmental effect parameters, operation process parameters and maintenance management parameters. The specific test parameters are shown in Table 5.12.

A total of 108 samples were collected for the validation test, and a total of 358 valid data were obtained. The samples were collected and measured in accordance with the methods specified in the relevant national standards (GB) and environmental protection standards (HJ). The statistics of the number of samples and the number of valid data for the validation test are shown in Table 5.13.

4) Validation of evaluation results

(1) The average logarithmic values of killing Bacillus subtilis black variant bacilli and Bacillus thermophilus bacilli were >4 for the medical waste treated with the hot-melt disinfection and solidification molding treatment technology, which met the disinfection effect requirements specified in the Technical Specification for Chemical Disinfection of Medical Waste Centralized Treatment Project.

(2) The emission concentration of VOCs from the exhaust vent can stably meet the relevant requirements of the “Medical Waste Pollution Control Standard” (reference implementation, under preparation, not released); whether the disorganized emission of VOCs can stably meet the relevant requirements of the “Comprehensive Emission Standards for Air Pollutants” for VOCs (reference implementation); the concentration of Hg and particulate matter from the exhaust vent and disorganized emission can stably meet the relevant requirements of the “Comprehensive Emission Standards for Air Pollutants Comprehensive Emission Standards for Air Pollutants”; whether the concentration of odor gas from the exhaust vent and disorganized emissions can stably meet the requirements of “Comprehensive Emission Standards for Odor Pollutants”.

Table 5.12 Test parameters of medical waste hot melt disinfection curing molding treatment technology

Parameter category	Test subjects		Specific parameters
Environmental effect parameters	Disinfection effect (bioassay)		Bacillus subtilis black variant bacteriophage killing log values; Bacillus thermophilus bacteriophage killing log values
	Pollutant emissions	Atmospheric pollutants	VOCs, Hg, particulate matter, malodor; Total number of airborne bacteria at the discharge port of the treatment equipment
		Water pollutants	pH, COD_{Cr}, BOD_5, SS, Hg, total residual chlorine, ammonia nitrogen, volatile phenols, fecal coliform count
		Solid waste	Emission of solid waste after medical waste treatment, emission of waste activated carbon, amount of calcium oxide dust
	Noise		Continuous Equivalent A Sound Level
Process operating parameters	Processing systems		Disinfection time
			Disinfection temperature
	Processing scale		Processing Capacity
Maintenance management parameters	Energy consumption		Water quantity, electricity
	Raw material consumption		Hydraulic oil dosage, lime powder dosage, chlorine dioxide dosage, activated carbon powder dosage and replacement time

Table 5.13 Statistics on the number of validation test samples and the number of valid data in medical waste hot melt disinfection curing molding treatment technology

Test subjects		Number of samples/pc	Number of valid data/pc
Disinfection effect (bioassay)		60	60
Atmospheric pollutants	Exhaust emission port	21	147
	Unorganized emissions	14	112
	Total number of airborne bacteria	6	6
Water pollutants		7	21
Noise		–	12
Total		108	358

(3) Before the treatment of medical waste, the bacterial test results were 5.8~8.4 CFU/dish, and after the treatment of medical waste, the bacterial test results were 13.5~14.8 CFU/dish, and the number of bacteria at the outlet of the treated material would increase.
(4) The treatment of 1 t of medical waste generates about 0.74 t of harmless solids. The activated carbon is replaced regularly, once every 3 months, and the annual usage of a single line (36.5 t/d output) is 1.2t. The amount of calcium oxide dust generated per unit weight of medical waste treated is 0.077 kg/t.
(5) The operation of the system equipment will not affect the regional sound environment. The noise at the plant boundary can meet the requirements of the Class II standard of the Environmental Noise Emission Standard for Industrial Enterprises (GB 12,348-2008), which is 60 dB (A) during daytime.
(6) The system disinfection temperature stability is about 160~180 °C, and have a stable operation of equipment, the processing time required to deal with 1t of medical waste is about 80 min. 80 min is required to treat 1t of medical waste, 24 h of operation, and a single production line processing size of 36.5 t.
(7) The consumption of water is 0.18 t/t, electricity consumption is 476.85 kW–h/t, lime powder consumption is 18.43 kg/t, chlorine dioxide consumption is 0.21 kg/t. The amount of circulating water in the technology application is 1 t. The hydraulic oil required in the technology application is replaced periodically, basically once a year, and the annual usage is 4000~5000 L. The activated carbon is replaced regularly, once every 3 months, with an annual usage of 1.2 t for a single line (36.5 t/d output).
(8) The cost of this technology to treat 1 t of medical waste is $315.

5.7.3.1 Technology Application and Promotion

Medical waste hot melt disinfection curing molding treatment technology is a new technology developed by a medical waste treatment and disposal enterprise in China, the intellectual property rights are unique, and there are already successful applications.

Given that the solid material formed by this technology has reusable resources, such as traffic isolation piers, grape pan piles, industrial bases, construction materials, etc., it is recommended to actively explore multiple ways of resource utilization and broaden the downstream industrial chain of the technology under the condition that the policy allows.

5.7.4 *Medical Waste Mobile Treatment Technology*

5.7.4.1 Technology Overview

Mobile medical waste disposal facilities single treatment scale is generally not greater than 2 t/d, mainly using high-temperature steam treatment technology, more suitable for small-scale local treatment, in Germany, France and other European countries have been used for many years, but also in China's Ya'an earthquake disaster areas of medical waste emergency treatment played an important role. After the New Crown epidemic, WHO proposed in the *Water, Sanitation, Hygiene and Waste Management for the COVID-19 Virus* that it is recommended to dispose of medical waste involved in the epidemic in situ (on-site) under the premise of ensuring safe collection and classification. China's Ministry of Ecology and Environment (MOE) urgently deployed 46 sets of mobile equipment to Wuhan for emergency medical waste treatment, and disposed of the medical waste generated by a few designated hospitals for centralized treatment of confirmed patients with New Guan Pneumonia in Huangpi District, Jiangxia District and Xinzhou District, effectively relieving the pressure of medical waste transportation and disposal (Ministry of Ecology and Environment 2020).

(1) Mobile dry chemical disinfection treatment technology

Medical waste is automatically transported to the automatic lifting tumbler through the conveying track of the feeding system, and after detection and weighing, the medical waste in the tumbler is lifted and turned into the primary crushing system, which is equipped with a low-speed, high-torque crushing device, during which the net weight of the feed is weighed out and stored by the automatic weighing device. When entering the primary crushing system, the bagged medical waste is crushed while automatically spraying water to humidify and add a certain proportion of dry alkaline disinfectant, medical waste is crushed, mixed with pharmaceuticals and disinfection treatment in the primary crushing system, so that microbial organisms and germs are killed.

After the primary crushing and grinding, the medical waste automatically enters the secondary crushing system to be crushed and ground again into fine particles to achieve further volume reduction and disinfection. After secondary crushing and disinfection, the volume percentage of discharged residue is reduced by more than 70%, and the residue is automatically transported to the enclosed residue transfer vehicle through the discharge winch. A pH monitoring system is set up at the outlet of the discharge winch to monitor the pH value in real time, and only when the requirement is reached can the disposed residue be transported to the enclosed residue vehicle, which means that the disposed residue is thoroughly disinfected and sterilized, and can be directly transported to the general sanitary waste landfill for landfill or incineration. If the pH value does not meet the requirements, the equipment system will automatically protect, stop feeding and alarm the staff for troubleshooting, and the residue that does not meet the requirements will need to be treated again before the next step of treatment.

The process flow diagram is shown in Fig. 5.25.

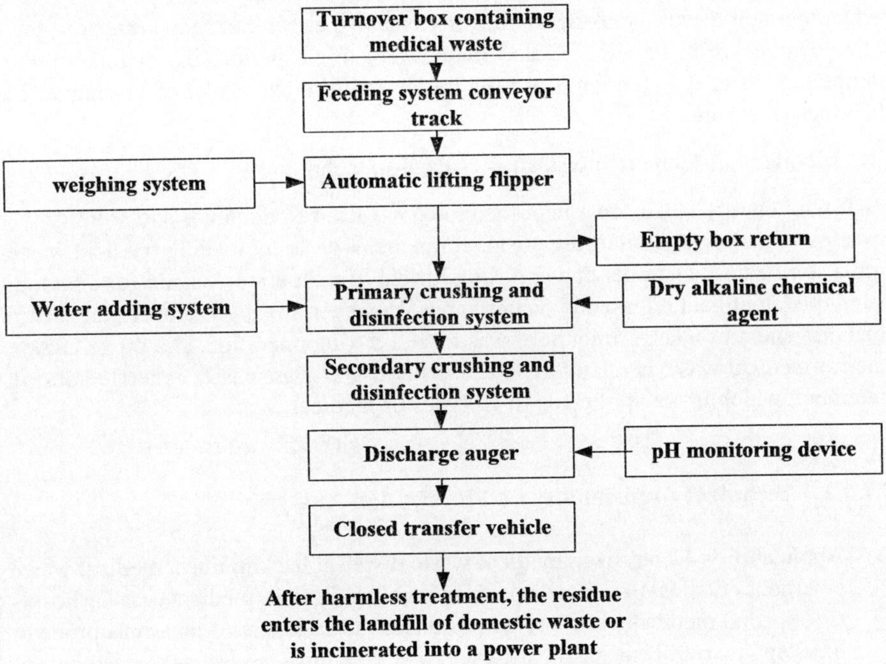

Fig. 5.25 Flow chart of mobile dry chemical disinfection treatment process

(2) Mobile microwave disinfection treatment technology

Microwave disinfection equipment crushes, disinfects and dries medical waste to some extent. It includes feeding unit, crushing unit, microwave disinfection unit, discharging unit, heat dissipation unit and exhaust gas treatment system. This medical waste treatment system can handle various model series such as 12, 25 and 50 kg waste per hour. It allows small institutions (such as laboratories, small clinics and hospitals) to safely process up to 250 L of biomedical waste per hour (about 25 kg/h) on site. Its innovative concept uses a specially designed medical waste vortex tumbling mechanism to uniformly sterilize and convert medical waste into domestic waste with a volume reduction of more than 60% and weight reduction of 20% (drying). In the premise of ensuring the safety of operators, the environmental impact and operating costs are reduced. Because of the pure microwave technology, no steam generator is needed, so there is no risk of excessive pressure and no liquid effluent is generated. Its ultra-compact size makes it an easy to use and maintain solution with only 10m^2 of installation space, the lowest operating costs on the market, and a sterilization log value of up to 6lg for the biological indicator (ATCC9372).

The process flow is roughly as follows: The material is crushed to the disinfection bin, and the discharge port and the inlet port are in the closed state. There are several microwave generating units around the disinfection bin, which output microwaves to irradiate the medical waste to raise the temperature for disinfection. The speed reducing motor drives the blades in the bin to make the material keep vortex turning. After heating up to 95~100 °C and maintaining for a period, the disinfection is completed. After disinfection, open the outlet and turn the blades to discharge the disinfected residue.

(3) Mobile high-temperature steam sterilization technology.

High-temperature steam treatment technology is used to denature and solidify the proteins of pathogenic microorganisms in medical waste by loading medical waste into a sterilization chamber and exposing it to high-temperature steam for a period, using the latent heat released by water vapor. However, the high-temperature steam method generates a large amount of wastewater during operation, and the reduction rate for medical waste is relatively low, and toxic VOCs are easily generated during treatment, which is not applicable to all types of medical waste.

5.7.4.2 Technical Applicability and Innovation Analysis

(1) Applicability: Centralized medical waste disposal backup lines, medical waste emergency disposal, on-site disposal of highly infectious medical waste in hospitals, special medical waste pre-disposal (for pretreatment of materials prone to blockage by pyrolysis gasification and other technical solutions), remote, rural areas. Especially suitable for emergency disposal, pre-treatment and disposal in remote areas.
(2) Innovation: The whole process of processing all-in-one machine, feeding to discharging one key intelligent operation, mobile, plug and play, and disinfectant can be recycled, disposal data can be monitored. Easy to install, plug and play, easy to operate, energy saving and environmental protection.

5.7.4.3 Technology Application Case Introduction

Take chemical disinfection method mobile treatment technology as an example. A company in Shenzhen introduced Israeli medical waste treatment technology, developed intelligent mobile medical waste on-site disposal equipment, providing an effective solution for emergency disposal, local disposal in remote areas and pretreatment of some medical waste, with a disposal capacity of 100 kg/h, 200 kg/h. Equipment parameters are shown in Table 5.14, and the main equipment is shown in Fig. 5.26.

Table 5.14 Main equipment names, specifications, design capacity, quantity, technical parameters

Model	Processing capacity/(200 kg/m)^3	Size/cm	Energy consumption/kW	Disinfectant consumption/(L/t)
Field type 700	960 L/h, 190 kg/h	250 × 210 × 225	5/18	9~11

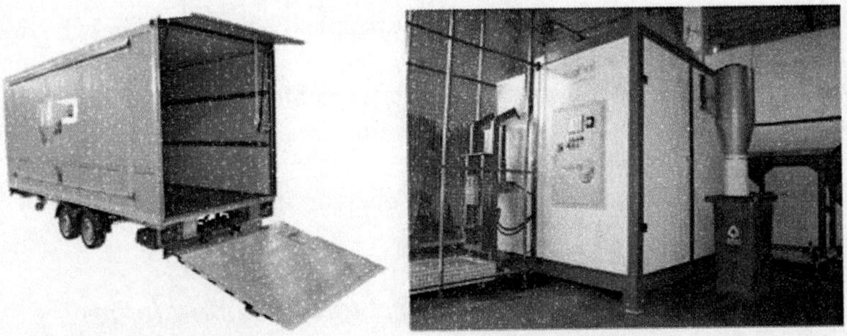

Fig. 5.26 Chemical disinfection method mobile treatment equipment diagram

The whole treatment system consists of feeding unit, crushing unit, pharmaceutical supply unit, chemical disinfection treatment unit, drying unit, discharging unit, disinfection solution automatic proportioning unit, control unit, etc.

(1) Feeding unit: 240 L standard drum, automatic loading, automatic weighing and feeding. Automatically open the bin door when feeding, and automatically close the bin door after feeding.
(2) Crushing unit: After closing the bin door, the crushing starts automatically, using a four-shaft crusher, the size of crushed particles is not larger than 5 m, and the crushing unit is in a closed state.
(3) Pharmacy supply unit: Disinfectant solution: 0.05~0.1% active ingredient of Sigemet disinfectant solution. Two-stage disinfection mode is adopted:

① Primary disinfection: Spraying disinfectant solution during crushing to disinfect the crusher and the dust generated during crushing at the same time, while removing the dust from falling into the chemical disinfection treatment unit.
② Secondary disinfection: After the crushing of medical waste directly into the chemical disinfection treatment unit, mixing and stirring disinfection takes about 10~12 min.

Chemical disinfection treatment unit: The feed hopper and disinfection treatment container are made of 304 stainless steel. The crushed medical waste falls directly into the hopper and is mixed and disinfected for 10~12 min. The disinfection solution storage tank and proportioning box are made of stainless steel. The pH value is monitored in real time, and the disinfection solution is automatically replenished to ensure disinfection efficacy. At the end of each treatment cycle, a small amount of waste liquid is discharged from the waste liquid discharge pipe. The amount of waste liquid discharged is the difference between the original water content of the raw material and the water content after squeezing and drying.

(5) Drying unit: After disinfection, the material is conveyed to the squeezing and drying equipment through the screw conveyor.

(6) Discharge unit: The dried material is automatically discharged into the material drum.
(7) Disinfectant automatic proportioning unit: Disinfection solution proportioning box, according to the pH value of real-time detection, automatically add the allocation of disinfection solution.
(8) Control unit, PLC control, and one key operation, fully intelligent.

5.7.5 *Small-Scale Decentralized Treatment Technologies for Medical Waste*

5.7.5.1 Technology Overview

The use of small decentralized disposal facilities for the local disposal of medical waste was once more common in Europe and the U.S. Since 2000, in the "Stockholm Convention on Persistent Organic Pollutants" and other international conventions such as external pressure for compliance and internal pressure to improve environmental quality, countries and regions have introduced more stringent air pollutant emission standards for incineration facilities. The original small incineration facilities are basically replaced by non-incineration facilities, new facilities also mainly use non-incineration technology without the emission of toxic and hazardous substances. China's centralized medical waste disposal model, "SARS" epidemic occurred more than 10 years after the basic did not build such facilities, only a very small number of the original facilities retained. During the New Crown epidemic, Wuhan's new Vulcan Hill, Thunder God Hill Hospital construction process synchronized with the configuration of a total of 24 t/d of medical waste incineration facilities. Other designated hospitals were equipped with 10.9 t/d medical waste disposal capacity to ensure the safe disposal of some of the epidemic-related medical waste.

Small decentralized treatment technology is a closed integrated medical waste disposal system, including feeding unit, crushing unit, pharmaceutical supply unit, chemical disinfection treatment unit, drying unit, discharge unit, automatic control unit, disinfection solution automatic proportioning unit and other auxiliary equipment. The disposal system is operated by one key with full intelligence, realizing the integration of disinfection treatment, crushing and drying equipment, avoiding the possibility of manual contact between the entrance of medical waste from the system and the exit of the material.

The process flow diagram is shown in Fig. 5.27.

(1) Non-incineration processing and crushing integrated machine

This technology is suitable for projects with a daily treatment capacity of less than 2 t, as well as the local treatment of medical waste generated by medical institutions. Its main features are: the treatment process (from feeding, disinfection to crushing) is completely closed, no odor discharge, and low impact on the surrounding environment. The treatment process is highly automated and requires low professionalism

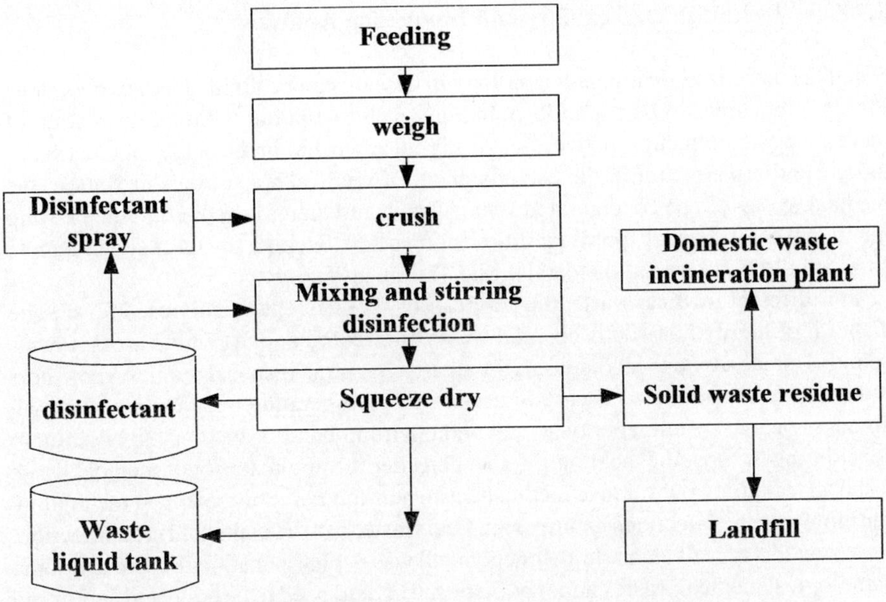

Fig. 5.27 Flow chart of small decentralized medical waste treatment process

of operators. The technology is adopted in the demonstration project of medical waste treatment in Shennongjia, China, and the equipment has been imported in large quantities from Europe by developing countries such as Indonesia since more than 10 years ago, and has become the mainstream equipment for medical waste treatment in these countries.

(2) Small non-incineration processing equipment (daily capacity less than 100 kg) + small crusher

This technology is a scaled-down version of centralized high-temperature steam, chemical, microwave, and dry heat disinfection treatment systems, and is characterized by low equipment cost, small footprint, usually self-contained disinfection generators, low operator expertise requirements, and high operational reliability. Many developing countries have adopted this technology in the United States, and small high-temperature steam processing equipment with a daily capacity of less than 100 kg is widely used in remote areas and small medical institutions.

(3) Small mobile medical waste treatment equipment

This technology is mainly applicable to emergency treatment and in situ disposal of medical waste on islands. The safe transfer of medical waste from islands to off-islands is costly and difficult to achieve, while local disposal is more feasible. Greece has more than 2000 islands, where mobile miniaturized non-incineration medical waste treatment facilities have been activated for medical waste disposal and sanitary landfills for the disposed medical waste.

5.7.5.2 Technical Applicability and Innovation Analysis

Small decentralized treatment technology in general can be divided into two models: First, remote areas to take a multi-point approach to a point, the implementation of medical waste collection network covering all township and village medical institutions and private clinics, the establishment of medical waste transfer station, the medical waste will be concentrated at each level, and then transported to the existing municipal centralized disposal facilities for disposal. Second, the local site in remote areas should build a small medical waste Treatment facility.

For different medical waste, the applicability is different, mostly used for on-site disposal of highly hazardous hospital waste, pretreatment and remote areas, towns, counties, small-scale local disposal. Small area but large disposal volume, the whole process of a key operation is easy to use, normal temperature and pressure disposal, no odor, easy to install, energy saving and environmental protection. In addition to considering the specific technology, also consider the applicability of medical waste disposal technology, the above medical waste non-incineration treatment technology can eliminate the infectious nature of medical waste, and the main goal is disinfection. However, this type of technology for chemical waste, pharmaceutical waste and some pathological medical waste cannot be disposed of, and need to be collected and stored centrally by qualified units for recycling and disposal. Using the National Health Service in the UK as a case study, it was found that the "best" treatment technology is still a mixed technology, and using hierarchical analysis as a research tool, it was suggested that deep landfills (safe segregated landfills) could be suitable for treating waste that is not infectious, but causes discomfort in terms of odor and appearance (e.g. urine). It is proposed that deep landfills (safe segregated landfills) can be used to treat wastes that are not infectious but cause discomfort to the human body in terms of odor and appearance (e.g. diapers, feminine hygiene products, etc.).

5.7.5.3 Technology Application Case Introduction

A company in Zhejiang introduced NW series on-site medical waste disposal system based on "frictional heat treatment (FHT)" patented technology from Italy. "FHT" medical waste treatment technology has several invention patents, and recognized by the WHO, in 2014, 2019. It is widely used in Italy, Britain, Germany, Sweden, Switzerland, Russia and more than 50 countries. This technology is published by WHO in "Safe Management of Medical Waste" as one of the best available technologies for solid medical waste.

The core of the NW series of processes is the thermal decomposition of the proteins that make up living cells (Fig. 5.28). In a closed sterilization vessel, a powerful rotor equipped with blades crushes the medical waste by impacting and rubbing it under shear stress for a period of time so that the heat generated by the friction is sufficient to reach and maintain a sterilization or disinfection temperature in the medical waste pile. The temperature of the waste is measured in real time and with high accuracy by a special sensor (patent number: PCT/EP94/02357) under constant and vigorous agitation. When the temperature reaches a predetermined level of 150 °C, water is automatically sprayed into the treated waste to cool it to 95 °C, followed by automatic unloading.

(1) Main equipment (Table 5.15, Fig. 5.29).

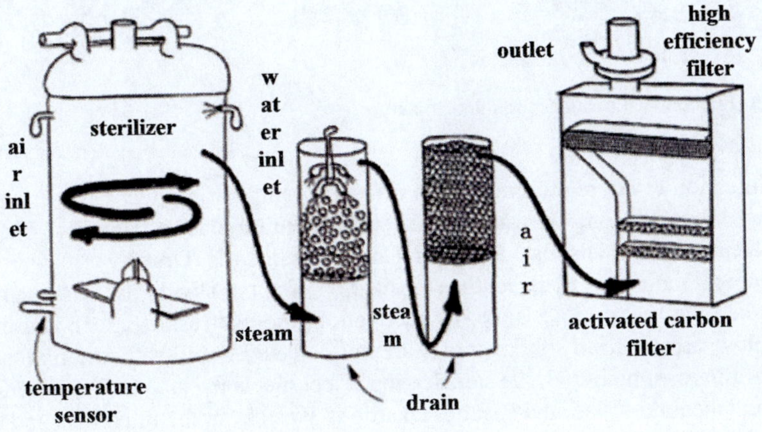

Fig. 5.28 Frictional heat treatment (FHT) process flow chart

Table 5.15 Technical parameters of NW series equipment

Item	NW5	NW15	NW50
Sterilization method	Frictional heat	Frictional heat	Frictional heat
Processing capacity/(kg/h)	15; 240	30; 320	90; 840
Sterilizer volume/L	100	175	460
Post-treatment waste form	Homogenized small particles	Homogenized small particles	Homogenized small particles
Volume of waste after treatment	Initial volume of 20~25%	Initial volume of 20~25%	Initial volume of 20~25%
Weight of treated waste	Initial weight of 70~75%	Initial weight of 70~75%	Initial weight of 70~75%
Energy consumption/(kW/h)	Up to 20; Average 13	Up to 30; Average 18	Up to 90
Water consumption (using the water cycle system)/(L/d)	50	Approx. 75; 50	About 50
Size/mm	1600 × 800 × 1300	1700 × 1000 × 1600	2500 × 1200 × 2100
Weight/kg	About 740	Approx. 980	About 2670

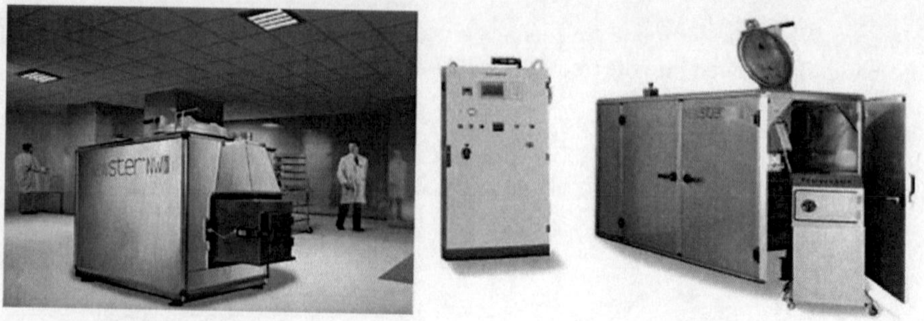

Fig. 5.29 Friction heat treatment equipment picture

Disinfection effect requirements and effects: Disinfection effect has been tested for many times for long-term use, to meet the requirements of China's "Medical Waste Treatment and Disposal Pollution Control Standards (Draft)".

Biological indicators to detect the sterilization effect of standards, waste samples to calculate the logarithmic value of inactivation, the test results are shown in the table below, all sterilized solid samples are not detected Bacillus thermophilus, kill the logarithmic value of ≥4, the sterilization effect meets the standard requirements value; and meet the "National List of Hazardous Wastes" in the infectious and injurious waste exemption conditions, can enter the landfill disposal or into the domestic waste incineration plant incineration disposal, disposal process is not managed by hazardous waste.

(2) Secondary pollution control equipment

The exhaust gas treatment equipment of the NW series medical waste disposal system consists of a two-stage heat exchanger, as well as two layers of different types of activated granular carbon and a HEPA absolute filter, with the activated carbon adsorbent and absolute filter being replaced periodically. The activated carbon adsorbent and absolute filter are replaced periodically. The two-stage heat exchanger contains a homogeneous Lacy ring packing and a spraying system, and the absolute filter is mainly composed of glass fibers.

5.7.6 Other New Non-incineration Technologies for Medical Waste

In recent years, with the continuous maturation of related technologies, medical waste non-incineration new technologies are also concentrated in a variety of technology composite integrated treatment and new secondary pollution control technology.

(1) Composite integrated treatment

High-temperature steam and microwave disinfection technology: such as the Italian company ECONOS developed a "microwave + high-temperature and high-pressure steam" composite enhanced treatment process (Gang et al. 2003). This technology integrates the advantages of microwave disinfection technology and high-temperature steam disinfection technology, no dioxin emissions during operation, to achieve the safe destruction of medical waste, volume reduction and harmless, suitable for installation in areas with high population density. At present, the United Kingdom, France and other developed countries in the West have successfully used microwave technology-based medical waste disinfection and destruction of integrated small-scale equipment to achieve the harmless disposal of medical waste in situ, and the residue can be treated together with ordinary household waste for centralized treatment. With the continued maturity of related technologies, Russia, Brazil, Romania, Malaysia, Singapore, Vietnam and other countries have also begun to use the integration of composite technology integrated treatment of medical waste.

In 2014, a company in Henan Shangqiu manufactured China's first prototype of medical waste microwave disinfection equipment, and won the national certification of scientific and technological achievements, marking the microwave disinfection technology as a non-incineration technology for medical waste is gradually being pushed in the country, to further promote China's medical waste "January 2020, Jiaxing City, Zhejiang Province, adopted the first medical waste disposal project using the more mature and advanced "high-temperature steam disinfection + microwave disinfection" process, equipped with automated processing device, with only three workers to operate, and the average daily capacity of medical waste disposal is 30 t to 45 t.

(2) Exhaust gas treatment

The treatment of medical waste, whether by high-temperature incineration or by non-incineration technologies, is accompanied by the generation of toxic and hazardous off-gases. Due to the special nature of medical waste materials, some of the gases have bad odor, and a small number of VOCs and mercury vapour generated is also toxic and harmful, which can cause secondary pollution of the environment. If not treated, it will not be able to meet the emission standards, and cannot really achieve the rational disposal of medical waste treatment.

High-temperature steam treatment system of the exhaust gas cleaning process is mainly to remove malodors, small amounts of VOCs and mercury vapour. Common odor treatment technologies include absorption, adsorption, thermal destruction, biological methods, plasma decomposition, etc. Most of the VOCs are toxic, such as formaldehyde, benzene and polycyclic aromatic hydrocarbons known to the public. Among the many treatment methods, the biological method with low cost, low energy consumption and no secondary pollution is more widespread. At present, there are several new technologies for the treatment of secondary pollutant waste gas.

Taking a medical waste microwave disinfection treatment system as an example, the exhaust gas treatment measures of medical waste microwave disinfection process were investigated, focusing on the primary exhaust gas treatment process of "primary filter + high efficiency filter + activated carbon adsorption" and the secondary exhaust gas treatment process of "cyclone tower + UV photocatalytic oxidation +

exhaust gas filter + activated carbon adsorption". Hence, the feasibility of microwave disinfection of medical waste was explored.

Medical waste microwave disinfection process, contains dust, NH_3, H_2S, volatile organic compounds and pathogenic microorganisms, bad odor and other gases. The primary waste gas is treated in the microwave disinfection equipment with the temperature not exceeding 50 °C, and the treatment of "primary filter + high efficiency filter + activated carbon adsorption" is adopted. The secondary exhaust gas treatment is collected centrally from the inlet and outlet as well as the exhaust gas after primary treatment, and then purified by the microwave disinfection exhaust gas treatment system of "cyclone tower + UV photocatalytic oxidation + tail gas filter + activated carbon adsorption", and then discharged through a 15 m high exhaust pipe. The process is in line with the requirements of the "Medical Waste Microwave Disinfection Centralized Treatment Engineering Technical Specification" for microwave disinfection waste gas treatment process.

Medical waste microwave disinfection exhaust gas disposal measures using "primary filter + high efficiency filter + activated carbon adsorption" and "cyclone spray tower + UV photocatalytic oxidation + tail gas filter + activated carbon adsorption" process, can effectively remove particulate matter and volatile organic. The process can effectively remove particulate matter, volatile organic pollutants, ammonia (NMHC), hydrogen sulfide and microorganisms, and the exhaust gas can meet the emission standards and the disposal measures are feasible.

(3) Wastewater treatment

Wastewater is also an inevitable product of medical waste disposal treatment process, especially in the non-incineration high temperature steam treatment process. The wastewater generated in the process of medical waste treatment is mostly treated with a combination of biochemical and physical treatment process. The process residue is first mixed with domestic wastewater, and after dilution, it enters the anaerobic tank for anaerobic treatment. The effluent first enters the sewage collection well, and then enters the contact oxidation tank for biochemical treatment under the action of lifting pump. Then it flows through the sedimentation tank, and the effluent from the sedimentation tank is discharged into the conditioning tank and mixed with flushing wastewater and initial rainwater to form a comprehensive wastewater for deep treatment. The operation practice of a disposal centre in Xiaogan (Shuangliu et al. 2018) shows that the high-temperature steam process waste gas treated by the biological filtration system and the high-temperature steam process wastewater treated by the combination of biochemical and physical treatment can meet the requirements of the corresponding national emission standards, which can provide reference for the construction of other high-temperature steam treatment systems in China.

References

Chen Y, Liu F Q, Shao CH Y. Chemical disinfection and medical waste treatment [J]. China Environmental Protection Industry, 2005c (10): 18-19.

Cheng L, Wu SHUNZE, Hou GUIGUANG, et al. Discussion on the application of high-temperature steam treatment technology for medical waste [J]. Non-ferrous metallurgy design and research, 2007(2–3): 35–42.

Jin Dengchao, Fang Boguan, Liu Xinyuan, et al. Factors affecting thermal penetration time in high-temperature steam treatment process of medical waste[J]. Journal of Environmental Engineering, 2021, 15(2). doi: https://doi.org/10.12030/j.cjee.202008217.

Zhang Feng'e. An investigation on the exhaust gas treatment measures of medical waste microwave disinfection disposal process [J]. Shanxi Chemical Industry, 2019, 184(6): 153-154.

YANG Gang, YUAN Xingzhong,ZENG Guangming. Design of high-pressure steam sterilizer for medical waste [J]. Environmental pollution control technology and equipment, 2003(1): 69-72.

Cai Ling, Jeng Pesong, Yang Jing. Application of microwave disinfection technology in medical waste treatment [J]. China Environmental Protection Industry, 2007(9): 40-43.

Luo JC, Li SY, Zhou Q, et al. A review of non-incineration treatment technologies for infectious medical waste - Microwave disinfection technology as an example [J]. Environment and Sustainable Development, 2020(4): 136-140.

Ministry of Ecology and Environment. Letter on soliciting comments on four national environmental protection standards, including the Technical Specification for Chemical Disinfection of Medical Waste Centralized Treatment Engineering (Draft for Comments) [Z/OL]. [2020–06–03]. http://www.mee.gov.cn/xxgk2018/xxgk/xxgk06/202006/t20200603_782480.html

Liu Shuangliu, Zhang Zheng, Sun, Ning. Secondary pollution control of high-temperature steam treatment process for medical waste--a case study of Xiaogan Medical Waste Disposal Center [J]. Environmental Protection Science, 2018, 42(4): 86–89.

Pei Zhaotang, Bai Xuemei,Xiu Fei. Process design for disposal of medical waste by high temperature steam sterilization treatment technology [J]. Environmental Engineering, 2009(1): 58-60.

Chapter 6
Medical Waste Emergency Disposal Management and Practice During an Epidemic

Since the "Thirteenth Five-Year Plan for Ecological Protection", "Guidance on Enhancing the Capacity of Hazardous Waste Environmental Supervision, Utilization and Disposal and Environmental Risk Prevention" (Environmental Solids [2019] No. 92), and the "Thirteenth Five-Year Plan for Medical Waste Disposal" all require the establishment of medical waste management practices. However, in 2020, the COVID-19 pandemic exposed the lack of emergency management and disposal capacity for medical waste generated by public health emergencies in China. The "Implementation Plan for Capacity Building of Centralized Medical Waste Disposal Facilities" was issued on April 30, 2020 (Development and Reform, Environmental Resources [2020] No. 696), which requires the promotion of emergency backup capacity building for centralized medical waste disposal facilities in large cities. This chapter aims to introduce medical waste emergency management and practices during an epidemic.

6.1 Medical Waste Emergency Disposal Management

6.1.1 Foreign Experience

For the management of emergency disposal of medical waste in sudden large-scale epidemics, some developed countries and regions have developed corresponding emergency plans and supporting mechanisms and present different characteristics.

(1) United States

In the United States, the National Center for Disease Control and Prevention (NCDC) is primarily responsible for developing guidelines for actions related to public health emergencies, including guidelines for the emergency disposal of medical waste. During the Ebola epidemic, the NCDC released a series of documents, including the

Y. Chen et al., *Innovative Disposal Technology and Management Practice for Medical Waste*, https://doi.org/10.1007/978-981-99-6786-5_6

"Ebola-Related Waste Management Guidelines" (hereinafter referred to as "Guidelines"), to guide medical institutions in managing and disposing of infectious medical waste related to the epidemic. The Guidelines classify waste contaminated (or potentially contaminated) by the Ebola virus as Class ''A" hazardous infectious waste, which must be disinfected or incinerated in accordance with relevant regulations. At the same time, local government regulations also play an important role in the actual implementation (U.S. National Environmental Protection Agency., 2016; National Safety Council, U.S.A., 2019; U.S. National Safety Council., 2017; Marshall Pandemic., 2020).

(2) United Kingdom

In the United Kingdom, the Department of Public Health (DPH) is primarily responsible for outbreak prevention and control and works in collaboration with the environmental protection department to manage medical waste (Tudor et al., 2005; UK Department of Health., 2013). The technical guide "Environmental and Sustainable Health Technical Memorandum 07–01: Safe Management of Healthcare Waste," published in 2013, details the process of medical waste generation and disposal. During the Ebola and COVID-19 outbreaks, emergency response plans in the UK followed the guidelines outlined in this document.

(3) Germany has made a detailed distinction between the various types of infectious medical waste in the Enforcement Assistance for Waste Disposal in Medical Institutions, and generally, it is not issued medical waste disposal guidelines for specific epidemics. When an epidemic occurs, the responsible institution can quickly find the corresponding disposal process and disposal requirements in the article, and carry out the collection and disposal of infectious medical waste. During the COVID-19 pandemic, the German federal government, on the advice of the Robert Koch Institute, an advisory body to the German Federal Ministry of Health, declared that that relevant waste should be disposed off according to the standards for medical waste of category 18 01 03* in the Enforcement Assistance for Medical Facility Waste Disposal document.

(4) Japan

The Ministry of the Environment (MOE) was responsible for establishing an emergency mechanism for the disposal of medical waste in the event of an outbreak of avian influenza in 2009 and the Ebola epidemic in 2019, prepared and issued the "Guidelines for Waste Disposal Response to Novel Influenza" and the "Notice on Waste Disposal for Ebola Hemorrhagic Fever", respectively, to guide the emergency disposal of medical waste. During the COVID-19 epidemic, the Ministry of the Environment issued documents such as "Measures on Waste Disposal in Response to the COVID-19" and established a special department, named " Novel Coronavirus Infectious Disease Control Headquarters COVID-19" to oversee the response efforts (Li et al., 2003; Ministry of the Environment of Japan., 2018).

(5) South Korea

Korea has established a contagious disease crisis warning mechanism, which is divided into four levels: "concern," "attention," "alert," and "serious" from slow to urgent. "During the Middle East respiratory syndrome coronavirus outbreak in 2015, South Korea formulated the Special Measures for the Safe Management of Medical Waste Isolated from the Middle East Respiratory Syndrome Coronavirus, which imposed stricter management standards on medical waste than before, such as requiring double sealed containers for medical waste. During the COVID-19 pandemic, South Korea formulated the Special Measures for Medical Waste Management of COVID-19 on January 28, 2020, and raised the outbreak alert to the highest level of "severe" on February 23, 2020, and formulated the Special Measures for Medical Waste Management of COVID-19 (second edition), adding specific measures to enhance the safety management of medical waste. The second edition of the "Special Measures for the Management of Novel Coronavirus Medical Waste" was formulated, adding specific measures to enhance the safe management of medical waste (Jang et al., 2006; Liu et al., 2021).

(6) Singapore

In Singapore, the National Infection Prevention and Control Committee (NIPCC) is responsible for outbreak prevention and control. In 2017, the NIPCC developed the "Guidelines for Infection Control in Healthcare Facilities in Singapore" based on the experience of the Ebola outbreak to provide guidance on national healthcare waste disposal. The National Environment Agency (NEA) issued Interim Guidelines for Environmental Cleaning and Disinfection in Non-Medical Commercial Settings Exposed to Confirmed COVID-19 Cases during the COVID-19 Outbreak, and made relevant recommendations.

During major infectious disease outbreaks, some countries and international organizations have further clarified the coverage of medical waste management to help prevent spread of pathogens. France stipulates that medical waste during an epidemic should include three sources: (1) waste produced by medical institutions; (2) waste generated by dispersed epidemic prevention and medical departments, and (3) medical waste generated by patients receiving self-treatment at home. The UK specifies that the targets for medical waste management during an outbreak are: (1) waste generated from patients with confirmed or suspected disease (regardless of whether the causative agent is known), and waste that may contain pathogens, and (2) waste generated by patients who are suspected to be infected or at potential risk of infection. The EU provides for the management of medical waste during the Ebola epidemic: (1) in the care of suspected or confirmed Ebola patients, after the use of sharps, dressings and other supplies, (2) suspected or confirmed Ebola patient samples in clinical laboratory testing of discarded supplies, (3) suspected or confirmed Ebola contaminated space (such as wards, aircraft, ambulances and other vehicles, airports and other transportation facilities, and (4) disposable personal protective equipment dismantled and discarded after working in a suspected or confirmed Ebola-contaminated environment.

6.1.2 Domestic Experience

6.1.2.1 Management Documents

At present, China has not yet incorporated the emergency disposal of medical waste related to public health emergencies in the design of the environmental emergency system at the national level, and the emergency disposal of medical waste during epidemics is mainly guided by temporary documents. Since the atypical pneumonia (hereinafter referred to as "SARS") epidemic, the Wenchuan earthquake, the H1N1 epidemic and the COVID-19 epidemic, the relevant state departments have issued a series of emergency management documents for the disposal of medical waste (Table 6.1), which are used to guide the emergency disposal of medical waste. Especially after the COVID-19 epidemic, the management of emergency disposal of medical waste has received high attention. The revised "Solid Waste Pollution Prevention and Control Law of the People's Republic of China", approved on April 29, 2020, added the responsibility for emergency disposal of medical waste in the case of an epidemic.

At the same time, some provinces in China also have issued a number of operational guidelines for emergency management of medical waste during the COVID-19 epidemic (Table 6.2). Zhejiang Province clearly emphasized the need to strengthen emergency management and closely monitor the medical waste disposal situation in each area. If the disposal load of the medical waste disposal units exceeds 60%, emergency preparation work should be carried out.

6.1.2.2 Management Requirements

According to Article 91 of "The People's Republic of China Solid Waste Pollution Prevention and Control Law," when a major infectious disease epidemic and other emergencies occur, the people's government at or above the country level should coordinate the collection, storage, transportation and disposal of medical waste and other hazardous wastes. China has issued the relevant standard documents, medical waste collection, storage, transportation, disposal and other aspects of standardized management during an epidemic has put forward specific requirements.

(1) Collection

Classification of medical waste management during an epidemic. "Medical Waste Management Regulations," "Medical Waste Classification Catalog" and other documents have put forward the following requirements: (1) Medical health institutions

Table 6.1 National-level medical waste emergency disposal-related management documents

Serial number	Management and technical documentation	Release Time	Department
1	Technical program of emergency response to wastes polluted by SARS virus	May 2003	Former environmental protection administration
2	Emergency treatment technology plan for sewage contaminated by SARS virus	May 2003	
3	Technical guidelines for the safe disposal of medical waste in earthquake areas (Interim)	June 2008	Former ministry of environmental protection
4	Preliminary plan for medical waste management in response to the influenza A (H1N1) epidemic	May 2009	
5	Notice on the environmental management of medical waste from the novel coronavirus-infected pneumonia outbreak	January 2020	Ministry of ecology and environment
6	Medical waste emergency disposal management and technical guidelines for novel coronavirus infected pneumonia outbreaks (Trial)	January 2020	
7	Technical program for emergency treatment of medical wastewater contaminated with novel coronavirus (Trial)	February 2020	Ministry of ecology and environment
8	Questions and answers related to the work of emergency disposal of medical waste from pneumonia epidemic at domestic waste incineration facilities	February 2020	
9	Notice on the management of medical waste in medical institutions during the pneumonia outbreak of novel coronavirus infection	January 2020	National health and wellness commission

should manage and dispose off domestic waste generated by patients with infectious diseases or suspected infectious diseases in accordance with the management and disposal of medical waste, (2) High-risk medical waste, such as waste generated from the pathogen culture medium, specimens and strains of bacteria, venom preservation fluid, and other items, should be disinfected in situ before delivery to the centralized disposal unit of medical waste. The "Notice on the Management of Medical Waste in Medical Institutions during the COVID-19 Infected Pneumonia Epidemic" issued during the COVID-19 Epidemic (National Health Office Medical Letter [2020] No. 81) refined the classification and collection of medical waste as follows: (1) Medical waste generated from fever outpatient clinics and wards (rooms) of patients and suspected patients with COVID-19 Infected Pneumonia, which are potentially contaminated areas, is medical waste related to the epidemic. Before leaving the contaminated area, disinfection must be carried out by spraying the surface of the bag with chlorine containing 1000 mg/L disinfectant (pay attention to spraying evenly) or add a layer of medical waste bags outside, (2) medical waste generated in the clean area should be collected in accordance with the conventional medical waste.

Table 6.2 Medical waste emergency disposal related management documents issued by the local government during the COVID-19 epidemic

Serial number	Management and technical documentation	Release time	Department
1	Urgent notice on effective response to the pneumonia outbreak of novel coronavirus infection-strengthening environmental management of medical waste	January 2020	Hubei department of ecology and environment
2	Technical guidelines for the prevention and control of contamination of medical waste for emergency disposal of novel coronavirus-infected pneumonia outbreaks (for trial implementation)	January 2020	Sichuan department of ecology and environment
3	Urgent notice on the identification of alternate facilities for emergency disposal of medical waste in the province for pneumonia outbreaks of novel coronavirus infections	January 2020	
4	Notice on the environmental control of medical waste in the novel coronavirus infection pneumonia outbreak	January 2020	Hunan provincial department of ecology and environment
5	Notice on effective strengthening of medical waste wastewater and special waste management for pneumonia epidemic of COVID-19	January 2020	Hunan provincial department of ecology and environment, Hunan provincial health and health commission, Hunan provincial department of housing and urban–rural development
6	Notice on the environmental management of medical waste from the novel coronavirus-infected pneumonia outbreak	January 2020	Department of ecology and environment of Henan province
7	Notice on the emergency disposal of medical waste during the pneumonia outbreak of novel coronavirus infection	February 2020	Eastern province department of ecology and environment
8	Workbook for the emergency environmental management of medical waste in the pneumonia outbreak of novel coronavirus infection in Hebei Province (for trial implementation)	February 2020	Hebei department of ecology and environment
9	Guidance manual for medical institutions	February 2020	
10	Law enforcement guide for healthcare facilities	February 2020	
11	Guidance manual for medical waste disposal units	February 2020	
12	Enforcement guide for medical waste disposal units	February 2020	
13	Guidelines for environmental emergency monitoring of pneumonia outbreaks in response to novel coronavirus infections	February 2020	

(continued)

Table 6.2 (continued)

Serial number	Management and technical documentation	Release time	Department
14	Notice on further improving the emergency disposal of medical waste in pneumonia outbreaks of novel coronavirus infections	February 2020	Zhejiang ecological environment department
15	Guidance on further improving the disposal of medical waste and waste masks during the prevention and control of the pneumonia epidemic with novel coronavirus infection	February 2020	Jilin provincial department of ecology and environment, Jilin Provincial department of housing and urban–rural development

The "COVID-19 Infected Pneumonia Epidemic Medical Waste Emergency Disposal Management and Technical Guidelines (for Trial Implementation)" provides guidance on the destination of medical waste after sorting and collection during an epidemic according to the degree of hazard of medical waste. It recommends that the infectious medical waste generated during the prevention and treatment of pneumonia epidemic and other medical waste be managed in a separate stream, with priority given to centralized medical waste disposal facilities and mobile medical waste disposal facilities to protect the disposal of medical waste involved in the epidemic.

(2) Storage

The "Medical Waste Management Regulations" require that the normal medical waste centralized disposal unit should collect and transport medical waste from medical health institutions at least every 2 days, in other words, the medical waste in the medical institution's temporary storage should not exceed 48 h. According to the "SARS" virus contaminated waste emergency treatment and disposal technology program" requirements, SARS virus contaminated waste should not be stored for more than 24 h, and should be disposed off on the same day it is generated. The "Medical Waste Emergency Disposal Management and Technical Guidelines for COVID-19 Infected Pneumonia Epidemic (for Trial Implementation)" requires that the storage site of medical waste in medical institutions should be disinfected in accordance with the methods and frequency required by the competent health authorities, and the temporary storage time should not exceed 24 h. Medical waste arriving at the disposal site should be disposed of as soon as possible, and the temporary storage time in the disposal unit should not exceed 12 h. Thus, it can be seen, during the epidemic that the temporary storage time frame of medical waste in medical institutions was shortened compared to the norm.

(3) Shipping

Medical waste disposal should still follow the principle of disposal in the vicinity during the epidemic, "the COVID-19 infection pneumonia epidemic medical waste emergency disposal management and technical guidelines (for trial implementation)" requires that the emergency disposal of medical waste should be given priority to the use of centralized medical waste disposal facilities within the administrative region. When the existing disposal capacity in the region cannot meet the emergency disposal needs, medical waste can be transferred to nearby centralized medical waste disposal facilities for disposal. Infectious medical waste transport during the epidemic can use special medical waste transport vehicles, or medical waste transport vehicles that have been temporarily modified to meet the requirements. The guidelines encourage the arrangement of fixed special vehicles to transport the infectious medical waste generated by the pneumonia epidemic prevention and control process alone. These vehicles should not mix infectious medical waste with other medical waste during transport. At the same time, the "National Hazardous Waste List (2021 Edition)" exempts the transport of medical waste during major infectious disease epidemics from management. The disposal program for transport should be determined by the people's government at the county level or above where the incident occurred, and hazardous waste transportation may not be used.

(4) Disposal

According to the "SARS" virus contaminated waste emergency treatment and disposal technology program, if there is no dedicated local incineration equipment, it is recommended to use the following equipment in order to temporarily replace the treatment of SARS virus contaminated waste: simple medical waste incinerator, hazardous waste incinerator, waste incinerator or industrial waste incinerator, cement kiln crematorium, crematorium incinerator, coal-fired boilers for heating (Science Times., 2003). According to the "response to the influenza A (H1N1) epidemic medical waste management plan", if the medical waste disposal capacity is insufficient in the region, you can choose to send it to neighboring areas of centralized medical waste disposal facilities for disposal, or use alternative facilities for disposal locally. Alternative medical waste disposal facilities includes: mobile medical waste disposal facilities, hazardous waste incineration facilities, domestic waste incinerators, and industrial kilns, etc. According to the "COVID-19 infection of pneumonia epidemic medical waste emergency disposal management and technical guidelines (for trial implementation)", mobile medical waste disposal facilities, hazardous waste incineration facilities, domestic waste incineration facilities, and industrial kilns, etc., are included in the pneumonia epidemic medical waste emergency disposal resources list.

From the requirements of the relevant management documents, medical waste emergency disposal technology has evolved from simple incineration during the "SARS" epidemic to a variety of forms of synergy during the COVID-19 epidemic. At the same time, the "National Hazardous Waste List (2021 Edition)" exempts the disposal of medical waste generated during major infectious disease epidemics from management. According to the disposal program determined by the people's government at the county level or above where the incident occurred, and it may not be disposed of as hazardous waste.

6.2 Medical Waste Emergency Disposal Practices

China's centralized medical waste disposal system was established as a result of the SARS epidemic, which built a solid defense line for medical waste emergency disposal during the COVID-19 epidemic, with nationwide coverage of centralized disposal capacity making a decisive contribution to medical waste risk prevention and control. Meanwhile, in the practice of medical waste emergency disposal, it has made significant progress from simple incineration during the "SARS" epidemic to the coordinated application of various forms during the COVID-19 epidemic. However, prior to the COVID-19 epidemic, the centralized medical waste disposal system had not experienced a large-scale epidemic impact. The current relevant guidance documents in China mainly set the scale of centralized medical waste disposal facilities based on the production of medical waste in normal service areas, without systematically designing emergency disposal capacity reserves and emergency material reserves for medical waste disposal under sudden epidemic situations. The COVID-19 epidemic has highlighted the existing problems in China's current emergency medical waste disposal system (Guo et al., 2020).

6.2.1 SARS Outbreak

Before the outbreak of SARS in 2003, only a few cities such as Shenyang, Taiyuan, and other cities in China had implemented centralized disposal of medical waste. Most cities dispose of medical waste in a decentralized manner, with large hospitals using small incinerators for on-site disposal, or outsourcing waste collection and disposal to the sanitation department. About 90% of medical waste was disposed of in landfills along with domestic waste, which did not only cause a number of pollution accidents, but also posed long-term potential harm to the water and soil environment in many areas of China. The remaining 10% of medical waste was disposed off by incineration, but most of the toxic and harmful substances produced by incineration due to backward technology and equipment, the lack of treatment facilities, resulting in certain risks to both the ecological environment and human health.

"After the outbreak of SARS, the timely treatment of medical waste involved in the epidemic received high attention from governments at all levels. Waste incinerators, medical waste treatment capacity is significantly improved, but there is still a shortage of treatment capacity for a certain period of time. Medical waste from the SARS epidemic that could not be handled temporarily was collected and sealed in multi-layer plastic bags for temporary storage. "During the SARS epidemic, both centralized disposal facilities and small incineration facilities were relatively simple incineration devices, which had no process control measures or tail gas treatment facilities, resulting in significant air pollution.

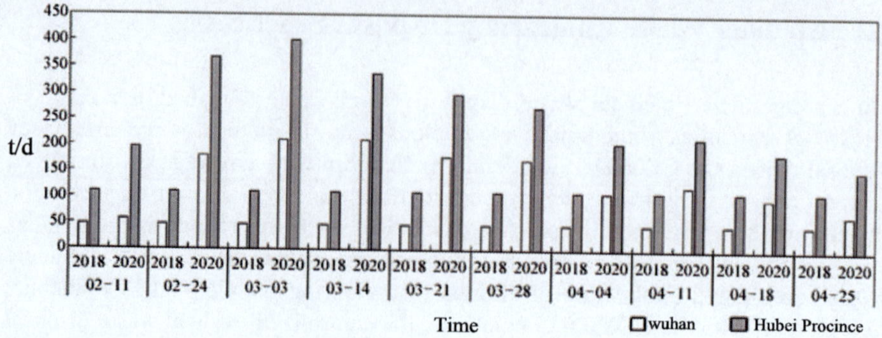

Fig. 6.1 Medical waste generation and disposal before and after the outbreak in Wuhan and Hubei Province

6.2.2 *COVID-19 Pneumonia Outbreak*

6.2.2.1 Medical Waste Generation

Due to the addition of a large amount of infectious household waste and the rapid consumption of protective equipment such as protective clothing, goggles, masks, and other protective supplies, the amount of medical waste generated during the epidemic increased compared to normal (Fig. 6.1). During the peak period of the epidemic in Wuhan, where the situation was most severe, the daily generation of medical waste reached 247.3t, which was 5–6 times of the normal level. The generation in Xiaogan, Huanggang and other neighboring cities also increased rapidly to 4–5 times of that before the epidemic. At the same time, the changes in the composition of medical waste caused a difference in its density compared to normal times. Normally, the density of medical waste in China is 100–120 kg/m^3, during the COVID-19 epidemic, the density of medical waste in Wuhan was 67–85 kg/m^3.

6.2.2.2 Medical Waste Disposal

After the COVID-19 epidemic, there were significant gaps in Wuhan, Xiaogan, Huanggang and other areas in China due to China's relatively weak medical waste emergency disposal capacity.

In terms of centralized medical waste disposal facilities, the emergency disposal capacity reserve is not sufficient. Most of the existing medical waste disposal facilities in China were put into operation before 2010. Compared with the rapid increase in medical waste generation, the improvement of medical waste disposal capacity is relatively lagging behind. In 2018, the overall load rate of medical waste disposal facilities in China was as high as 76.0%, with four provinces having a facility load rate of over 100% (Fig. 6.2), 13 provinces having a load rate over 80%, and more than one-third of cities having a load rate of over 90%. In addition, some centralized disposal facilities have poor operational conditions and may face the need for upgrading, making it difficult to achieve approved operating scales.

In terms of medical waste emergency disposal facilities, the capacity for direct deployment is extremely limited. Before the COVID-19 epidemic, there were 24 hazardous waste disposal facilities nationwide that could co-dispose the medical waste. These facilities can supplement the emergency medical waste disposal capacity of the region by adjusting the ratio of waste feeding. At the same time, the short-term deployment of 48 mobile medical waste disposal facilities can provide an emergency disposal capacity of 122t/d. Other hazardous waste incineration, domestic waste incineration, cement kilns and other disposal facilities basically do not have a special discharge area for medical waste disposal, feeding equipment, cleaning and disinfection equipment and other supporting conditions. Therefore, they cannot be used directly for medical waste co-disposal.

Within 2 months after the COVID-19 epidemic, the national medical waste disposal capacity increased rapidly from 4902.8 t/d before the epidemic to 6074.0 t/d on April 10, 2020, with a growth rate of 23.9%.The disposal capacity of Hubei Province increased from 180 t/d to 667.4 t/d before the epidemic, and the disposal capacity of Wuhan City increased from 50 t/d to 265.6 t/d before the epidemic. The national medical waste emergency disposal capacity has been able to match the disposal needs, medical waste basically achieve "daily production and daily cleaning". It should be clear that the new medical waste disposal capacity during the new epidemic is mostly temporary. In Wuhan City, for example, according to public information, the majority of the 215.6 t/d of new disposal capacity added during the period from the COVID-19 epidemic to April 10, 2020, came from new emergency disposal facilities, which could hardly meet the technical requirements under normal conditions (Table 6.3).

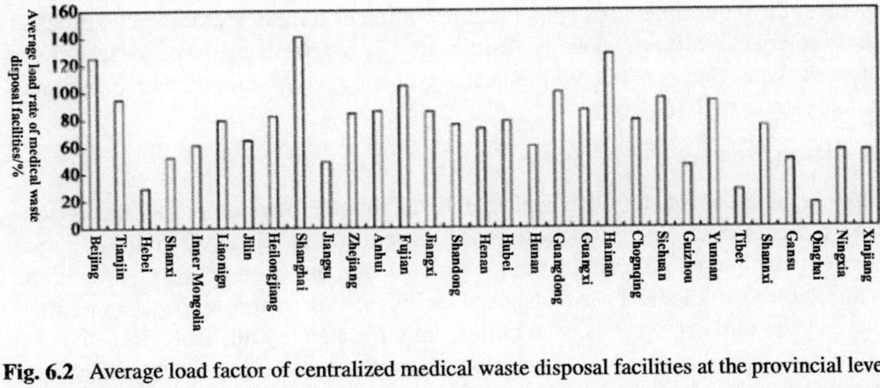

Fig. 6.2 Average load factor of centralized medical waste disposal facilities at the provincial level in China in 2018

Table 6.3 Composition of medical waste disposal capacity in Wuhan during the COVID-19 epidemic

Serial number	Disposal facility type	Disposal capacity/(t/d)
1	Original centralized disposal facilities	50
2	Newly built centralized disposal facilities	30
3	Co-disposal of hazardous waste incineration facilities	15
4	Co-disposal of domestic waste incineration facilities	100
5	Vulcan Mountain, Thor Mountain and other hospital-owned facilities, mobile treatment facilities	70.6
Total		265.6

6.2.2.3 Local Experience

(1) Hubei Province

During the COVID-19 epidemic, in order to ensure timely, orderly, efficient and harmless disposal of medical waste, Hubei Province established the overall principles of "local proximity, timely priority, centralized safety, and guaranteed capacity," as well as "a team, double the capacity, a program, a temporary storage "specific requirements: the establishment of a task force, point-to-point verification of the implementation of emergency disposal capacity one by one; determine emergency disposal units or implement mobile disposal facilities based on the principle of expanding the existing disposal capacity by 1:1; local ecological environment, health and other departments quickly Organized the preparation of the local centralized medical waste disposal emergency capacity preparation program; and determine suitable locations that meet the requirements of epidemic prevention as temporary storage facilities to ensure emergency storage capacity. As of March 21, 2020, all cities and prefectures in Hubei Province had constructed medical waste emergency storage, with emergency storage capacity of 1552.6 t.

(2) Henan Province

After the outbreak of the COVID-19, Henan Province planned ahead and coordinated the province's medical waste emergency disposal capacity, dividing it into four tiers to cope with the epidemic: the first echelon has been approved around the commissioning of medical waste disposal facilities, with a disposal capacity of 307t/d, and the second echelon has been built around the standby medical waste disposal equipment, as well as part of the centralized hazardous waste disposal facilities, with an emergency disposal capacity of 184t/d, the third echelon of mobile medical waste disposal facilities, municipal solid waste incineration power plants, biomass power plants, etc., with an emergency disposal capacity of 73 t/d, the fourth echelon of cement kiln co-disposal enterprises, medical waste emergency disposal capacity of 370t/d. Except in extreme circumstance, the first to third echelon A total of 564t/d of medical waste disposal capacity, which can fully meet the province's medical waste disposal needs in a timely, efficient and safe disposal.

Henan Province scientifically guided the province's medical waste disposal work, and the Department of Ecology and Environment fought for 16 days to compile and publish the "Reader's Digest of Medical Waste Emergency Disposal Knowledge During the Epidemic", which became a powerful tool and operational guide for medical waste generation and disposal units and industry regulators on the "epidemic". At present, "medical waste during an epidemic emergency disposal knowledge book" has been translated into 8 languages and 15 versions, and transferred the copyright to multiple countries, serving as a "Chinese program" to support epidemic prevention and control work in 199 countries overseas. Henan Provincial Health Committee released "100 Questionnaires on Integrated Medical Waste Management in Medical Institutions in Henan Province", which provides an authoritative, accurate, and scientific technical guide for front-line operators to prevent and control medical waste risks.

(3) Chongqing

The Ecology and Environment Bureau of Chongqing Municipality promptly issued the "Medical Waste Management Plan for Responding to the COVID-19 Epidemic", requiring all district and county ecology and environment departments and 15 medical waste disposal units to develop corresponding work plans, improve mechanisms for managing epidemic medical waste and emergency response, clarify specific requirements for medical waste disposal units in the collection, transportation, storage, and disposal of epidemic medical waste, and defines the supervisory responsibilities of district and county ecology and environment departments, forming a three-level emergency management system of "one municipal government, 41 district and county governments, and 15 disposal units".

The centralized medical waste disposal facilities in Chongqing all adopt high-temperature steam treatment technology. During the COVID-19 epidemic, Chongqing used three hazardous waste incineration facilities, including the Changshou hazardous waste disposal site and the Bishan hazardous waste disposal site belonging to Chongqing Zhongminggangqiao Environmental Protection Co., Ltd., and Chongqing Herun Zhongtian Environmental Protection Technology Co., Ltd., as emergency disposal facilities for the city's epidemic medical waste. The emergency facilities bear the original centralized disposal facilities difficult to effectively deal with medical waste, mainly including the diagnosis and treatment of pneumonia patients generated by the process of discarded uniforms, sheets and quilts and other large pieces of medical waste, as well as other medical waste requiring emergency disposal.

6.3 Types of Medical Waste Emergency Disposal Facilities

During the COVID-19 epidemic in early 2020, China launched a large-scale exploration of medical waste emergency disposal, and achieved the safe disposal of all medical waste under the combined effect of different types of disposal facilities, such as centralized medical waste disposal facilities, hazardous waste incineration facilities, domestic waste incineration facilities, cement kiln facilities, mobile medical waste disposal facilities, and medical institutions' own disposal facilities. The use of other solid waste disposal facilities for co-disposal is the main form of achieving rapid improvement in the emergency disposal capacity of medical waste.

6.3.1 Centralized Medical Waste Disposal Facilities

The use of centralized disposal facilities for medical waste emergency disposal is the fastest response and lowest risk approach. Emergency disposal capacity can be quickly increased by activating the backup production line, and backup facilities, or extending the operating time of existing facilities. Additionally, centralized disposal facilities in neighboring cities can supplement each other's emergency response capabilities. However, before the COVID-19 epidemic, many major cities in China had only one medical waste disposal facilities, and had not equipped themselves with backup production lines or facilities. Taking Wuhan as an example, the only medical waste incineration facility had a disposal capacity of 50t/d and a loading rate of more than 90%, was almost unable to meet the demand for disposal of medical waste generated by the unexpected epidemic before the COVID-19 epidemic. In this situation, some cities fully exploit the disposal capacity of centralized disposal facilities to increase their emergency disposal capabilities, either by building new production lines or extending the operating time of their facilities. For example, Wuhan quickly built a new 30t/d medical waste emergency treatment center in February 2020, greatly relieving the pressure of medical waste disposal during the epidemic. Some cities with insufficient emergency disposal capacity also sent their medical waste to centralized disposal facilities in neighboring cities for disposal. For example, at the beginning of the outbreak of the COVID-19 epidemic, some medical waste from Wuhan was transported across the city to Xiangyang for disposal in the face of an extreme shortage of medical waste disposal capacity in Wuhan.

Case: In early February 2020, medical waste began to accumulate rapidly in major hospitals in Wuhan, creating an extremely critical situation. On February 7, the Wuhan Municipal Government and China Energy Conservation and Environmental Protection Group Co., normally takes about 10 months to build a 30-ton-per-day medical waste disposal facility, but the construction team of Wuhan Qian zishan Medical Waste Emergency Disposal Center completed the construction plan within 10 h, the engineering design within 3d, and the main equipment lifting and pipeline

Fig. 6.3 Wuhan Qianzishan medical waste emergency disposal center

laying within 8d (Fig. 6.3). In the afternoon on February 22, the three 10t/d high-temperature steam treatment lines were officially put into operation, 31 h ahead of schedule, fully completing the construction tasks originally planned to be completed in 15d, and put into operation.

6.3.2 Hazardous Waste Incineration Co-Disposal Facilities

The co-disposal of medical waste in hazardous waste incineration facilities has been practiced for many years both at home and abroad. Since the operation and management requirements of hazardous waste incineration facilities are like those of medical waste incineration facilities, they are suitable as the preferred alternative facilities for medical waste co-disposal. During the COVID-19 epidemic, the relevant units in China were equipped hazardous waste disposal facilities with emergency medical waste disposal capacity, which to some extent relieved the pressure of emergency medical waste disposal.

Case: During the COVID-19 epidemic, under the coordination of Wuhan Ecological Environment Bureau, Wuhan Beihu Yunfeng Environmental Protection Technology Co., Ltd., transformed the existing hazardous waste treatment facility for medical waste disposal and process technical modification, and completed the configuration of 12t/d of medical waste disposal capacity within a few days (Fig. 6.4). From the first barrel of medical waste into the plant to February 18, 26d time, and the disposal of medical waste in Wuhan City nearly 200t.

Fig. 6.4 Hazardous waste disposal facility co-disposal of medical waste site

6.3.3 Co-disposal Facilities for Domestic Waste Incineration

The Basel Convention Technical Guidelines for the Environmentally Sound Management of Biomedical and Health Care Waste states that infectious waste can be disposed off in the same way as domestic waste after sterilization treatment. Prior to the COVID-19 epidemic, mechanical grate waste incinerators for the disposal of infectious medical waste had been successfully used in practice at home and abroad. Klametsrud (Klemetsrud) domestic waste incineration plant in Oslo, Norway, using mechanical grate incinerator disposal of infectious medical waste, with a medical waste blending ratio of no more than 5% of domestic waste. During the 2009 H1N1 influenza virus epidemic in Japan, the Guidelines for New Influenza Waste Management Measures issued pointed out that the possibility of receiving municipal waste incineration facilities could be discussed with the municipal authorities if the quantity of infectious waste was feared to exceed the disposal capacity. Shanghai, China, has been using domestic waste incineration facilities for emergency disposal of medical waste since 2014. During the COVID-19 epidemic, Shanwei, Dongying, Putian, Zhuhai, and Xiantao began to use domestic waste incineration facilities to dispose off medical waste since late January and early February 2020 (Fig. 6.5). In mid-February, Wuhan dynamically opened two grate waste incineration facilities for co-disposal of medical waste, providing a sizable medical waste disposal capacity and achieving 70 consecutive d of co-disposal of about 7000t.

The treatment temperature requirements of the domestic waste incineration process are similar to those of medical waste incineration, and there is potential for synergy between the two in terms of the treatment process in the furnace. However, there are significant differences in the pre-furnace processing procedures and methods, which result in significant differences in the ability to control the spread of pathogens during the treatment process. Domestic and foreign practice

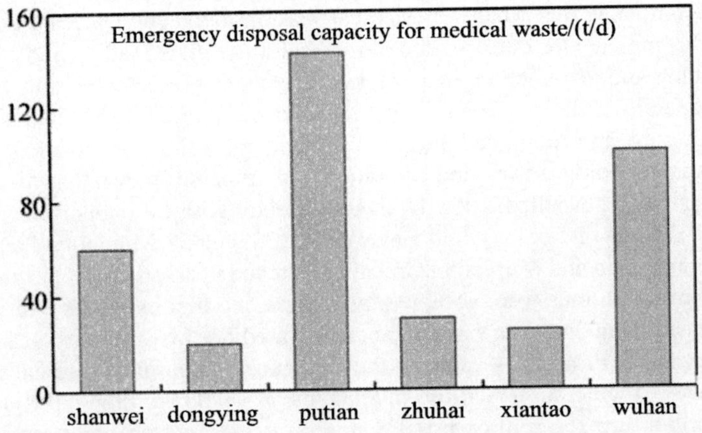

Fig. 6.5 Some of China's municipal domestic waste incineration facilities in January–February 2020 medical waste emergency disposal capacity comparison

has shown that through six layers of barriers such as "separation-disinfection-enhanced packaging-improved feeding process-negative pressure-sanitary protection", the infectious risk of emergency medical waste disposal in household waste incineration facilities can be effectively prevented by: (1) Diversion: Priority is given to the disposal of epidemic medical waste and other infectious waste in medical waste centralized disposal facilities, and infectious medical waste is not sent to household waste incineration facilities for disposal as much as possible. (2) Disinfection: It is sent to the domestic waste incineration facilities for emergency disposal of infectious medical waste, in the packaging, transfer, disposal of the relevant links to strengthen the disinfection process, and to reduce the risk of infection. (3) Strengthen packaging: It can reduce the probability of breakage in the process of grabbing material by strengthening the packaging (such as double plastic bag packaging and then disposable carton packaging, tape wrapping. (4) Improved the feeding process: through the establishment of a separate discharge port (such as direct hopper feed through the maintenance elevator), switch to the use of curved hooks and mesh pockets supporting the grab (no grab), video surveillance to guide the refinement of feeding, shorten the time spent in the waste pit and other operations to reduce the possibility of medical waste packaging breakage. (5) Negative pressure: Strictly implement a "micro-negative pressure" environment for the waste pit. (6) Sanitary protection: Strengthen the hygiene and protection measures of operation workers and training, and select professional personnel with medical waste disposal experience to participate in the disposal when conditions permit.

Case: In response to the emergencies such as medical waste incineration facility maintenance shutdown and public health emergencies. Shanghai issued the "Domestic Waste Incineration Air Pollutant Emission Standards" (DB 31/768) in 2013, which stipulates the entry requirements for co-disposal of medical waste in domestic waste incineration facilities in emergency situations. A company in

Shanghai completed the "Shanghai Medical Waste Emergency Disposal Protection Plan", and completed the construction of medical waste crate loading and unloading area, cleaning and disinfection area, stacking area and sewage collection and other supporting facilities in the general industrial solid waste landfill pretreatment workshop in Laogang, and completed the medical waste unloading area, loading elevator, medical waste feeding device and sanitation and epidemic prevention and control in the Laogang Renewable Energy Utilization Center with the support of Shanghai Municipal Bureau of Ecology and Environment, Shanghai Municipal Health and Health Commission and Municipal Greening Bureau in early 2014. After the emergency disposal facilities were accepted by experts, the first emergency disposal of medical waste using domestic waste incineration facilities in Shanghai was launched in February 2014. From 2014 to 2019, the emergency disposal of medical waste in domestic waste incineration facilities in Shanghai was 0.085 million t, 0.193 million t, 0.463 million t, 0.756 million t, 0.931 million t, and 1.44 million t respectively. The emergency disposal process in accordance with the "Shanghai Medical Waste Health Management Code" and "medical waste management related items disinfection methods" for sanitation and epidemic control, medical waste and domestic waste mixed incineration rate per hour is controlled within 5%.

6.3.4 Cement Kiln Co-disposal Facilities

In 2014, WHO issued a revised version of the "Medical Waste Sound Management Manual", which states that the emergency disposal of medical waste can be disposed of by incineration in high-temperature industrial kilns under the condition of ensuring safe transfer. Before the COVID-19 epidemic, there was little domestic and foreign experience in the co-disposal of medical waste in industrial kilns such as cement kilns. During the COVID-19 epidemic, number of domestic cement enterprises made the first attempt to co-dispose of medical waste in cement kilns, and a total of 143.7t of medical waste was disposed of by February 24, 2020.

Medical waste has the characteristics of high calorific value and high chlorine content. Two technical points should be noted in the co-processing of medical waste in cement kilns: the renovation connection should be done well. The packaging size of medical waste is different from that of general hazardous waste. Cement enterprises should ensure that the feeding system and inlet are smoothly connected with the medical waste container during technological transformation to reduce the risk of infection. Second, it is necessary to match with cement production. Due to the high chlorine content (3–5%) and high calorific value of medical waste, it is necessary to reasonably determine process parameters such as the amount of addition and residence time of the decomposition furnace to ensure stable operation of cement production.

Case: Yichang City Epidemic Prevention and Control Command issued instructions to strengthen the emergency disposal capacity of medical waste and designated a certain cement production enterprise as a backup unit for the emergency disposal of special hazardous waste and medical waste in the city On February 2, 2020. On February 10, Huaxin Cement started the co-disposal transformation of the second production line, and completed the trial burn on February 14, achieving an emergency co-disposal capacity of 10t/d of medical waste.

6.3.5 *Mobile Medical Waste Disposal Facilities*

In the guidance document for personal and public health protection and medical waste management for COVID-19 (*Water, Sanitation, Hygiene and Waste Management for the COVID-19 Virus*), on-site disposal of medical waste related to the outbreak, while ensuring safe collection and sorting. Mobile medical waste disposal facilities are suitable for small-scale on-site treatment, and have been used in Germany, France and other European countries for many years, and have also played an important role in the emergency treatment of medical waste in the Ya'an earthquake area in China. After the outbreak of the COVID-19 epidemic, the Ministry of Ecology and the Environment deployed 46 mobile units to Wuhan for emergency treatment of medical waste, which effectively alleviated the pressure of medical waste transportation and disposal by on-site treating the medical waste produced by designated hospitals in Huangpi, Jiangxia, and Xinzhou districts with confirmed COVID-19 patients.

Mobile medical waste disposal facilities is generally small, mostly less than 2 t/d, usually using high-temperature steam, microwave disinfection and other non-incineration treatment technology. The main advantages of mobile medical waste disposal facilities are: adaptability, the operating site conditions are not high, the rapid formation of emergency disposal capacity; stable operation, the emergency state can be 24 h non-stop operation; simple operation, 1~2 people can be controlled after simple training.

Case: The medical waste disposal centre in Jingzhou City, Hubei Province had two high-temperature steam production lines with a total capacity of 8t/d and a total capacity of 16t/d. After the outbreak of the COVID-19 epidemic, the amount of medical waste generated in Jingzhou City increased exponentially. Jingzhou City Medical Waste Disposal Center purchased a 1.5t/d scale of mobile high-temperature steam treatment equipment for emergency disposal. The schematic diagram of the mobile medical waste high-temperature steam disposal facility is shown in Fig. 6.6.

6.3.6 *Medical Institutions to Provide Their Own Disposal Facilities*

Since 2000, under the combined pressure of external compliance with international conventions such as the Stockholm Convention on Persistent Organic Pollutants and internal pressure to improve environmental quality, countries and regions have

Fig. 6.6 Schematic diagram of mobile medical waste high-temperature steam disposal facility

successively implemented more stringent emission standards for air pollutants from incinerators. Small-scale incinerators have been largely replaced by non-incineration facilities, and new facilities mainly use non-incineration technologies that emit no toxic or hazardous substances. However, due to the centralized disposal model for medical waste in China, few new facilities of this kind have been built in the more than 10 years since the outbreak of SARS. During the COVID-19 epidemic, Wuhan's new Vulcan Mountain, Thunder God Mountain Hospital construction process was synchronized with the configuration of a total of 24t/d of medical waste incineration facilities, other designated hospitals configured 10.9t/d of medical waste disposal capacity, to ensure the safe disposal of some of the epidemic-related medical waste.

Case: Huo Shenshan Hospital in Wuhan is a designated hospital for the treatment of COVID-19 pneumonia. The construction process of Huo Shenshan Hospital was supported by the construction of two medical waste incinerators with a design capacity of 4t/d and 5t/d respectively for the temporary disposal of medical waste generated in the hospital. The two emergency incinerators were officially put into operation on February 10, 2020. During the operation, the medical waste basically cleared on a daily basis. After the Qianzishan Medical Waste Emergency Disposal Centre in Wuhan was completed, the medical waste from Huoshenshan Hospital was transported to the center for disposal. The medical waste disposal facilities at Huoshenshan Hospital stopped operating on March 6, 2020.

6.4 Medical Waste Emergency Disposal System Construction Path (Cheng et al. 2020; Zhang et al. 2020; Shao et al. 2020)

6.4.1 *Analysis of the Shortcomings of the Emergency Disposal System of Medical Waste*

(1) Lack of top-level design for medical waste emergency disposal

The National Emergency Response Plan for Environmental Emergencies issued in 2014 provides for the emergency operation procedures for environmental emergencies, but there is no description of the emergency disposal of medical waste, and the relevant provisions are not applicable to the emergency disposal of medical waste. The relevant provisions are also not applicable to the emergency disposal of medical waste. The current "Medical Waste Management Regulations" and other medical waste regulations do not specify the requirements for the environmental emergency management and disposal of medical waste. The Ministry of Ecology and Environment issued the "Thirteenth Five-Year" Ecological Environmental Protection Plan, "Hazardous Waste "Three Capabilities" Construction" and other documents on the establishment of medical waste coordination and emergency disposal mechanism requirements are also relatively general and lack of operability.

(2) Inadequate emergency disposal capacity for medical waste

A considerable number of mega-cities have only one centralized medical waste disposal facility, which is already at full capacity or overloaded. Hazardous waste incineration, domestic waste incineration, cement kilns and other co-disposal facilities basically do not have the supporting conditions for direct emergency disposal. The centralized disposal facilities of emergency capacity reserves and can be directly deployed emergency disposal capacity are not optimistic, directly leading to serious lack of medical waste disposal capacity during the COVID-19 epidemic in Wuhan and other places. In addition to the emergency state of disposal capacity, the shortage of medical waste transportation capacity is more prominent. Some areas have sufficient disposal capacity, but there are not enough vehicles and personnel for timely collection of medical waste. Most of the temporary transformation of transfer vehicles cannot meet the technical requirements. In addition, medical waste collection bags, transfer barrels and other large gaps, protective supplies, and disinfection supplies are also very scarce, resulting in many problems faced in the timely and safe disposal of medical waste during the epidemic.

(3) Shortage of medical waste emergency disposal technology

During the epidemic, on the one hand, there were significant changes in the composition and infectious characteristics of medical waste, and timely adjustments to technical parameters were needed in the pre-processing and treatment stages of centralized medical waste disposal facilities. On the other hand, there were significant differences in the technical requirements between emergency and regular disposal of medical waste, and corresponding technical expertise, such as expert teams and specialized operational teams, were particularly important. During the epidemic, many new temporary and unscheduled disposal capacity of the technical reserves is extremely lacking, secondary pollution control and secondary environmental risk prevention is not enough. Due to the large amount of domestic waste generated by infected persons in epidemic-related medical waste, some centralized incineration facilities were unable to operate normally in the early stages of the epidemic; most hazardous waste incineration facilities and domestic waste incineration facilities were urgently retrofitted for collaborative disposal during the epidemic, and there were risks in the storage and addition stages. The overall pollution control level of the incineration facilities urgently built for Huoshenshan and Leishenshan Hospitals was not high, and there was a lack of relevant experience in the proportion and method of waste addition for co-disposal of medical waste in cement kilns.

(4) Emergency disposal supervision capacity needs to be strengthened

The supervision system for the whole process of medical waste disposal in China is not yet perfect, and modern technologies such as remote monitoring, real-time positioning, and electronic transfer notes are still in their infancy in medical waste supervision. During the epidemic, in addition to the regular regulatory tasks, there will be many new medical institutions wastewater monitoring, medical waste incineration gas monitoring, medical wastewater treatment sludge supervision, water quality monitoring and other emergency tasks, at the same time, a large number of temporary, decentralized emergency disposal of medical waste, but also to environmental supervision, law enforcement brought greater pressure. The insufficient capacity of environmental supervision was further highlighted during the COVID-19 epidemic, and there is still a need for further strengthening of personnel allocation and supervision methods.

6.4.2 Medical Waste Emergency Disposal System to Build Ideas

Based on the several outbreaks of epidemics both domestically and internationally, it can be observed that densely populated large cities such as Beijing during the SARS outbreak, Wuhan during the COVID-19 pandemic, and New York City in the United States have all been heavily affected. Large cities have a concentration of medical resources and high population density with frequent movement, which leads to a significant increase in medical waste generation once an epidemic occurs. Therefore, large cities are critical for emergency medical waste management, and it is essential to accelerate the development of a complete emergency management mechanism and capabilities.

According to the "Notice on the Adjustment of the Criteria for the Classification of City Size" (Guo Fa [2014] No. 51), cities are classified into five categories and seven

Table 6.4 China's city size classification standards

Serial number	City type	Number of permanent residents in urban areas
1	Mega city	10 million or more
2	Megacities	5~10 million
3	Large cities	Type I large cities: 3~5 million
		Type II large cities: 1~3 million
4	Medium-sized cities	500,000~1 million
5	Small cities	Type I small cities: 200,000~500,000
		Type II small cities: under 200,000

blocks according to the number of urban residents (Table 6.4). In 2018, 28 super-large, mega-cities, and type I megacities produced 477,000 t of medical waste, accounting for 48.7% of the total amount of medical waste produced nationwide. Based on the population data and medical waste generation in large cities, mega-cities, and first-tier large cities in 2018, the medical waste generation intensity per million people was compared. The results show that the range of medical waste generation intensity per million population in large cities, megacities, and type I large cities is 3–9, 4–10, and 5–10 t/d. If the emergency disposal capacity is configured according to the scale of 1 times the normal generation of medical waste, the emergency disposal capacity of medical waste per million population in large cities should be not less than 10 t/d.

From the perspective of emergency disposal of medical waste, emergency disposal capacity can be composed of centralized medical waste disposal facilities, hazardous waste incineration facilities, domestic waste incineration facilities, cement kiln facilities, mobile medical waste disposal facilities, medical institutions own disposal facilities and other integrated composition, which are coordinated and used in stages. Among them, municipal solid waste incineration facilities, as an essential urban infrastructure, can coordinate and cooperate with centralized disposal facilities to form a considerable scale and controllable risk emergency disposal capacity.

6.4.3 *Main Tasks of Building Medical Waste Emergency Disposal System*

(1) Build medical waste emergency disposal management system

Connect with the emergency management system for public health emergencies, carry out environmental risk assessment of medical waste under different event levels at the national level, and clarify the corresponding medical waste emergency plans to develop the main body and content. Implementation of the "Prevention of Solid Waste Pollution of the Environment Act" requirements, clarify the responsibilities of the central, provincial, and municipal governments, as well as the competent departments of health, ecology and environment, environmental health, transportation, etc.

Accelerate the improvement of medical waste regulations and standards system, the revised version of the "Medical Waste Management Regulations" has added relevant requirements for medical waste emergency disposal, and the medical waste-related standards that are being developed and revised propose targeted requirements for the operation parameters and pollutant discharge indicators of medical waste centralized disposal facilities under the situation of sudden outbreaks, and ensure the connection and unity of relevant requirements. Promote the establishment of regional joint prevention and control mechanism for medical waste emergencies in urban agglomerations among the Yangtze River Delta, the Guangdong-Hong Kong-Macao Greater Bay Area, and the Beijing-Tianjin-Hebei region.

(2) Enhance the medical waste emergency disposal capacity reserve

In terms of centralized disposal facilities, the national ecological and environmental departments jointly with the relevant departments to introduce medical waste disposal facilities construction planning, taking into account the reserve of emergency disposal capacity, the technical route, capacity configuration and other principle requirements: each prefecture-level cities above the design scale of centralized medical waste disposal facilities should maintain sufficient margin, mega-city centralized disposal facilities should be automated. A higher degree of production line; for rural and remote areas of medical waste to take the "small box into the big box" collected by level after centralized disposal, the construction of their own disposal facilities, the configuration of mobile disposal facilities and other medical waste disposal model, the corresponding emergency disposal capacity reserve requirements; around the actual demand, in the public health emergencies centralized rescue designated hospitals and other hospitals inside, set up small medical waste treatment facilities or mobile disposal facilities, as a supplement to the capacity of the emergency state. In the emergency disposal facilities, the city can establish a list of emergency medical waste disposal resources for the unit, included in the list of collaborative emergency disposal facilities, and should set up a special unloading area, and the configuration of special feeding equipment, cleaning and disinfection equipment, etc. Medical waste disposal facilities should be with the subsequent disposal facilities and emergency disposal facilities to play a synergistic layout to maximize effectiveness.

(3) Implementation of safe economic emergency disposal technology route

The technical route for emergency disposal of medical waste during an epidemic outbreak is shown in Fig. 6.7. After the centralized medical waste disposal facility is operating at full capacity, the backup production line or backup facility can be activated, and the mobile medical waste disposal facility or the hospital's own disposal facility can be activated simultaneously used for on-site disinfection of epidemic-related waste, followed by subsequent disposal. In case of insufficient centralized medical waste disposal capacity, centralized disposal facilities should be given priority to protect the disposal of epidemic-related medical waste. Emergency disposal of medical waste should be used as far as possible within the city and surrounding areas of the facility, giving priority to the regional process like the hazardous waste incineration facilities, followed by domestic waste incineration

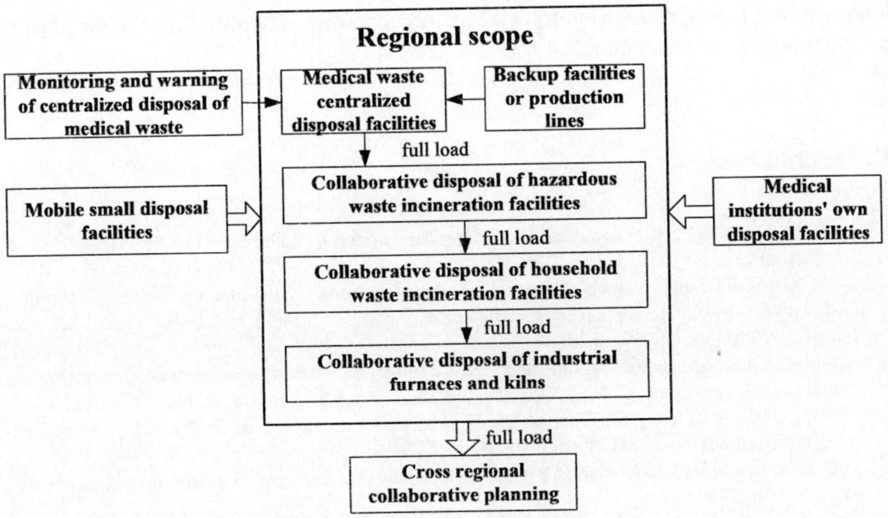

Fig. 6.7 Medical waste emergency disposal technology route

facilities (grate type); if still cannot meet the demand, then the use of industrial kilns and other facilities for collaborative emergency disposal. The Ministry of Ecology and Environment should organize a comprehensive assessment of the economic, technical, and risk aspects of emergency disposal of medical waste during the COVID-19 epidemic as soon as possible, providing applicable technologies and equipment for the construction of the subsequent medical waste emergency disposal system. At the same time, the rapid response technology and application development for temporary storage, transfer and disposal facility site selection, equipment configuration, disposal capacity design, etc., should also be quickly followed up.

(4) Improve the emergency disposal of medical waste supporting measures

The establishment of medical waste emergency disposal capacity and material stockpile system in response to an epidemic or disaster, which will be integrated into the local emergency security system planning and construction. Establish a regional medical waste emergency disposal team composed of leading companies in the medical waste disposal industry and regional hazardous waste centralized disposal facilities, as well as a team of experts composed of environmental management experts, disposal technology experts, and disposal facility operation experts. Relying on national, provincial and municipal solid waste management information system, accelerate the construction of medical waste information supervision platform, covering medical and health institutions, centralized medical waste disposal units and medical waste emergency disposal units, realizing information exchange and sharing, making full use of modern means such as the Internet of things, big data, cloud computing, the establishment of centralized disposal facilities disposal load warning mechanism, to guide emergency response scheduling in the event of a sudden epidemic (Almuneef and Memish, 2003).

References

Almuneef M, Memish Z. Effective medical waste management: it can be done [J]. American Journal of Infection Control, 2003, 31(3): 188-192.

Cheng L, Zhang Z, Sun N, et al. Reflections and suggestions on making up for the shortcomings of medical waste and hazardous waste collection and treatment[J]. Environmental Science Research, 2020, 33(7): 1698-1704.

Guo Chunxia, Chen Yang, Shi Xiangming, et al. Readings on medical waste knowledge during an epidemic [M]. Zhengzhou: Henan Science and Technology Press, 2020: 64-65.

Jang Y C, Lee Cargro, Yoon Oh-Sub, et al. Medical waste management in Korea [J]. Journal of Environmental Management, 2006(80): 107–115.

Li Q X, et al. Medical waste disposal system in Japan [J]. Shanghai Environmental Sciences, 2003(7): 508–511, 518.

Liu Zhaoxiang, Chen Yang, Wang Qin, et al. Analysis and inspiration of Korean medical waste management and disposal experience [J]. Journal of Environmental Engineering, 2021, 15(2). doi: https://doi.org/10.12030/j.cjee.202008215.

Marshall J. Pandemic Could the U.S., like China, face a medical waste crisis? [N/OL]. (2020–03–24) [2020–09–26]. Washington D. C.: E&E News. https://www.eenews.net/stories/1062690625.

Ministry of the Environment of Japan. Guidelines for disposal of infectious waste based on the Waste Disposal Act (2018 revision) [Z]. 2018.

National Safety Council, U.S.A. Guidance for the management of Class A infectious solid waste [Z], 2019.

Science Times. Best disposal for SARS medical waste: Burn! [EB/OL]. [2003–05–14]. http://www.cas.cn/zt/jzt/kpzt/zkywfzfdzgx/zlff/200305/t20030514_2666762.shtml.

Shao L M, Lv F, Peng W, et al. Exploration of methods and technologies for emergency management of domestic source waste during major infectious disease outbreaks [J]. Environmental Health Engineering, 2020, 28(2): 1-5.

Tudor T L, Noonan C L, Jenkins L E T. Healthcare waste management: a case study from the national health service in Cornwall, United Kingdom [J]. Waste Management, 2005, 25(6): 606-615.

U.S. National Safety Council. Planning guidance for the disposal of solid waste contaminated with Class A infectious substances [Z], 2017.

U.S. National Environmental Protection Agency. Introduction to medical waste-related [EB/OL]. [2016–02–17]. www.epa.gov/rcra/medical-waste.

UK Department of Health. Environment and Sustainable Health Technical Memorandum 07–01: Safe Management of Medical Waste [EB/OL]. (2013–03–20). www.gov.uk/government/publications/guidance-on-the-safe-management-of-healthcare-waste.

Zhang Zheng, Cheng Liang, Wang Xiahui, et al. Ideas for building an emergency disposal system for medical waste in China [J]. Environmental Science Research, 2020, 33(7): 1683-1690.

Chapter 7
Medical Waste Information Regulation

In recent years, with the rapid development of the Internet and information technology, medical waste management technology and supervision have also developed and progressed accordingly. This chapter introduces the need for medical waste informatization and application of key technologies, and gives an analysis of the regulatory process; based on the top-level design of cloud services and big data, the functional design of the medical waste regulatory system is explained one by one for reference of relevant organizations. Finally, the application cases of medical waste informatization supervision and future development prospects are given.

7.1 The Need for Information-Based Regulation of Medical Waste (Zhang G. P., Zhou G. M., 2003; Cao Yunxiao et al. 2021)

Medical waste (HW01) is highly infectious and contagious and poses a great threat to human health and environmental safety. Due to the special characteristics of the medical industry, medical waste has strong characteristics of space pollution, acute transmission and latent pollution. It will have a great threat to the environment and human health if we fail to management medical waste. Therefore, the collection and disposal of medical waste is an important work related to the environmental safety of a city, especially in the current stage of the strictest environmental protection policy situation. However, there are still some problems and difficulties in the collection, disposal, and supervision of medical waste, and some cases of loss and dumping have occurred. Most medical institutions and disposal units, such as medical waste classification collection, handover link using manual registration, manual handover, time-consuming and inefficient, resulting in statistical management is not timely, the loss is not easy to trace; supervision done by the supervision department include the

Y. Chen et al., *Innovative Disposal Technology and Management Practice for Medical Waste*, https://doi.org/10.1007/978-981-99-6786-5_7

use of on-site inspection, manual statistics. However, its management is difficult, and there are loopholes in the supervision.

In recent years, the national management departments at all levels of medical waste management have become increasingly important, the National Health Commission issued a notice to further standardize the management of medical waste and audit department continuous special audit, but also stressed that the ecological and environmental management departments should increase efforts to regulate medical waste, to strictly prevent the loss of medical waste. With the rapid development of China's economy and comprehensive national strength, the rapid maturation of medical technology, medical institutions are also increasing, accompanied by the generation of more and more medical waste, and supervision is increasingly difficult.

To this end, the National Health and Health Commission, the Ministry of Ecology and Environment, the Development and Reform Commission, and other 10 ministries and commissions in February 24, 2020 issued a "comprehensive management of medical institution waste work program" to strengthen the comprehensive management of medical institution waste, to achieve waste reduction, resource, and harmless operations. Establishing the medical waste information technology supervision platform, covering medical institutions, medical waste centralized storage sites, and medical waste centralized disposal units, is important to achieve the sharing of information. The health and health departments can promptly report to the ecological and environmental departments of medical institutions to generate, transfer, or self-disposal of medical waste. The ecological and environmental departments send to the health sector in a timely manner to announce the administrative approval of the centralized medical waste disposal units, the community to disclose the list of centralized medical waste disposal units, disposal types and contact information. Through the above-mentioned programmes, build up a closed-loop system of medical waste data supervision, medical waste online traceability, and monitoring of the trace of the connection, as well as the whole process of waste, without dead-end supervision (Ni Chuanming et al., 2018).

7.2 Key Technology for Medical Waste Information Technology Application

The key technology for the application of medical waste information technology is to take advantage of the opportunity to develop big data, through "innovative means of supervision. The full use of modern Internet of things technology, improvement of management efficiency" as the spirit of guidance, to "vigorously develop medical waste information technology, enhance the level of key environmental prevention and emergency response capabilities" as the goal, to promote medical waste from the source of classification and quantification of management systems and regulatory means of innovation.

(1) Internet of Things Technology (Zhang Hongfang, 2012)

The Internet of things can connect things with things and people with things. The main function of medical waste informatization is to use IOT technology to obtain accurate perception of medical waste source classification and reduction through IOT devices. Combined with the business scenario of medical waste and user needs, the entire medical waste information collection programme, control functions, communication mode and other targeted design and development, customized offline business workflow and information systems, so that the system can be connected in accordance with the process of sensors and Internet of Things facilities, and real-time collection of important data information such as weighing, transfer, in and out of storage. Through the integration of medical waste supervision business scenario model, medical waste, transport vehicles, operators and other monitoring target behaviour to track, detect, determine, and alarm.

The application of Internet of things technology makes medical waste from the generation, handover, measurement, collection, bulk storage, and other aspects of intelligent dynamic supervision, reduce the errors of manual entry, the use of electronic induction, and other means of Internet of things to make all the processes into data entry system, so as to achieve the construction of the terminal perception layer, can effectively prevent the loss of medical waste and outflow. Medical safety plays a great role in the protection.

(2) Cloud Computing Platform (Li Guojuan, 2020)

Cloud computing integrates distributed computing, parallel computing, network storage, load balancing, virtualization technology, and other computer and network technologies. Cloud computing is a large number of distributed computers to complete computing tasks, rather than local computers, which makes cloud computing have the characteristics of ultra-large scale, high reliability, high scalability, and on-demand services. The application of the cloud platform allows the unification of services, allowing a closer connection between services, and can significantly improve management efficiency. Medical waste informatization applications can be used to unify resource scheduling and unified release management of medical waste numerical modelling, traceability modelling and other computing resource-consuming applications with the cloud computing platform. At the same time, the technology includes both multi-application intelligent desktop integration technology and intelligent management technology; the former can isolate the physical collection of applications, to ensure the security performance of the system, and the latter can effectively control the long-term occupation of the license, and can effectively improve the utilization of resources.

(3) Big Data Technology

The big data applications of medical waste information technology. Big data applications cover mainly the entire process of medical waste generation, classification, collection, storage, transfer, disposal, and other business data. These medical waste production data are of various types and frequent data transactions, so the performance of database storage and data mining analysis has raised higher requirements. In the data processing and application layer, we need to analyze the correlation of different types of data set, aggregate multiple data sets, form a fusion of different types of data integration, and correlation analysis.

(4) Intelligent AI technology

Intelligent AI technology is the basis for intelligent decision-making on medical waste. Through artificial intelligence technology, deep machine learning technology, and expert database technology, the value of data can be extracted faster and the value laws can be identified. Database information is modeled for different data levels to carry out medical waste traceability, early warning, and trend analysis applications, providing managers with decision-making information on medical waste reduction, harmlessness, and resource utilization.

7.3 Analysis of the Entire Medical Waste Supervision Process (Zhou W.-L., 2016)

According to the distribution of medical waste collection points, the number of large changes, collection and transportation network routes complex and other characteristics, the entire medical waste supervision process can be divided into three main stages: front-end management, transfer process, and end management, focusing on the transfer process stage. Back to the front-end management is mainly for the production of medical institutions, transfer, and temporary storage management; back to the end of the management is mainly for the disposal of units in and out of storage, disposal and utilization management. The use of Internet of Things technology to build the whole process of monitoring the elements of perception network, and actively promote the whole process of medical waste supervision, do the whole process of information technology intelligent monitoring, to ensure that medical waste from temporary storage, transfer, transport, disposal of each key link is without dead-end monitoring and perception.

The chain of medical institutions' medical waste treatment industry generally consists of three parts: the collection, storage and transportation of medical waste within the hospital; medical waste from the medical institution to the destruction of companies or reuse enterprises in transit transport; medical waste destruction or reuse of the destruction of reuse institutions. The whole process of medical waste information technology supervision is the use of Internet of things, Internet technology for medical waste collection and disposal of the whole process of monitoring and management, for medical institutions, centralized disposal units, environmental protection, and health regulatory agencies to provide the whole process of information management system, medical waste collection and disposal business operation standardization, automation and standardization, traceability, online supervision, to eliminate the loss of medical waste.

The basis of the whole process of information-based supervision of medical waste is to contain medical waste packaging containers, installed electronic tags, thus achieving the digitisation of medical waste, the realisation of the whole process of flow data information, easy data sharing, real-time query, and online supervision.

The key to the entire process of information-based medical waste is the generation and collection of medical waste data within the medical institution and the centralised medical waste disposal unit collection, transfer, interoperability of the disposal data, and then with the environmental protection and health supervision departments network.

According to the existing division of administrative functions, medical institutions belong to the health department in charge, the centralized disposal units belong to the ecological and environmental management department in charge, resulting in difficulties in the construction of the whole process of medical waste information technology process, the need for two departments to build their own and then the two departments of the data sharing.

At present, most of the medical waste information systems have been built around the medical institutions to build a set of internal, centralized medical waste disposal units, to build a set of two systems are independent, and their data is difficult to do consistent. According to the national requirements for the modernization of medical waste management, the construction of medical waste information-based supervision of the whole process shall be carried out by the provincial and municipal administrative units to carry out the top-level design, the health and health management department in charge of medical institutions and the ecological environment management department in charge of the overall design of the centralized medical waste disposal unit, electronic tags for the packaging container, identification standards for terminal equipment, unified data processes, unified system. So that health, environmental protection supervision to achieve a network, consistent data, and the whole process is clear and traceable. The medical waste classification and disposal technology process is shown in Fig. 7.1.

7.4 Cloud-Based Services and Top-Level Big Data Design (Zhao Hui, Hu Shuyu, 2016; Xuan Junfang et al. 2019; Geng W., Wu Xiaoyan, 2009)

The application of a medical waste information management system will shift medical waste disposal management from equipment-centric to information-centric, avoiding regulatory agencies from investing a lot of IT resources and human resources in the operation and management of information systems, and will focus on medical waste information technology data analysis and decision analysis. Provide on-demand allocation and quantifiable computing and storage resources through

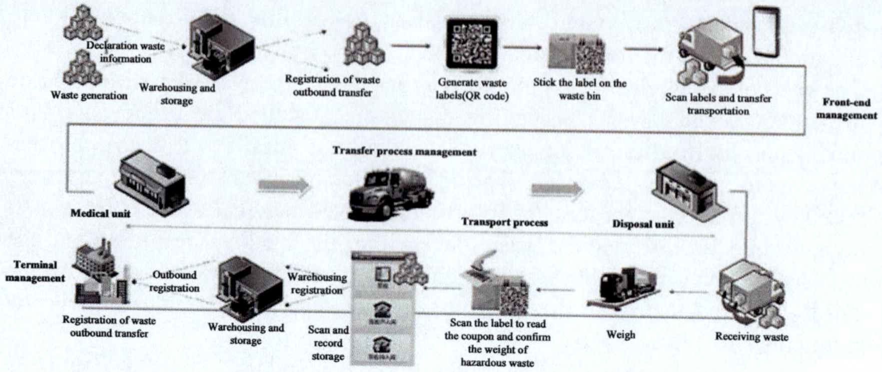

Fig. 7.1 Medical waste classification and disposal technology process

Infrastructure-as-a-Service (IaaS) of cloud service architecture; Using software development platforms or middleware as a service to build platform-as-a-service (PaaS) can solve the phenomenon of inconsistent technical specifications, inconsistent data standards, and various reports, GIS, and workflows caused by the proliferation of suppliers of front-end IoT devices. Information security and other services; SOA concept for a variety of application layer service development to build customizable software-as-a-service (SaaS) applications to meet the health and health commission, the ecological environment and other departments of the medical waste from the department to put, classification and collection, classification and packaging, timing and route delivery, temporary storage and handover of centralized disposal and other aspects of the entire process of information supervision. The top-level design of big data based on cloud services is shown in Fig. 7.2.

At the same time, the medical waste information management system uses big data, machine learning, intelligent AI, and other technologies to build an intelligent platform from data collection, data storage, data mining, data analysis, and information display. Data algorithms and numerical models are used to help managers identify hidden patterns and correlations from data, make classifications and predictions, allow automatic pattern finding and interactive analysis, and analyze correlations between the same or different monitoring elements. The back-end management platform, in addition to the conventional forms of reports and charts, can also be used for different applications, using a variety of tools for a more intuitive presentation of data. For example, combined with geographic information data in the form of maps to show the data. The technical process of the big data application is shown in Fig. 7.3.

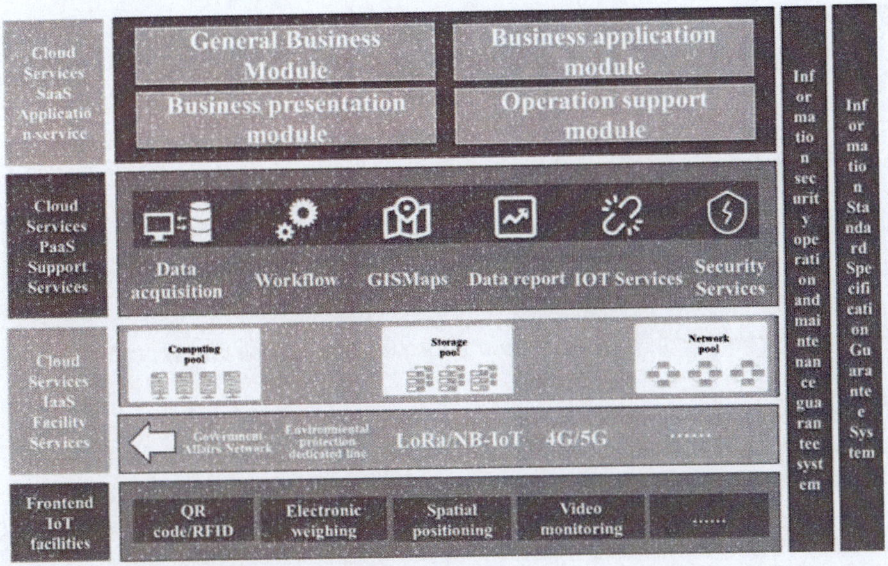

Fig. 7.2 Top-level design of big data based on cloud services

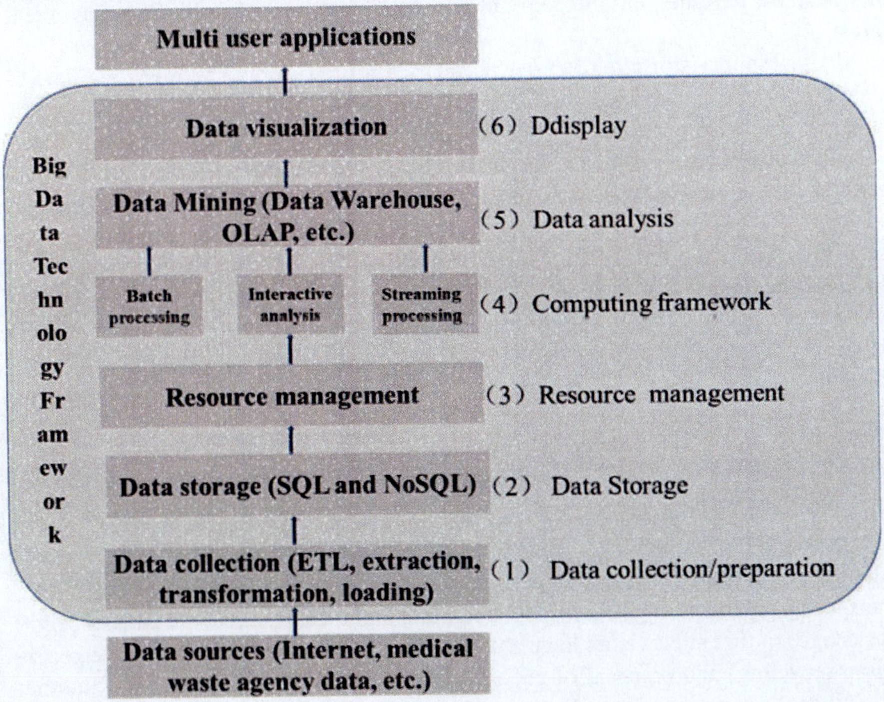

Fig. 7.3 Medical waste big data application technology process

7.5 Functional Design of the Medical Waste Supervision System (Zhou W.-L., 2016)

The construction of the system software platform is based on the open software system framework, and the application system is designed according to the ideas of standardization, structuring, hierarchy, platform, and modularization. The design of the main functional modules includes the following content.

7.5.1 Portal for Medical Waste Regulatory Cloud Service

The medical waste information management system mainly involves regulatory departments, waste production units, collection units, transport units, and business units 5 types of users, different application objects with different business, process, and functional requirements; therefore, for the different users of the medical waste management process, respectively, build targeted application portals: waste production unit portal, collection unit portal, business unit portal, transport unit portal, and regulatory department portal to achieve the declaration of the medical waste regulatory business extranet, intranet audit of advanced, fast, and convenient management mode.

Medical supervision cloud service portal uses unified account permissions, unified message service, and unified application integration to provide a unified and centralized portal service for system users. Production waste units, collection units, transport units and business units can be unified through the cloud services portal for the maintenance of basic information management units, as well as related medical waste business processing, transfer management, information declaration and filling, notification notices received, and for other matters. Supervisory departments can view different medical waste regulatory business data, review and approve the medical waste administrative licensing business according to the authority of each functional department.

7.5.2 Dynamic Surveillance of Medical Waste Internet of Things

All-weather data surveillance of medical waste, real-time medical waste collection, transportation, and storage equipment operating parameters sent to the control centre backstage, so that supervisors have a comprehensive grasp of medical waste generation, recycling, and disposal of each health care institution. The use of "Internet of Things+" mode, in the medical waste collection, transportation, and storage of medical waste, covers the whole process of tracking management and real-time monitoring. The main functions include IOT equipment management, IOT equipment collection management, key link video monitoring management, online monitoring data management, IOT facility operation management, etc.

7.5.3 *Integrated Regulation of Medical Waste Operations*

This function provides health and environmental management departments with a comprehensive integrated regulatory service function for the medical waste supervision business. For example, the management of the medical waste transfer approval process, the medical waste administrative approval business, the generation and disposal of medical waste statistics and decision-making, as well as the daily supervision and management of medical waste-related enterprises. Specifically to achieve: ① intelligent supervision: collection of medical waste classification, temporary storage, transfer, disposal of the whole process of supervision, sharing of medical waste data sharing to prevent illegal dumping and trading; ② accurate traceability: through the process of medical waste flow of information technology to achieve each link, you can accurately trace the type of medical waste, source, quantity, weight, handover personnel, handover time, handover location and final destination, etc. Important information; ③ scheduling arrangements: according to the amount of medical waste generated to arrange the emergency disposal capacity scheduling, medical waste should be received, with unified collection; ④ statistical decision making: shows the total data of medical waste, total real-time data of medical waste, panoramic map, video surveillance and other functions, online law enforcement, supervision and management, and AI decision analysis of integration, intelligence, visualization, and early warning.

7.5.4 *Emergency Disposal of Medical Waste and Prevention and Control*

According to the requirements of "COVD-19 infection of the pneumonia epidemic" and technical guidelines (for trial implementation), the establishment of medical waste emergency disposal and unified prevention and a control system. According to the principle of "unified management and hierarchical management, combined with the division of labour and joint prevention and control, combined with centralized disposal and disposal in the vicinity", construction of medical waste in health, ecological environment, housing, urban and rural construction, transportation and other competent departments under joint coordination of the standard emergency disposal activities of medical waste process.

In accordance with the technical route of emergency disposal and technical requirements, the construction of medical waste risk sources for a source of a file management, medical waste disposal materials unified scheduling management, medical waste risk sources of early warning management, medical waste emergency collaborative scheduling management, medical waste disposal after the management and other functions, guidance around the timely, orderly, efficient, harmless disposal of medical waste, standardised medical waste emergency disposal activities, to prevent disease transmission and environmental pollution, the timely release of emergency disposal information. Medical waste management from "previous approval management" to "after the fact process supervision", from "passive management" gradually to "active prevention type management" change, timely warning to remind the rapid investigation and supervision, play a macro-guidance role of the government.

7.5.5 *Decision on One Map Decision*

With the help of big data management mode, the management of medical waste visualization, standardization, data more accurate, detailed, unified, but also online for effective supervision, the receipt and confiscation, who is responsible for the handover, big data all have records, all based on data speak. BI tools and business reports and other data analysis techniques and means of thematic and comprehensive analysis of various business data, combined with tables, charts, rich text, and other rich display methods to present the results of the analysis to the user, for viewing and query, and to assist decision-making. Big data-based correlation analysis of conventional monitoring data, nonconventional monitoring data, business data correlation analysis summarizes the laws between data, better data traceability, finds the essence of the factors that cause environmental problems, and their potential changes in the relationship, to help managers find hidden patterns and associations from the data, classification and prediction, to provide a reliable basis for waste medical waste management decisions.

7.5.6 *Mobile Application for Integrated Medical Waste Communication*

Combine mobile applications with medical waste supervision business, through smart phones, tablets, and other smart terminals, to build medical waste APP and WeChat app to meet the business needs of the different users of medical waste information management system. Users can realize medical waste in and out of management, transfer orders, declaration, and approval of mobile application scenarios, electronic operation, convenient for users to handle related business at anytime and anywhere.

7.6 Applications for the Regulation of Medical Waste Information (Chen W., 2020)

The supervision of medical waste information technology is a closed-loop management system for the entire medical waste from in-hospital classification and collection to centralized disposal. According to the application of information technology scenarios, it is divided into three categories of medical institutions, disposal institutions and regulatory bodies, the following will be applied from these three types of objects to explain the application of medical waste information technology to regulate the whole process of closed-loop management applications.

7.6.1 Application of Information Technology in Medical Institutions

7.6.1.1 Application Units

Select two hospitals in the South to carry out medical waste informatization application cases.

7.6.1.2 Disposal Scale

A hospital in South I is a tertiary level A general hospital. The medical area of the hospital covers an area of nearly 50,000 m^2, with a total construction area of nearly 170,000 m^2 and 2,300 beds, with 65 specialized departments and 35 teaching and research departments. The daily generation of medical waste is about 200 kg for injury, 1800 kg for infection, 20 kg for pathology, and 10 kg for medicine.

A hospital in the South II was founded in 1941, a large comprehensive tertiary hospital, the first batch of the top 100 hospitals in China. The hospital has established 111 diagnosis and treatment subjects, has 14 national key clinical specialty construction projects, and 4 national key disciplines of traditional Chinese medicine. A hospital in the south which generates 2 medical wastes a day is about 1000 kg of injury, 3300 kg of infection, 40 kg of pathology, and 20 kg of medicine.

7.6.1.3 Content of Information Construction

1. Main Body Equipment

A hospital in South I Medical Waste Classification Collection and disposal main equipment list and Operation Schematic Diagram are shown in Table 7.1 and Fig. 7.4a, respectively.

Table 7.1 A hospital in the South I medical waste classification collection and disposal of the main equipment list

Serial number	Equipment name	Equipment technical parameters	Quantity	Unit
1	Portable information collection terminals (with PDA)	It is applied to the whole hospital for intelligent cleaning management and can be used with the original medical waste cleaning truck of the hospital 1. Two-dimensional code scanning, full-touch operation screen 2. Thermal printing, automatic code output at the weigh end 3. GPS positioning, wireless transmission, display track 4. Intelligent weighing, data acquisition, real-time monitoring 5. Portable, handheld, integrated design 6. Length, width, and height: 50 cm × 30 cm × 10 cm	5	Taiwan
2	Medical waste traceability source calibrator	Used in medical waste temporary storage, hospital medical waste in medical waste temporary storage for review into the storage and transfer out of the storage use 1. 15-inch smart touch screen 2. Scan the code to identify, out of the warehouse, into the warehouse, check the data 3. Voice prompts, automatic broadcast 4. Intelligent weighing, data upload, platform monitoring 5. Thermal printing, automatic code 6. Length, width, and height: 70 cm × 50 cm × 90 cm	1	Taiwan
3	Medical waste wise energy supervision platform	Includes PC terminal, mobile APP, and WeChat mini-programme. The main functions include management of medical waste inlet and outlet, management of query statistics, hospital information management, monitoring of waste handover management, equipment status, management of GIS maps, etc.	1	Set

A hospital in South II, Medical Waste Sorting and disposal main equipment list and Operations Schematic Diagram are shown in Table 7.2 and Fig. 7.4b, respectively.

2. Specifications of the main equipment parameters

(1) Portable information collection terminal. The terminal is applied to some buildings without special elevators for collecting and transporting medical waste, and this portable collection device is placed in the dirt room of these buildings for the collection of medical waste in the department, with compact appearance, convenient movement, and complete functions, which is convenient for

Fig. 7.4 Collection and disposal photos

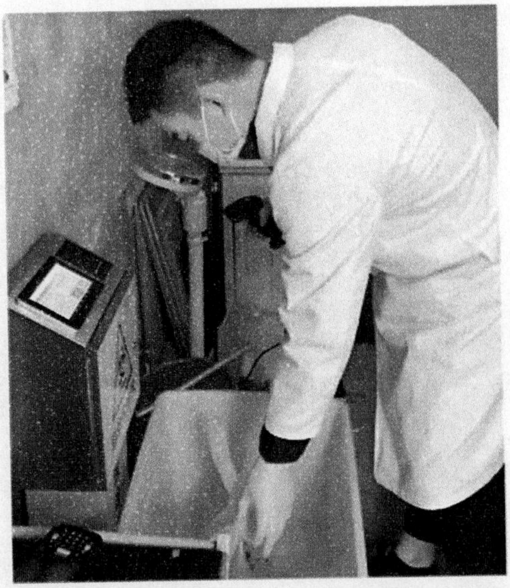

Medical waste staging room traceability calibration

Handover of medical waste scanning

(a) A hospital in the South I

Fig. 7.4 (continued)

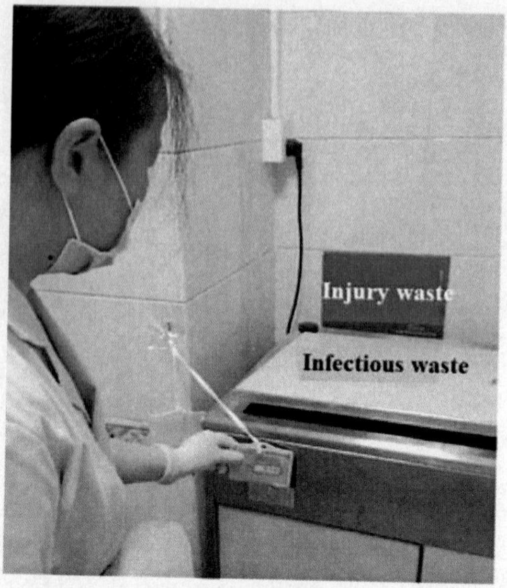

Medical waste swipe card drop-off

Medical waste staging room traceability calibration

(b) A hospital in the south-II

Table 7.2 List of the main equipment of a hospital in the South II medical waste collection and disposal

Serial number	Equipment name	Equipment technical parameters	Quantity	Unit
1	Medical waste wise can collect terminals (box type)	It should be used in temporary storage rooms within departments with limited space or on hospital floors 1. Opening of the sensor door, only in and out 2. Full overflow alarm to prevent secondary pollution 3. The nurse swipes the card and authorizes cleaning 4. Automatic weighing, automatic code, to prevent loss 5. GPS positioning, voice transmission 6. Automatic upload of information, platform monitoring	4	Taiwan
2	Portable information collection terminals (with PDA)	It is applied to the whole hospital for intelligent cleaning management and can be used with the original medical waste cleaning truck of the hospital 1. Two-dimensional code scanning, full-touch operation screen 2. Thermal printing, automatic code output at the weigh end 3. GPS positioning, wireless transmission, display track 4. Intelligent weighing, data acquisition, real-time monitoring 5. Portable, handheld, integrated design 6. Matching the existing 660L medical waste collection vehicle	4	Taiwan
3	Medical waste traceability source calibrator	Used in medical waste temporary storage, hospital medical waste in medical waste temporary storage for review into the storage and transfer out of the storage use 1. 15-inch smart touch screen 2. Scan the code to identify, out of the warehouse, into the warehouse, check the data 3. Voice prompts, automatic broadcast 4. Intelligent weighing, data upload, platform monitoring 5. Thermal printing, automatic code	1	Taiwan

(continued)

Table 7.2 (continued)

Serial number	Equipment name	Equipment technical parameters	Quantity	Unit
4	Smart handheld tablet devices	Screen size: 10.8 inches Running memory (RAM): 4 GB 1. Storage capacity (ROM): 128 GB 2. Front camera: 8-megapixel, fixed focus 3. Rear camera: 13 megapixels, autofocus 4. Resolution: 2560 × 1600 5. Extension support: micro SD (TF) card 6. Expandable capacity: 512 GB 7. CPU cores: eight cores 8. Touch screen: capacitive ten-point touch 9. Bluetooth transmission: Bluetooth 5.0 10. USB interface: USB Type-C interface, support and PC data synchronisation, fast charging, and other functions 11. APP for intelligent medical waste intelligent supervision APP	10	Taiwan
5	Medical waste wise energy supervision platform	Includes PC terminal, mobile APP, and WeChat mini-programme. The main functions include management of medical waste inlet and outlet, management of query statistics, hospital information management, monitoring of waste handover management, equipment status, management of GIS maps, etc.	1	Set

the scavengers to collect medical waste. It is composed of a handheld PDA with a Bluetooth portable scale. The handheld device is mainly used for medical waste collection personnel in the medical waste collection to do personnel identification, access to Bluetooth scales medical waste weight, in and out of the warehouse, and traceability enquiries. The portable scale mainly functions as a weighing tool.

Medical waste data collected at the front end is transmitted to the medical waste management platform in real time via a handheld PDA. This device is suitable for all types of medical institutions and is used with existing carts. The portable information collection terminal device is shown in Fig. 7.5.

The device has the following features: ① PDA scanning identification authentication, full touch screen operation; ② GPS positioning, real-time data collection; ③ intelligent weighing, automatic code; ④ portable hand-held, Bluetooth data transmission; ⑤ automatic upload of information, platform monitoring.

(2) Medical waste collection and information transportation terminal. The terminal is applied to the hospital building with a special medical waste collection and transportation elevator, equipped with electric power trolley, cleaners use this equipment to collect medical waste for the whole building, the equipment comes with weighing, coding, medical waste handover and other functions in one (Fig. 7.6), the information of recovered medical waste is automatically uploaded

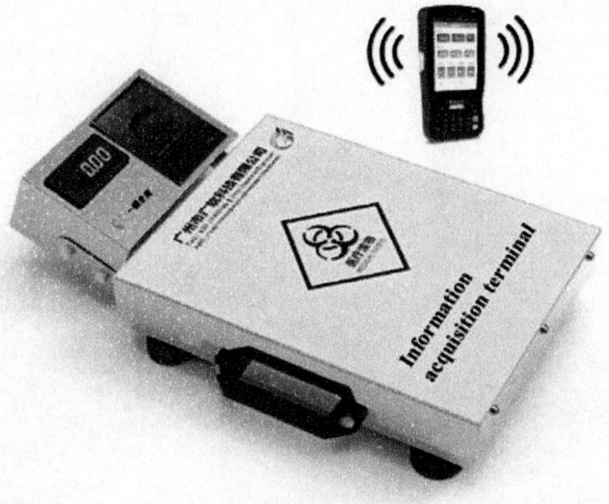

Fig. 7.5 Portable information collection terminal equipment

to the regulatory platform by this equipment. Equipment using a fully enclosed design, the cart set information collection and weighing in one, generally a building (medical waste elevator) equipped with one such cart can meet the entire floor of medical waste collection and transportation work, and can be used for inbound, outbound, and other operations. With electric power, the operator can freely match the speed, which is convenient for medical waste collection personnel.

The device has the following features: ① sweep code authentication, full touch screen operation; ② GPS positioning, real-time transmission of data collection; ③ thermal printing, intelligent weighing, automatic code; ④ automatic upload of information, platform monitoring.

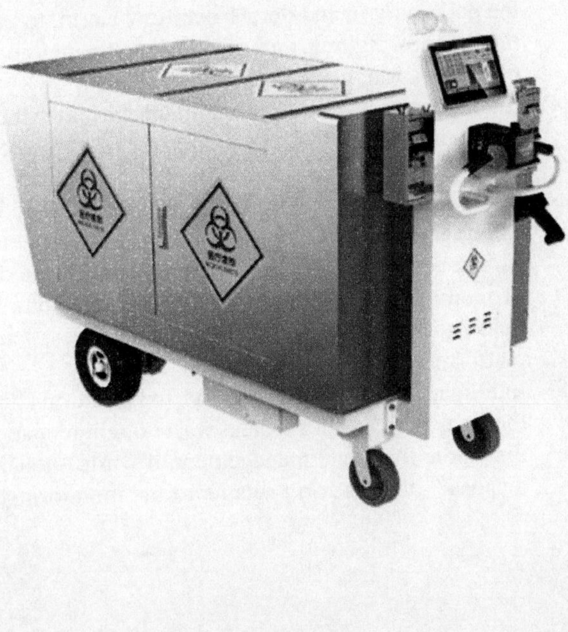

Fig. 7.6 Medical waste collection information transport terminal

Fig. 7.7 Airtight medical waste source separation box

(3) Airtight source separation bin for medical waste. Operating rooms, surgical departments, and other infectious risk departments are equipped with this source-separated collection box with closed design, which can prevent the spread of infection. Medical waste delivery and cleaning through the card access to open the door, to avoid bystanders contact with medical waste, automatic weighing out of the code. Automatic weighing out code function, reduce the collection link, reduce the frequency of manual errors, improve the accuracy of data, and effectively enhance the supervision. The closed medical waste source separation box is shown in Fig. 7.7.

The equipment has the following features: ① swipe card authorization to open the door, only in and out; ② overflow alarm, to prevent secondary pollution; ③ intelligent weighing, authorized cleaning, automatic code; ④ automatic upload of information, platform monitoring.

(4) Intelligent traceability calibrator. It is placed in the hospital's main staging room and is used to manage the entry and exit of medical waste collected in portable carts and sorting bins and traceability enquiries, while the weight of medical waste collected at the source can be calibrated and reviewed, and medical waste can be traced back to whether it is lost, leaked, etc.

The equipment has the following features: ① data verification, management of medical waste inlet and outlet management; ② functional weighing, data upload; ③ thermal printing, automatic code; ④ automatic information upload, platform monitoring.

(5) Intelligent medical waste supervision platform (Yang Chunmei, 2015). Through the establishment of medical waste intelligent supervision platform to achieve the whole life cycle management of waste "generation - collection - transfer - disposal", the platform sets real-time monitoring, business flow, data sharing,

prediction and early warning, scientific decision-making and service management and other functions are integrated. According to different application scenarios, the platform is divided into three parts. PC desktop, mobile APP, and mobile WeChat app, and main functions include medical waste inlet and outlet management, query statistics management, hospital information management, waste handover management, equipment status monitoring, GIS map management, etc. The PC desktop platform software interface is shown in Fig. 7.8 and the mobile APP interface is shown in Fig. 7.9.

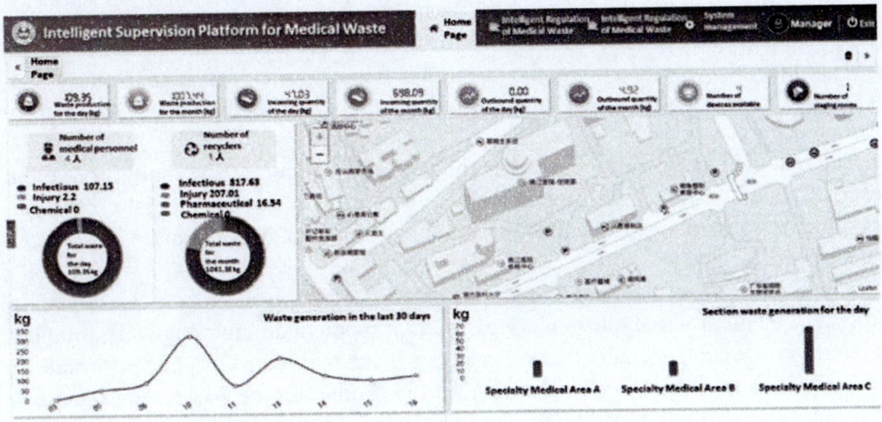

Fig. 7.8 PC desktop side platform software interface

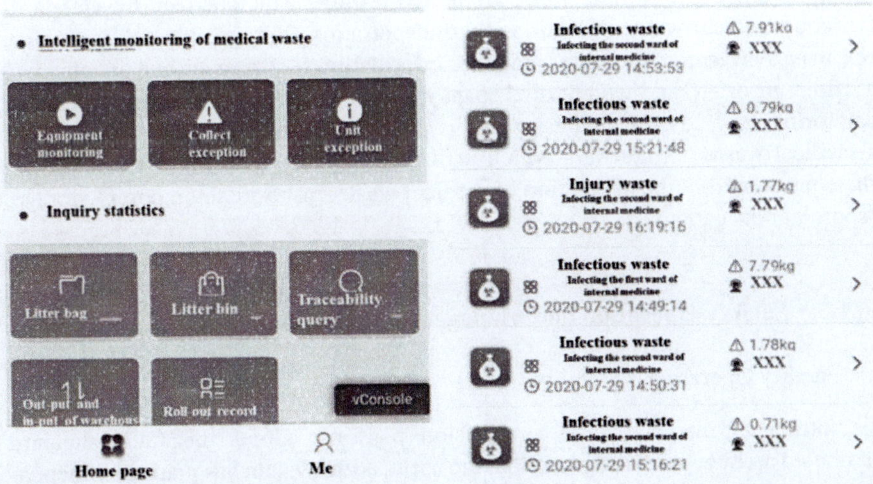

Fig. 7.9 Mobile APP interface

7.6.1.4 Effectiveness of Information Technology Operation

A hospital in South I and a hospital in the south II are equipped with different models of special medical waste collection equipment according to the actual situation of the hospital. The medical waste information collected through these devices is transmitted in real time to the information management software platform, from medical waste, temporary storage into the warehouse, out of the warehouse handover, disposal, and transport links to establish a medical waste prevention and control of the physical network, forming a collection, storage, transportation in one, and information interoperability of medical waste management "a network".

Based on the data collection of the front-end hardware facilities, the software management platform can query the hospital's medical waste data in real-time statistics and data reports exported, leaving behind the traditional manual registration, manual accumulation.

The data management mode makes the data more accurate and real; through the reserved data interface, real-time and health supervision, environmental protection, and other regulatory departments can be obtained for data docking and reporting can be obtained.

The entire process management and intelligent control process is shown in Fig. 7.10, medical waste cleaners, departmental medical and nursing staff, inbound and outbound managers and other roles are assigned their own QR code information, from cleaners to scan the code cleaning, to the production of waste departments to scan the code and nurse handover, the handover information is printed and generated QR code, after reaching the staging room to scan the QR code in the staging room for inbound and outbound storage, the whole process is through the scan code. The entire process is carried out by scanning the code, which realizes the standardization of medical waste collection and transportation, storage, and out of storage, etc.; an early warning forecast is made for information such as overtime uncollected, overtime unsorted, and weight discrepancy, etc. Meanwhile, combined with video monitoring, etc., loss of medical waste is prevented and traceability management of medical waste is improved, which realizes the closed-loop management of the whole process from the generation of medical waste, in-hospital temporary storage, disposal, and cleaning.

7.6.1.5 Facility Operation and Management Programme

(1) Facility Operational Stability

The intelligent medical waste supervision platform adopts the J2EE standard distributed architectural design, so that the application system has platform independence and can be deployed on any J2EE compliant application server, improving the deploy ability of the system and reducing maintenance and management costs. The

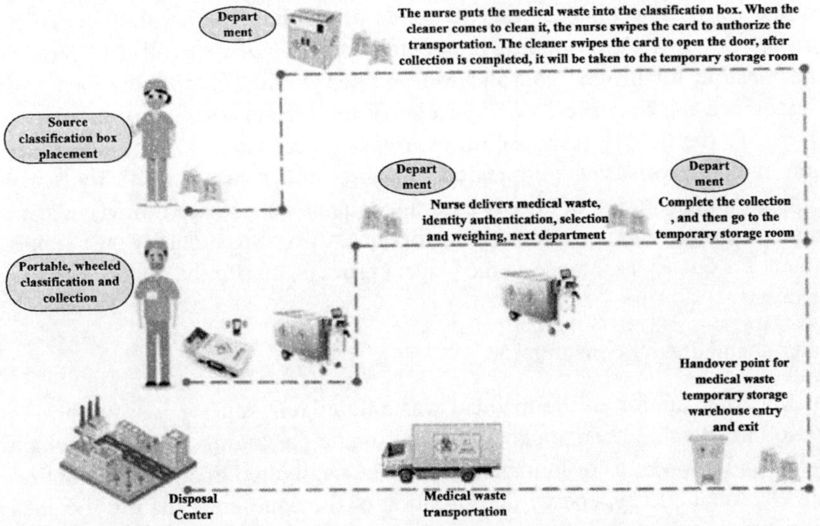

Fig. 7.10 Total process management and intelligent control process

operation of the overall project is combined with the information system project operation and maintenance management system standards, and the operation and maintenance management system is established, and the hardware and software facilities are regularly maintained and checked, including checking the data status, checking the application configuration, the operation of the facilities, and the necessary patch upgrade, so as to nip the failure in the bud in advance. For systems that cannot be repaired or are not functioning properly, replacement equipment is provided to ensure normal operation of the system.

(2) Maintenance management requirements

According to the daily management needs to arrange at least one full-time maintenance management personnel, mainly responsible for the daily operation and maintenance of the system, its main responsibility is to ensure that the hardware and software facilities in a fault-free operation of the state, should have a certain degree of computer, network knowledge of the staff to serve. Network, operating system and business application system installation and maintenance work, system functions, real-time performance monitoring and the necessary status checks; should be regular backup data, in the case of system interruptions, operational paralysis, timely troubleshooting the corresponding fault to avoid or minimize system losses; in addition, should also analyze the operation of the system, to provide reasonable expansion, transformation recommendations to ensure The system's sustainable development.

(3) Implementation of quality control

In the project implementation, the project must be standardized management, project implementation organization, project implementation management plan, project progress plan, project acceptance plan and other programmes to ensure the quality of the project. Set up a corresponding project team and designate a full-time project leader responsible for project coordination and scheduling. Clarify the needs of quality assurance in the project process; clarify the content of quality assurance (object); clarify the quality responsibilities in the project team; clarify the implementation of quality assurance strategies and ways and means; clarify the entry point for the implementation of quality assurance (quality assurance timing); clarify the material resources and personnel inputs for quality assurance; clarify the support environment for quality assurance in the project process; clarify the quality process submissions.

(4) Material and energy consumption

The material consumption of the medical waste intelligent supervision platform is mainly two-dimensional code stickers, medical waste packaging bags, etc., which are low-cost and low-value consumables that can be negligible. Energy consumption is mainly electrical energy, energy consumption of the equipment mainly includes servers and server room network equipment, according to the 24 h work is expected to consume about 20 kW-h, portable information collection terminal, medical waste traceability calibrator, and other low power consumption equipment, according to the daily work 8 h is expected to consume about 0.5 kW-h.

(5) Application of the evaluation results

Medical waste source classification collection and disposal technology with innovative Internet of Things management means, from medical waste, temporary storage into the warehouse, out of the warehouse handover, disposal, and transportation to establish a hospital internal medical waste control network, leaving behind manual registration, manual accumulation of the traditional mode of data are management, so that the data can be more accurate and more real. To achieve automated, quantifiable, visualization, remote closed-loop management of the whole process, so that the level of medical waste management has a qualitative development.

The medical waste intelligent supervision platform from the perspective of business processing improves the efficiency of medical waste management, is more scientific, from the perspective of monitoring, standardizes the medical waste management business processes, to a large extent controls the emergence of human or illegal behaviour, from the perspective of automation. It is the starting point and key point to become the ultimate intelligent management.

The results of the evaluation of effects of the traditional and information-based approaches are shown in Fig. 7.11.

Content	Conventional methods	Informationization methods
Record Mode	Paper (manual)	Electronic (automatic)
Save Mode	Paper	Electronic
Data Error	Yes	No
Information sharing	No	Yes
Statistical methods	Post event manual statistics	Real time upload
Weighing method	Manual scale	Bluetooth electronic scale

Collection Mode	Imperfect	Perfect
Early warning and forecasting	No	Yes
Traceability	No	Yes
Data statistics detection	Manual (with large errors)	Automatic (precise)
Visualization	No	Yes
Registration time	Casualness	Real time
Process supervision	No	Yes
Human factor	Uncontrollable	Controlled

Traditional manual weighing (kg)	Bluetooth electronic scale weighing (kg)
65	60.91
70	80.87
65	85.67
79	95.21
66	97.01

Conventional methods	Informationization methods
Collection	Collection
Department confirmation	Scan code handover confirmation
Transportation	Confirmation of medical waste types
Manual warehousing	Bluetooth electronic scale weighing
	Print label
	Scan code for storage
	Scan code for outbound delivery
	Statistical table

Fig. 7.11 Evaluation results of the effect of traditional and information-based methods

7.6.2 *Applications of Disposal Agency Information Technology*

7.6.2.1 Application Units

A medical waste safety disposal centre in a southern city was selected to carry out a medical waste information application case.

7.6.2.2 Disposal Scale

A city in the south of the operation of the medical waste safety disposal centre for many years, the company's disposal of medical waste generated year-by-year growth, the center's incinerator and its supporting treatment system is often overloaded with disease operation, resulting in the frequent failure of key equipment, and there is no emergency disposal capacity for unexpected epidemics. To better regulate the company's internal management, to meet the company's medical waste disposal business operation needs, and to effectively solve the problem of long-term overload operation of the centre. The company invested in a new incinerator to enhance treatment capacity, introduced an information management system for medical waste disposal, constructed an information-based platform for the medical waste safety disposal centre, strengthened basic management, improved process management and implementation supervision, comprehensively improved execution, and established a strict internal control system to provide support for scientific and agile decision making to ensure the sound operation of the medical waste disposal centre in a southern city.

7.6.2.3 Content of the Information Construction

A city in the south of the medical waste safety disposal centre information construction content, mainly to medical waste "two points a line" as the core, around the medical institutions to generate units, operating units "two points", transfer transport "a line" in the key links, key nodes, the use of intelligent video, RFID, and other technologies to achieve automated analysis of information flow, the formation of "collection - transport - receiving - re-weighing - storage - treatment" of medical waste throughout the information management (Fig. 7.12).

(1) Positioning Monitoring of medical waste transportation vehicles

The use of license plate recognition, GPS/Beidou positioning, vehicle video surveillance, and other technologies to achieve full supervision of medical waste transport vehicles. Not only can you view the online status of the vehicle equipment, but also you can display the real-time location of the vehicle and, at the same time, can view the geographic location information of the positioning point. When the vehicle is tracked in real time, the map is always centered on the target vehicle. Vehicle driving status can also be displayed, including normal, offline, and alarm three states. The vehicle history travel route for track playback, playback speed can be divided into low speed, medium speed and high speed, playback the maximum span of 24 h. Track playback display speed information, and in the interface to show the speed curve.

(2) Monitoring of medical waste vehicle weighing

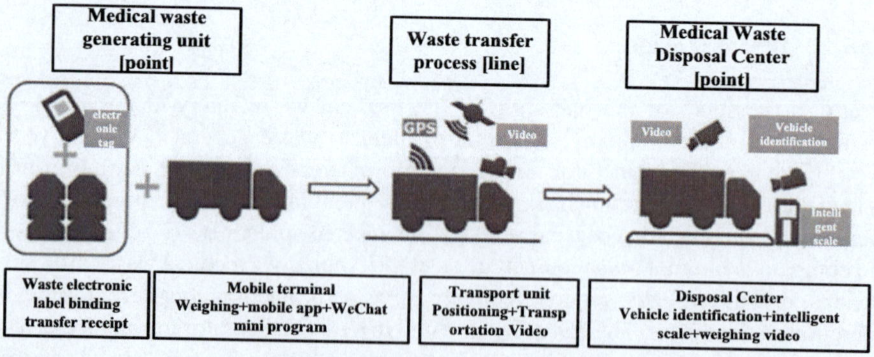

Fig. 7.12 Disposal centre "two points and one line" business process

The weighting of medical waste vehicles is a key link in the transfer of medical waste to the disposal centre, and all vehicles that transport solid waste must be strictly in accordance with the specifications for the weight machines and video recording. Weighing monitoring using high-definition entrance and exit video surveillance, gate control, weighing terminal to control the vehicle, to achieve the analysis of vehicle information in and out of the warehouse capture, the driver's face capture, etc. The main equipment list is shown in Table 7.3.

(3) Monitoring of the video surveillance network.

Video surveillance network monitoring includes medical waste in and out of the factory entrance, intelligent weighbridge, warehouse, and other key parts of the medical waste transfer vehicle license plate recognition, weighing process monitoring, weighing record monitoring, roof monitoring, import and export, loading and unloading areas, warehouse areas and other dead-end video surveillance, video surveillance information in the enterprise internal storage.

(4) Portable terminals

Disposal center according to the medical waste transportation, storage and disposal business management configuration of a certain number of mobile terminal equipment, mainly including handheld terminals and portable Bluetooth electronic scales. The main components of the portable terminal equipment are shown in Table 7.4.

(5) Process monitoring of disposal process conditions

Through the construction of the disposal process monitoring system, the process data focused on the medical waste from storage to disposal of all process management. Combined with the system using material balance, energy conservation of energy and other rules, it can automatically analyze the working conditions according to the rules of verification, accurate evaluation of the operational status of facilities, and the effectiveness of emission data. At the same time, to give alarm alerts, disposal centre managers can real-time control the state of the disposal process, and disposal emissions.

7.6.2.4 Effectiveness of Information Technology Operations

The information management of the medical waste collection, transportation and disposal process in the medical waste safety disposal centre of a southern city solved the technical backwardness problems in the medical waste collection, transportation and disposal network in the medical waste safety disposal centre. For example, the paper inefficiency of the collection and transportation process transfer record sheet, inaccurate collection quantity information and lagging feedback, inlet and outlet accounts, and disposal management information are not synchronized, disposal production management, and equipment operation cannot be real-time scheduling monitoring, low operating efficiency, and high operating costs. The following three

Table 7.3 List of main equipment for weighing and monitoring medical waste vehicles

Serial number	Product name	Technical parameters
1	5 series road gates	[Straight Rod] [Medium Speed] 5 Series Road Gates Includes 2 remote controls, 1 anti-smash radar; Gate type: straight bar; Gate direction: omni-directional; On the gate pole: 4 m; Running speed: 2.5–3 m/s; Features: anticollision, frequency, anti-condensation, manual lock gate function
2	Entrance and exit capture All-in-one	[Intelligent fill light capture one machine] [Deep learning] [Large angle recognition] Includes shield, lens, camera, 2 LED fill lights, etc. Video compression standard: H.264/H.265/MJPEG Storage function: built-in TF card slot Intelligent recognition: licence plate recognition, car model recognition, car logo recognition, body colour recognition; Fill light control: fill light automatic light control, time control Features: external gate, armed state can automatically control the external gate open/close according to the stored black and white list; vehicle capture: support licence plate, model, car tag, body colour recognition, electric zoom, automatic aperture, built-in LED fill light, synchronous fill light, synchronous video, black and white list control, video trigger
3	Column	[DS-TCG225 Matching Column] [DS-TCG227 Matching Column] Column height: 1.3 m; Column diameter: 60 mm; 1.3 m can be installed at the all-in-one machine, 0.5 m can be installed "four lines of LED display"

(continued)

Table 7.3 (continued)

Serial number	Product name	Technical parameters
4	Pan smart camera	Support three types of intelligent resources switching: face capture (default), road monitoring, and event analysis Face capture: support detecting, tracking, capturing, scoring and filtering moving faces, outputting the best face capture image, and detecting up to 30 faces at the same time; Road monitoring: licence plate capture, vehicle color, type, machine nonhuman identification; Event analysis: crossing detection, area intrusion detection, enter/exit area detection, wandering detection, people gathering detection, fast motion detection, parking detection, item left/pickup detection; Support ONVIF (profile S/profile G), ISAPI, GB/T28181, and E-HOME access; Support five-stream technology, dual HD, support 20 streams at the same time; Video compression standard: H.265/H.264/MJPEG
5	Second generation control terminal	Entrance/exit control terminal [with management software Processor: Intel Bay Trail platform processor; Memory: 4 GB; Features: with entrance/exit management software, fanless design, integrated switch, 485 interfaces, alarm 4-in/4-out, microphone input, video HDMI interface, 1 TB hard drive, 22-inch 1080p display, configured with keyboard and mouse kit, genuine Windows system

parts from medical waste collection, transportation, and treatment and disposal illustrate the effectiveness of information technology construction.

(1) Collection and transportation of electronic medical waste

Disposal centre using the Internet of Things, mobile terminals, and other technologies to achieve information management of the medical waste collection and transportation process. When medical waste collectors and hospitals are handed over, through the mobile terminal can quickly identify the type of medical waste, weight, and other handover information, and can quickly fill out the medical waste transfer information. Similarly, data are also uploaded to the central back-end platform, and the disposal centre can track the "movement" of these medical wastes in real time (Fig. 7.13).

(2) Real-time monitoring of the transfer and transportation process

The disposal centre through the information technology platform can obtain GPS positioning information for medical waste transfer transport vehicles in real time, abnormal warning of abnormal location, abnormal alarm history, transfer node

Table 7.4 List of main components of portable terminal equipment

Serial number	Product name	Technical parameters
1	Handheld terminals	Operating system: Android 5.0 or above; Processor: quad-core 1.2 GHz; Memory: RAM 2 GB, ROM 16 GB; Display: 5-inch capacitive screen, resolution 720 × 1280, support glove touch, support multi-touch; Charging time: <4 h; Working time: 10–12 h; Bluetooth: Bluetooth 4.0; Mobile communication: 4G full network access; Camera: front 200 W, rear 800 W; IP: IP65
2	Portable Bluetooth Electronic weighing scale	Maximum range: 100 kg; Indexing value: 10 g; Unit of measurement: kg, g, pounds; Scale table size: 30 cm × 40 cm; Power supply: DC5V1A; Built-in rechargeable lithium battery power supply; Display resolution: (3–300) millionths; Accuracy class: OIML class III; Inductor: resistance strain type C3 class, 30,000 divisions; Sensitivity to wet drift: $\leq 12 \times 10^{-6}$/°C; Communication interface: RS-232 serial output interface, baud rate 9600; Bluetooth function: standard 2.0/4.0 optional

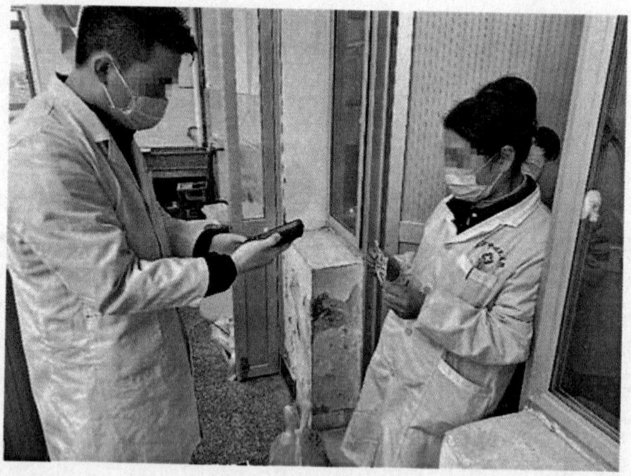

Fig. 7.13 Electronic Handover of medical waste collection and Transportation

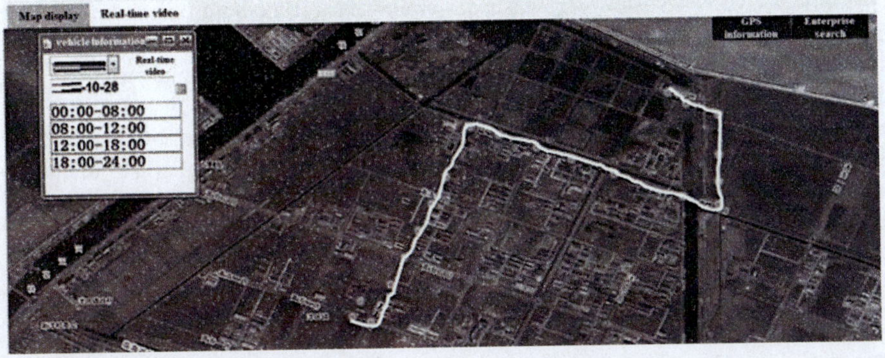

Fig. 7.14 Real-time monitoring of medical waste transport vehicles

information, etc., to help managers transport vehicles for real-time scheduling (Fig. 7.14).

(3) Weighing of medical waste into the plant

When medical waste enters the disposal center plant on the weighing scale, it can quickly identify the license plate number and correlate with the weight of the weighed waste through the camera, generating a list of vehicle record details ledger. Vehicle weighing records, video surveillance screenshots, and transfer orders and other information according to the time period, license plate number, etc. (Fig. 7.15).

(4) Regulation of medical waste disposal work conditions

Medical waste in the disposal process, through the work condition management real-time collection of process production equipment status, pollution control facilities

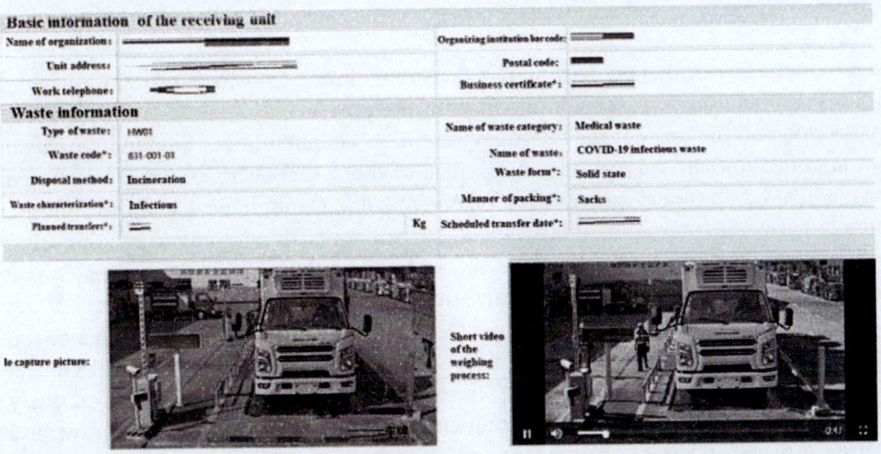

Fig. 7.15 Medical waste vehicle weighing record ledger

operation, as well as wastewater, waste gas emissions online monitoring, through the facility operation stop analysis, analysis of the limit, stop/limit analysis, process correlation analysis, etc., timely discovery of environmental management facilities, equipment, such as failure to open, idle, deceleration, frequency reduction, and abnormal shutdown and other cases of abnormal work. To provide data support for medical waste disposal management. For example, in the medical waste incineration process, through real-time data docking, the production of a working conditions map, a panoramic view of the medical incineration process, and various parameters recorded.

7.6.3 Application of Information Technology for Regulatory Agencies

7.6.3.1 Application Units

A provincial government department in the South was selected to carry out the application case for medical waste informatization supervision.

7.6.3.2 Construction Scale

A provincial government department in the south to carry out the province's medical waste disposal supervision, covering the province's prefecture-level cities, counties (districts) of medical institutions, medical waste transportation and disposal users, involving more than 50,000 medical and health institutions in the province.

7.6.3.3 Content of Information Construction

A provincial government department in the south of medical waste information technology applications closely combined with the province's medical waste disposal status quo and the current new coronavirus infection pneumonia epidemic medical waste collection and transportation disposal of the actual needs of the work, based on medical waste supervision and management work, through the use of the Internet, mobile network, Internet of things, video surveillance, cloud computing, and other information technology, comprehensive coverage of the province's medical waste from the generation, collection, transfer, treatment and disposal of all aspects of the collection, transfer, treatment and disposal of all aspects, and effectively enhance the regulatory capacity and efficiency of medical waste, the completion of the province's unified medical waste collection and transportation disposal system, and actively explore the province's medical waste refinement classification collection, control, and transport disposal services governance management pattern, the province's medical waste generation, utilization, transfer of dynamic visualization, refinement and intelligent management, and effectively enhance the medical waste regulatory capacity and efficiency.

(1) Cloud Service Infrastructure Construction

Under the overall structure of the "digital government" construction, the provincial medical waste informatization and supervision platform will fully rely on the "digital government" government affairs cloud in accordance with the principle of intensive construction and will migrate part of the external business to the public cloud, while the main core internal business platform may migrate to private clouds. Using the infrastructure and information security service capabilities provided by the government cloud, public and private cloud platforms can be consolidated, while ensuring the security and reliability of business data (Fig. 7.16).

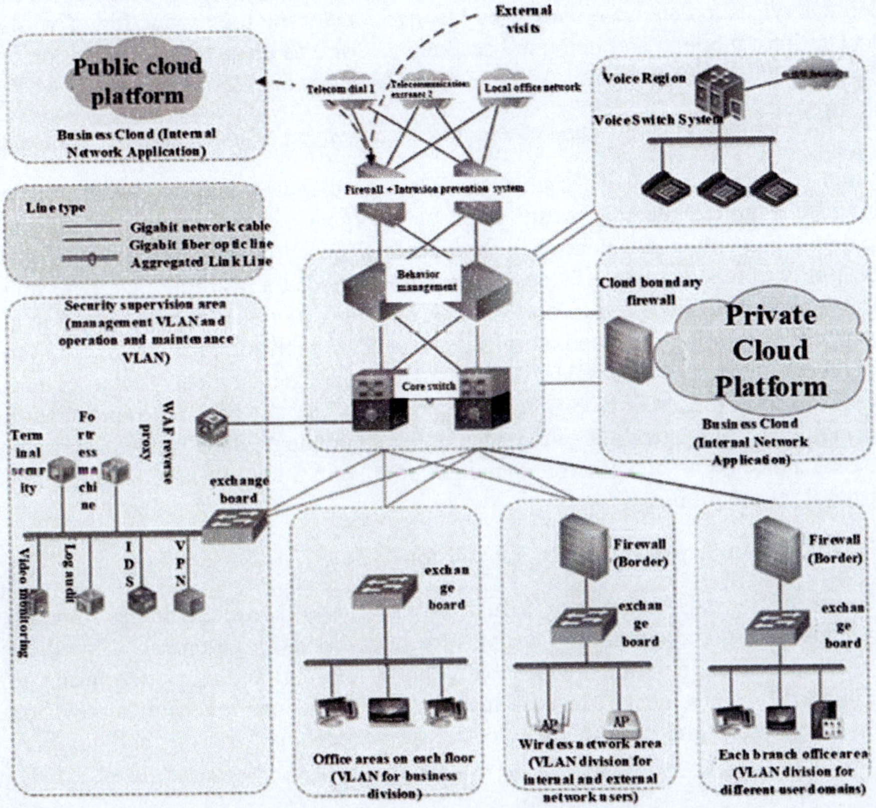

Fig. 7.16 Network topology diagram of the medical waste cloud platform

(2) Dynamic supervision of the whole process of medical waste Internet of Things

Make full use of the results of the construction of medical institutions and disposal institutions of information technology to build up a province-wide medical waste supervision “a network”. Full coverage of medical waste and all-weather data supervision, real-time medical waste collection, transportation, and storage, disposal of the entire process of data collection, dynamic tracking, and monitoring, so that government regulators can fully grasp the province’s medical health institutions of medical waste generation, recycling, and disposal.

(3) Refined management of medical waste prevention, control and disposal

According to the life cycle of medical waste supervision, the establishment of medical institutions, disposal institutions, archival management, comprehensive collection of medical waste generation, and disposal process of basic information, transfer information, ledger information. Medical waste-related institutions can be unified cloud portal services to achieve cloud-based business, including the production of waste declarations, inbound and outbound accounts, as well as disposal and utilization of the unit’s business declaration, and transfer of business one-door processing.

(4) Prevention and control and emergency management of medical waste disposal

Based on the principle of “visible, audible, commanding and cooperative”, the province’s unified medical waste disposal emergency management system and epidemic prevention and control of emergency disposal coordination mechanism, the ability to medical waste disposal threshold exceeds the warning, late reporting notice, and medical waste transport vehicles operating in violation of the warning to achieve risk warning information upload and send. Establishing the city as a unit to coordinate emergency disposal facilities resources, the establishment of a list of emergency disposal resources of pneumonia epidemic medical waste, with information technology to fully standardize the pneumonia epidemic medical waste, to prevent disease transmission and environmental pollution, and timely release of emergency disposal information.

(5) Big data, a map of integrated supervision

The use of big data technology, visual analysis, and other technologies, through in-depth mining and correlation analysis to show the various elements of medical waste supervision, combined with tables, charts, rich text and other rich forms, to facilitate the management of decision makers to carry out comprehensive supervision of medical waste and epidemic medical waste research and analysis and auxiliary decision making (Fig. 7.17).

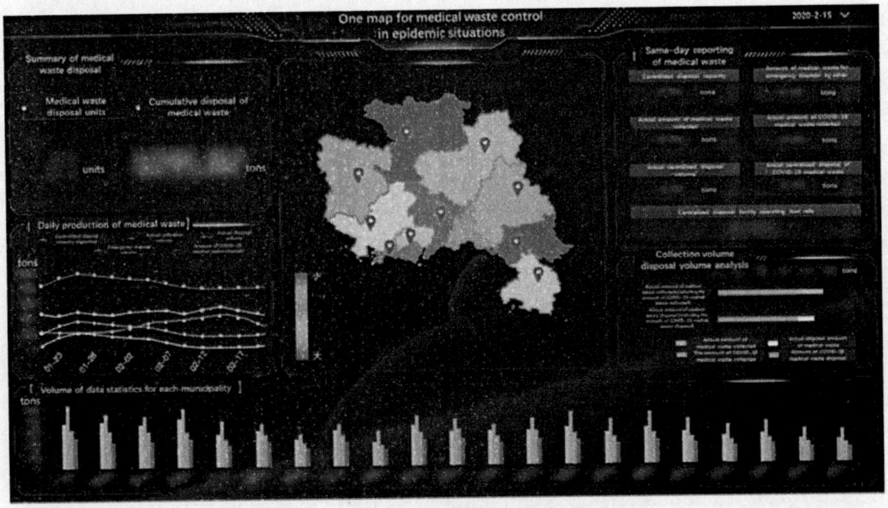

Fig. 7.17 Big data for medical waste a map of integrated supervision

(6) Medical waste prevention, control, and disposal "medical waste" mobile application

Based on the mobile terminal application, we have built the mobile application "Medical Waste", including two versions of WeChat applet and the APP application. There are various types of mobile applications integrated in a portal interface, single sign-on, unified user and permission management, to meet the medical waste in and out of storage management, transfer orders, and other application scenarios of mobile, electronic needs. Medical waste companies involved in the mobile terminal to complete the business declaration, the transfer of a single business function. Supervisory departments can complete business approval, daily supervision, and task handling functions in the mobile terminal, thereby improving the overall effectiveness of medical waste management.

7.6.3.4 Effectiveness of Information Technology Operations

A provincial regulatory authority through the information technology supervision of medical waste management, the realization of COVD-19 infection pneumonia provincial designated treatment hospitals, medical waste disposal units, emergency disposal units throughout the process, and real-time dynamic supervision, effective control of medical waste loss, leakage caused by human and environmental hazards. Through the information management system on line, real-time supervision and positioning of medical waste, paperless management, medical waste disposal throughout the traceable, to avoid the risk of handover errors and loss, leakage. All medical waste data to achieve information management, data uploaded to the regulatory platform, regardless of the link, its origin, weight, type, and handover personnel can be traced back to the source, effectively eliminating the risk of loss of medical waste. The use of cell phones, computers, "a map" to help fine supervision, the realization of data micro-collection, information management, decision support and many other functions effectively improve the regulatory data acquisition, analysis efficiency, and the response to intervene to guide the speed of service, to strengthen the collection, transfer, storage, disposal of medical waste supervision, and management of the whole chain. All data are integrated in a base map, visualized by spatial dimensions, a one-time centralized medical waste disposal capacity, the actual amount of medical waste collection, the actual amount of centralized medical waste disposal, the amount of COVD-19 medical waste disposal, other ways of emergency disposal of medical waste, etc., through the "one map" for medical waste during the epidemic.

During the epidemic through the "Medical Waste" WeChat app pointing fingers, medical waste-related enterprises can quickly upload the actual daily medical waste collection, epidemic waste collection, disposal facility load rate, as well as the emission port soot, sulphur dioxide, and other monitoring data. Reduce the variety of data and repeat the process. At the same time, the system background to achieve information query, data sharing, regulatory departments at all levels in the critical period can be the first to grasp the information, easy and fast to complete the statistical reporting and other work, timely intervention to do supervision, identify problems, and guidance services in place.

7.7 Future Prospects for the Development of Medical Waste Information Technology

In the current Internet era, the State Council is vigorously promoting the "Internet government services". In this context, relying on the Internet and big data, to strengthen the whole process management, in order to control the source, process supervision, end of the standard disposal, and other links to solve the problem of medical waste disposal management. In the future, the demand for the collection and disposal will grow more rapid growth. In this regard, the following recommendations are made for the next phase of the development prospects of medical waste:

(1) Establishing medical waste classification and collection of the Internet of Things standard specification system will be more conducive to the national push.

Collection and disposal using IoT technology to solve medical waste loss, false concealment, overtime collection and transportation, illegal transfer, data falsification and other regulatory issues. Considering that medical waste IoT must access complex heterogeneous related hardware and hardware access to a standard port protocol, to ensure the flexibility of the software system application to adapt to different front-end IoT facilities heterogeneous network integration requirements, which requires the establishment of a unified medical waste classification and collection of IoT standard specification system, with the help of IoT conversion module and conversion procedures, to provide a unified access standard, to achieve. The software system is scalable and reliable on IoT devices.

(2) To serve the society with the "Internet+" mode to realize the deep excavation and application of medical waste information.

The ultimate purpose of the management of medical waste data, and intelligent operation is to connect big data, and the use of big data, the region's medical waste for scientific, efficient, and environmentally friendly operation and disposal. The establishment of environmental protection information management system to adapt to the new era of ecological civilization, to further strengthen the value of ecological data mining and application, so that ecological and environmental supervision more accurate, ecological and environmental work more efficient, and ecological and environmental public services more convenient for the people.

(3) In-depth integration of the various sectors of medical waste regulatory model will be strict enforcement and innovative ideas at the same time.

The management of medical waste in China involves a number of departments, such as health and environmental protection. Horizontally, the health department focusses on medical waste management within medical and health institutions and the treatment, disposal of disease prevention and control; Environmental protection departments focus on medical waste treatment, disposal management, and related environmental pollution prevention. Health, ecological, and environmental departments at all levels should each do their own job, the formation of a clear division of labour linkage cooperation system, strengthening the attribution of responsibility, increase supervision, and inspection to ensure that medical waste is classified according to the provisions of classification, collection, transfer, storage, and harmless disposal. At the same time, actively innovate and explore new ideas of medical waste supervision and management, strengthen the classification of medical waste collection and disposal information sharing and coordination, comprehensively improve the standardized management of medical waste, and effectively prevent environmental risks.

(4) Strengthen emergency medical waste disposal capacity building and medical waste disposal.

At present, China's medical waste in the collection and disposal capacity, regulatory capacity and emergency response capacity, there are still shortcomings. To implement the requirements of the Party Central Committee and accelerate the shortcomings of medical waste collection and treatment, it is urgent to coordinate the construction of medical waste disposal facilities nationwide, establish a sound medical waste collection and disposal system that is compatible with the country's economic and social development, strengthen the collection and disposal of medical waste, key regional collaborative disposal, information technology to enhance the capacity of supervision and other aspects of capacity building. Strengthen the training and publicity of medical institutions, especially large grassroots medical institutions, and effectively improve the awareness of the main responsibility of medical health institutions medical waste pollution prevention, accurate dissemination of knowledge related to medical waste, the establishment of a training system including medical waste management personnel, clinical departments on duty, new staff in clinical departments and logistics cleaning staff positions, to strengthen medical staff's awareness of environmental protection and laws, consciously resist improper handling of medical waste, strengthen the management awareness of medical institutions, and help medical staff to consciously do a good job in the standardization of waste disposal. At the same time, through multiple channels and forms of publicity, to improve the environmental awareness of the public, the formation of the public to actively participate in the social synergy, common supervision, and management of medical waste.

References

Cao Yunxiao, Yu Xiaodong, Shan Shujuan, et al. Evolution of pollution prevention and control policies for medical waste treatment and disposal in China, analysis of existing problems and suggestions [J]. Journal of Environmental Engineering, 2021, 15(2). https://doi.org/10.12030/j.cjee.202012018.

Chen W. Application of medical waste management system based on cloud management--practice exploration in Yancheng First People's Hospital [J]. Science and Technology Communication, 2020, 12(14): 147–149

Geng W, Wu Xiaoyan. Requirements and authority division of security managers of classified information systems [J]. Information Security and Communication Security, 2009(7): 114–115.

Li Guojuan. Research on the application of mobile medical information system based on cloud computing [J]. Times of Agricultural Machinery, 2020, 47(6): 61–62.

Ni Chuanming, Lu Wenwei, Cai Ping, et al. Application of online traceability management system in medical waste supervision and management [J]. China Journal of Health Supervision, 2018, 25(2): 138–143.

Xuan Junfang, Chen Changgui, Li Ziming, et al. Design and application of medical waste information management system in our hospital [J]. China Modern Doctor, 2019(32): 134–137.

Yang Chunmei, Hai Ling, Chu Guiyang, et al. Design and application of intelligent medical waste management system [J]. China Digital Medicine, 2015, 10(5): 113–114.

Zhang Hongfang. Application of Internet technology in medical waste management [J]. Contemporary Nurses (Academic Edition), 2012(2): 184–185.

Zhang G. P., Zhou G. M. A review of hazardous waste management and its incineration treatment [J]. Energy Research and Information, 2003(3): 172–179.

Zhao Hui, Hu Shuyu. Research on the information management of the whole industrial chain of medical waste in the era of Internet of Things [J]. China Management Informatization, 2016, 19(5): 70–71.

Zhou W.-L. Construction and application of integrated management information platform system for environmental monitoring [J]. Green Technology,2016(22): 42–45.

Chapter 8
Preventive and Management of Pollution from Medical Waste Disposal

Nowadays, the standardized management of medical waste has become an important link in improving the environment and preventing pollution. This chapter will analyze the general idea of medical waste disposal in China and analyze the problems in the operation and management of disposal technology facilities. To eliminate the environmental safety hazards of the medical waste disposal process, we should be committed to strengthening the operation management and supervision of the medical waste treatment and disposal technology process, combining with the emergency needs of medical waste treatment and disposal in China, and improving the emergency management system for medical waste treatment and disposal.

8.1 The Starting Point and General Idea of Pollution Prevention and Control of Medical Waste Disposal

8.1.1 The Starting Point for Pollution Prevention and Control of Medical Waste Disposal

In the planning and construction of medical waste disposal facilities, China has implemented a centralized disposal, rational layout of the general idea (Zhang SC et al. 2003). China has implemented the idea of planning units for the construction of municipalities to build centralized medical waste disposal facilities to accept and dispose of medical waste in all counties within a reasonable transportation radius, and generally do not advocate or allow hospitals to disperse. In addition, on the basis of the construction of centralized medical waste disposal facilities to prefecture-level cities, cities in areas with developed transportation and dense towns are encouraged to jointly build and share centralized medical waste disposal facilities; at the same time, hazardous waste facilities and medical waste facilities should be integrated

Y. Chen et al., *Innovative Disposal Technology and Management Practice for Medical Waste*, https://doi.org/10.1007/978-981-99-6786-5_8

construction, hazardous waste centralized disposal facilities to deal with medical waste generated by the city; each county (city) have built medical waste collection transfer and disposal system to achieve full collection and treatment of medical waste above the county level, and gradually covered to the established towns, for medical waste in rural areas to get standardized disposal; population of more than 500,000 counties (cities) or medical waste collection and disposal of more than 5t county-level areas, the construction of incineration, high-temperature steam and other centralized medical waste disposal facilities; for remote grass-roots areas, the use of medical waste by mobile disposal and pre-treatment facilities is encouraged to achieve local disposal of medical waste. In terms of incineration technology, incineration has more advantages in the disposal of hazardous and medical waste (Chen Y et al. 2005). Although incineration can achieve harmless, reduction, and even resource recovery of waste, incineration also has its limitations, namely, emissions from the incineration process can produce environmental pollution, especially the incineration process produces dioxins as POPs substances in the category controlled by international conventions, and the incineration process is difficult to control, the operation and management process is complex, and the monitoring means are insufficient. These limitations bring more difficulties to the operation of the facilities and the implementation of supervision by environmental management departments.

In terms of non-incineration treatment technology, it was initially proposed as an alternative technology to incineration, and new medical waste disposal facilities were proposed to encourage the use of alternative technologies. However, the use of non-incineration treatment technology for the disposal of medical waste should also pay attention to pollution prevention issues, not only to consider the type of medical waste treatment, bacterial inactivation efficiency, technical reliability, automation level, environmental emissions and environmental impact, occupational safety and health and other basic factors, but also to consider the degree of recognition of policies and regulations, public acceptance, treatment space requirements and site selection, supporting public facilities and installation requirements and other basic factors to ensure the effectiveness of non-incineration treatment technology applications. Therefore, the following three aspects must be noted in the application of non-incineration treatment technologies: ① chemical waste, pharmaceutical waste and pathological waste that cannot be disposed of by non-incineration treatment technology should be separated out in the process of source classification, and other appropriate ways to manage and dispose off; ② strict control of non-incineration treatment process parameters control and treatment effect testing, to ensure that the non-incineration treatment of medical waste can reach the corresponding national standards below, no longer infectious; ③ non-incineration treatment of medical waste should also be properly managed, the implementation of standardised landfill or incineration disposal, not arbitrarily discarded; otherwise there is the possibility of reverting to the original characteristics of infectious waste.

After a comprehensive analysis, the domestic medical waste disposal facilities operation and management problems are mainly manifested in the end control, light process control; heavy administrative management, light technical management; heavy engineering construction, light capacity training. So that the operation of medical waste disposal facilities is difficult to achieve the technical performance of the operational management level of facilities; and the local environmental protection department to implement the supervision and management of the lack of knowledge and understanding of specific technologies and facilities, so that the supervision and management personnel difficult to understand the key aspects of supervision and management of medical waste disposal facilities, the content and the corresponding supervision and management methods. The existence of the above-mentioned problems brings hidden dangers to the environmental safety of medical waste incineration and disposal processes.

In fact, it is difficult for governmental regulatory authorities to achieve complete continuous monitoring of waste management facilities and to detect various violations at any time. However, a good medical waste management model can better promote the environmental behavior of enterprises operating the facilities and motivate them to actively participate in the management of hazardous waste and achieve the goal of safe medical waste treatment and disposal. That is, by taking priority measures, combined with specific process and technical characteristics, from the perspective of pollution control, a comprehensive clarification of the key issues in the process of medical waste disposal, comprehensive promotion of the concept of life cycle and whole process management, so that facility operators can understand more deeply; environmental supervision and management personnel to achieve supervision and management process and has rules to follow; so that the public know how the government and facility operators ensure that environmental protection facilities standardization and operation is of standard (Ministry of Ecology and Environment 2019). The realization of the above goals depends on the understanding of the key issues of medical waste disposal facilities operation and management and putting forward targeted solutions (Ministry of Ecology and Environment 2011).

8.1.2 General Idea of Pollution Prevention and Control of Medical Waste Disposal

(1) Emphasis on source classification, achieving source classification and end pollution control means to match.

The issue of medical waste management within medical waste institutions is crucial, It is related to the effective elimination of infectious medical waste infectious, reducing the harm to the human body, and the environment and different medical waste disposal technologies need to be implemented in medical institutions within the scientific classification. First of all, the implementation of medical waste reduction from the source, especially such as medical waste incineration and disposal of the source of chlorine source reduction control with foreign developed countries there is a certain gap, which is one of the main reasons why China's current hazardous waste incineration disposal tail gas management is more difficult to meet the standards. Although the standard level of medical waste bags and other materials put forward certain requirements, such as the use of polyvinyl chloride and other chlorine-containing materials to make medical waste bags. The field of hazardous waste management problems does not only exist in the medical waste of this waste, and on the other hand, due to China's slow progress in environmental law enforcement and material substitution, the lack of basic policy guidance, which in turn causes hazardous waste and medical waste and the lack of effective control of hazardous and medical waste reduction at the source. For example, the mixing of chemical and pharmaceutical medical waste into non-incineration facilities for treatment is outside the scope of non-incineration medical waste treatment facilities. As far as heavy metals are concerned, mercury is the most likely heavy metal in medical waste, and how to extract mercury in the process of source separation of medical waste is the key to promoting the end to achieving standard discharge.

(2) To process control as the core, promote process control and end control combined to promote the standardization of facility operation and management.

The core objective of the operation of medical waste disposal facilities is to implement medical waste disposal while preventing all types of pollutants generated during the disposal process, including air, wastewater and solid waste. For incineration technology, the top priority is the control of dioxins, and how to ensure that the corresponding process parameters, such as the incinerator temperature of 850 °C, the flue gas in the furnace residence time of not less than 1 s, combustion efficiency is greater than 99.99% and other critical parameters. For non-incineration technology, disinfection temperature, disinfection time, chemical dosage, etc., are essential to eliminate the infectious nature of medical waste. Medical waste incineration disposal facilities should be combined with their own technology and facility characteristics, process control as the core, taking into account the source classification and end of control measures, combined with the actual enterprise to establish and improve the rules and regulations, strengthen personnel training, do a good job of safety and emergency protection, do a good job of monitoring pollutant emissions at the disposal facility and environmental monitoring records around the disposal unit, and assessment work to ensure that medical waste disposal facilities standardise operation and management.

(3) Performance control as a guarantee to carry out performance evaluation of medical waste incineration and disposal facilities, to promote the high level of stable operation of disposal facilities.

International experience shows that to regulate the operation and management behaviour of medical waste disposal facilities, exploring a practical performance evaluation method has gradually become an inevitable choice for the management and disposal in China. To determine whether a facility meets emission standards and is suitable for a particular type of waste disposal, a systematic evaluation is required, taking into account the characteristics of the waste and the working conditions of the facility. Therefore, the scientific and systematic verification of the performance of a set of incineration facilities, from the waste characteristics, facility performance, pollutant emission performance and facility operating parameters of the whole process of facility performance testing and evaluation methods, taking into account the waste characteristics indicators, system performance indicators, flue gas emission indicators, equipment operating parameters of the organic combination of four types of indicators, taking into account the medical waste in the boundary conditions and normal conditions of both the situation, the standard waste as a reference for the performance testing and evaluation of disposal facilities to evaluate the disposal facilities to determine the specific performance of the disposal facilities. After evaluation, for facilities that meet the standards, the waste characteristics, working parameters, and main operating parameters should be clarified in the case of standard emissions; for the facilities that do not meet the standards, the facilities need to be modified and retested and evaluated. Therefore, different types of disposal processes should be combined with the implementation of targeted performance testing to provide a basis for mapping the performance of specific facilities, the issuance of permits, and the implementation of the corresponding environmental regulation.

(4) Supervision and management as a means to establish a standardized approach to promote the scientific and standardised supervision and management process.

Local ecological and environmental administrative authorities taking effective measures to promote the safe operation of the facility from the perspective of supervision and management is also the key to achieving pollution control. The competent ecological and environmental administrative departments can implement supervision and management of the operation of centralized medical waste disposal facilities through written inspections, on-site verification and remote monitoring; local ecological and environmental administrative departments should collect information according to actual needs before implementing supervision and inspection. It should be carefully studied in the centralized medical waste disposal unit to provide relevant materials or technical assessment reports on the basis of determining the focus of the inspection, content and evaluation criteria, and the preparation of supervision and inspection implementation plan, the establishment of a standardized system for the supervision of the operation of centralized medical waste disposal facilities file management; according to daily supervision and information reported by the public, for the basic operating conditions of medical waste disposal units, disposal facilities operational process, pollution prevention and control facilities configuration and operational effectiveness, as well as safety and labor protection measures to check. Among them, the most critical is to combine the characteristics of different medical waste disposal facilities, effectively around the life cycle and the concept of whole process management, from the perspective of risk control, pollution prevention, and clean production combined to promote the realisation of centralized medical waste disposal facility supervision, management standardisation, and institutionalization.

8.2 Transportation, Storage, and Receipt of Medical Waste

8.2.1 Baseline Management of Medical Waste Transportation, Storage and Reception Management Basis

Medical waste storage involves the storage of medical waste generating units and medical waste transported to the disposal of storage in place after two cases. Medical waste generation unit temporary storage facilities should be in accordance with the “Technical Specifications for Centralized Disposal of Medical Waste” requirements for construction, medical waste disposal unit storage facilities should be in accordance with the “Centralized Incineration of Medical Waste Disposal Engineering Construction Technical Specifications”, “Medical Waste High Temperature Steam Centralized Treatment Engineering Technical Specifications”, “Medical Waste Microwave Disinfection Centralized Treatment Engineering Technical Specifications”, “Medical Waste Chemical Disinfection Centralized Treatment Technical specifications”, “centralized medical waste disposal technical specifications” and other requirements for construction. The operation and management of medical waste storage facilities should be in accordance with the “Technical Specifications for the Centralized Disposal of Medical Waste” implementation. The operation and management of storage facilities for medical and health institutions should be in accordance with the implementation of the “Medical Waste Management Regulations” and the “Medical Waste Management in Medical and Health Institutions”. The above provisions provide a legal and technical basis for the implementation of the construction and operation of medical waste storage facilities.

Medical waste transportation should be fully considered in the risk avoidance delivery process, and the configuration of the medical waste transfer vehicle should be in accordance with the implementation of the “Medical Waste Transfer Vehicle Technical Requirements” (GB 19217–2003). Medical waste transport should be in accordance with the “rules for the transport of dangerous goods by car” (JT 617–2004) and “medical waste centralized disposal technical specifications” implementation. Medical waste transport should use special vehicles, and the vehicle compartment should be separated from the cab and sealed; compartment should meet the airtight requirements, smooth and flat interior, easy to clean and disinfect; compartment materials should be waterproof and corrosion-resistant; compartment bottoms to prevent liquid leakage, and set up a drainage collection device for cleaning sewage. Medical waste transport, transport vehicles and swing containers in accordance with the “Technical Specifications for the Centralized Disposal of Medical Waste” and the relevant provisions of the implementation.

8.2.2 *Medical Waste Transportation, Packaging Reception, and Storage Management Requirements*

8.2.2.1 Transportation Requirements

According to the "Medical Waste Management Regulations", the centralized medical waste treatment unit shall be responsible for the delivery process of medical waste. According to the 'Hazardous Waste Management Licence Management', China's hazardous waste management licence according to the mode of operation, is divided into hazardous waste collection, storage and disposal of the integrated business license and the hazardous waste collection business license, which received a hazardous waste collection business licence, units can only be engaged in maintenance activities generated in the mineral oil and residents of daily life generated in the waste cadmium nickel batteries Hazardous waste collection and operation. The transport of medical waste can only be carried out by a centralized disposal unit, which obtains a comprehensive business license. In general, medical waste is mainly transported by land vehicles, and the vehicle must comply with the "Medical Waste Transfer Vehicle Technical Requirements" of the special vehicle. The vehicle is in the stereotypical car two modified chassis insulation car, refrigerated car on the basis of the appropriate transformation, for the transfer of special medical waste special trucks. Delivery vehicles must be in front and rear of the vehicle, and both sides of the vehicle place special medical waste warning signs; cab both sides of the sprayed medical waste disposal unit and delivery vehicle number. The special vehicle should not transport other items, such as the need to change to other uses, should be thoroughly disinfected, and obtain the consent of the environmental protection department. If you cancel the vehicle delivery vehicle number, it should be in accordance with the provisions of public security traffic management to reapply for the vehicle use change procedures.

Medical waste disposal units according to the total disposal programme, equipped with a sufficient number of delivery vehicles and standby emergency vehicles. Transport units (for medical waste, that is the disposal units) equipped with delivery vehicles to meet the waste generation delivery cycle and reasonable spare emergency vehicles. Delivery vehicles should be equipped with: "hazardous waste transfer coupons" (medical waste special), "medical waste delivery registration card", delivery route map, communication equipment, medical waste generation units and their management personnel list and telephone, accident response plan and contact units and personnel list, telephone, tools for the collection of medical waste, disinfection equipment and drugs, spare medical waste special bags, sharps boxes and personnel protective equipment.

(1) Frequency of delivery

For medical and health institutions with inpatient beds, the disposal unit should send a car to collect medical waste daily, and medical waste should be daily production; for there cannot be daily production of medical and health institutions without inpatient beds, such as outpatient clinics, clinics, disposal units should collect medical waste at least once every two days (Ministry of Ecology and Environment 2020).

(2) Transportation routes

Meet the requirements of laws and regulations to ensure safety. Avoid peak periods of pedestrian traffic, busy areas, traffic congestion, and as far as possible at night. Avoid duplication of routes and use circular routes as much as possible to ensure the economy. The collection time and the collection volume are considered in an integrated way, and the specification of the transport vehicle is equipped with the amount of waste generated on the route. Transport processes, as far as possible, do not occur in the second transfer. The planning of medical waste route must be based on the geographic location of the disposal centre, the regional scope of services, the geographical distribution of health care units, the size of each health care unit and the amount of medical waste generated, the distribution of transport time, traffic routes, road conditions, and other circumstances to be considered.

(3) Transporters

Medical waste management in disposal plants involves medical waste transporters, operators, and managers. Due to the different nature of their work, the responsibilities assumed in the process of medical waste disposal are different, so the training of all types of personnel at all levels should be targeted, so that they are clear about the importance of medical waste disposal, master the correct treatment of medical waste, do a good job of self-protection, training of professional skills and safety protection of transport personnel, and to achieve a certain level of professional skills.

(4) Safety requirements

It requires a person in charge. Medical waste disposal units should be designated for each delivery vehicle responsible for the medical waste delivery process. The medical waste transfer vehicles should ensure that the car doors are locked to make sure that the transport process of medical waste will not be lost, spilt, or open the package to remove medical waste. Medical waste in transit is not allowed to be transferred, and medical waste and its crates are not allowed from a transfer vehicle to another transfer vehicle to prevent accidents. Prohibit the dumping and stacking of medical waste in non-storage locations. Medical waste transfer vehicles must be dedicated, not loaded or mixed with other goods and plants and animals in addition to medical waste, transported without other unrelated people. Prohibit mailing of medical waste. Prohibit the transport of medical waste by rail, air and land access and also, prohibit the transport of medical waste by water; no land access must be transported by water medical waste, and should be approved by the municipal government above the ecological environment administrative department, and take strict environmental protection measures before transport by water. Prohibit medical waste and passengers from using the same means of transport. Prohibit the transport of medical waste on the water bodies of protected drinking water sources.

(5) Execution of transfer coupons

Centralised medical waste disposal units and waste transfer from each medical and health institution should be implemented using the transfer coupon system. Medical and health institutions for the first time to the centralised disposal unit before the transfer of waste should be proposed to the municipal environmental protection department transfer plan; the municipal environmental protection department approved the transfer plan, medical waste units and disposal units of daily medical waste handover can be used to simplify the "hazardous waste transfer coupon" (medical waste specific) specific implementation. "Hazardous Waste Transfer Form" (for medical waste) in duplicate, one per month, by the disposal unit, medical waste delivery personnel and medical waste management personnel of medical and health institutions to fill out the handover, medical and health institutions and disposal units are kept, with the retention time of 5 years.

8.2.2.2 Packaging and Receiving Requirements

(1) Packaging requirements

According to the "medical waste management regulations", "medical waste management in medical and health institutions" and other provisions, medical waste classification and packaging should be completed by medical institutions. Packaging medical waste bags, sharp boxes, and crates (barrels), in terms of materials, appearance, specifications, physical and mechanical properties to meet the "medical waste special bags, containers, and warning signs standards" requirements. Packaging to use the appropriate packaging and containers, damaging waste to be packaged in sharp boxes, cannot be packaged in bags, nor allows the use of double bag packaging but other waste can be packaged in bags. When transporters accept medical waste, they have the right to request health care institutions to repackage, identification, and contain medical waste in the crate if medical waste does not meet the requirements of packaging and identification or is not contained in the crate.

As the mechanism of action of high temperature steam and other treatment technologies is to kill pathogenic bacteria through steam heat transfer, the presence of bags and their materials will, to some extent, affect the effect of steam treatment. China has introduced regulations on medical waste bags for "medical waste special packaging, container standards, and warning labels regulations". Provisions require the use of bags that can be made of polyethylene (PE), and the strength of the bag, volume and other requirements, but no specific treatment technology for the bag should have the corresponding requirements. However, it should be noted that the bags currently used in China are prone to adhesion to the cavity of the processing equipment (such as the use of high-temperature steam treatment technology), which usually requires anti-adhesive treatment of the cavity wall within the sterilization chamber of the processing equipment. In overseas, they are specifically for high temperature steam treatment technology and the use of bags, which can simultaneously meet two opposing requirements: In the waste transport and feeding process, the bag must be absolute leak-proof; in the steam treatment process, the bag must be permeable, so that steam can enter the bag and direct contact with the waste. In fact, this bag has two layers, respectively, composed of different densities of polypropylene material; the inner layer is not a heat-resistant sealing layer, which prevents the entry of gases and liquids; the outer layer is a porous, permeable layer, which can withstand high temperatures of 140 °C. At the beginning of the treatment cycle, the inner layer starts to melt and the steam penetrates directly into the waste through the porous outer layer. However, its high cost limits its application.

(2) Receiving requirements

The operating units of medical waste treatment facilities should check the waste received and the consistency of the scope of operation specified in the business licence to eliminate the receipt of hazardous waste beyond the scope of operation. Operating units of medical waste non-incineration treatment facilities should be configured to measure the system, and should have weighing, recording, transmission, printing, and data processing functions. Medical waste non-incineration treatment facilities should be equipped with radioactive waste detection and alarm equipment to prevent radioactive waste in non-incineration treatment facilities.

8.2.2.3 Storage Requirements

The temporary storage site for medical waste should be designed in the main plant and the waste contained in plastic bags and transfer boxes should be sorted and placed in a separate place. The design size of these storage places should be appropriate for the amount of waste generated and the frequency of collection. The main requirements include: ① medical waste unloading site, temporary storage, storage cold storage and other operations and management should meet the "hazardous waste storage pollution control standards" of the relevant requirements; ② should ensure that the medical waste storage temperature meets the relevant requirements of medical waste storage; ③ medical waste storage facilities should be closed and slightly negative pressure state, and ensure that the collected gas is purified and discharged; ④ medical waste storage facilities observation window should be used to regularly observe the waste storage conditions.

8.2.2.4 Cleaning and Disinfection Requirements

Medical waste disposal units should be set up to wash and disinfect the facilities for medical waste transfer vehicles, crates, and other medical waste delivery tools for cleaning and disinfection. Prohibit the washing of medical waste transfer vehicles in social vehicle washing places. Through strict disinfection and sterilisation measures and systems to achieve the purpose of disease control and environmental protection in all aspects, to ensure that no cross-contamination occurs.

Medical waste transfer vehicles should be disinfected using ultraviolet light, an ultraviolet disinfection intensity of 30 mJ/cm^2, irradiation time of not less than 30 s, or use the equivalent disinfectant disinfection capacity, disinfection after rinsing with water. The crates containing medical waste must be disinfected before reuse. If the use of chlorine disinfectants, the chlorine content of the disinfection solution should not be less than 500 mg/L, continuous contact or spraying time of not less than 30 min; if the use of ozone disinfection, the ozone content of the disinfection solution should not be less than 20 mg/L, continuous contact or spraying time of not less than 15 min. Crate disinfection is complete, it is also appropriate to use clean water for cleaning. The design of the cleaning and disinfection process should be based on the principles of easy operation, economics, and reasonable. Metal items such as vehicles should be cleaned with water after contact with disinfectant to avoid corrosion of equipment and vehicles.

8.3 Management of the Operation of Medical Waste Disposal Facilities

8.3.1 Medical Waste Pyrolysis Incineration Disposal Facilities Operation Management

8.3.1.1 Management of Feeding System

Since medical waste is sorted and packaged by the medical department, the incoming form of medical waste can be basically considered as solid. The incoming medical waste is solid and wrapped using sorted bags in crates. The packaging of incoming medical waste is not uniform, but has a difference in size. Also, because medical waste is not compacted when sorted and collected, the feed can appear puffy, and sometimes the feed needs to be pushed. Medical waste-feeding devices generally use two-stage feeding gate feeding devices. This device is airtight and it can reduce the instability of combustion caused by the large amount of air entering the furnace when feeding. The feed furnace has two doors, the first for the opening door and the second for the gate, also known as the fire door. General feeding, open the door, waste will be sent into the feed tank; at the end of feeding, close the open door, open the second gate, the push rod will push the waste into the furnace, and then the gate is closed, and the push rod is restored, then open the door, and start feeding.

Operators must wear the necessary protective equipment, such as helmets, gloves, protective shoes, and protective clothing, when feeding the material. Medical waste is bagged when it comes into the plant, so it can only be input in batch mode in drums, which makes it difficult to control the stability of the combustion situation.

As any object in the incinerator must go through the process of heating, volatilisation, combustion, etc., volatilisation and combustion rate directly affect the stability of the combustion chamber. If the content of volatile substances in the barrel is high, the material in the high-temperature furnace, will suddenly volatile combust in a short period of time, resulting in local overheating or the risk of rapid temperature rise. Due to the simultaneous combustion of excessive organic vapours, it is difficult to increase the oxygen in the furnace in a short period of time and the consequences of incomplete combustion can occur. Therefore, in the input of lump or barrel waste, the air input should be adjusted to increase to the maximum capacity, the input of liquid waste is reduced, while the temperature is reduced to the normal operating conditions or the low limit permitted by the specifications.

General incineration plants are based on the furnace type, exothermic rate, and their own experience, to establish a set of laboratory drums (or containers) with the waste calorific value, volatile content, and maximum limits of the flammable substances. At the same time, the containers are classified according to calorific value and flammability, and then based on experience to establish the input rate of different categories of waste specifications.

Medical waste incinerators must be equipped with an emergency feed cutoff system so that when operating conditions are not controlled, the feed can be automatically stopped to avoid equipment damage or environmental pollution. In addition to testing before operation, regular tests should be performed during normal operation to ensure proper operation of the system.

8.3.1.2 Operation and Management of the Incineration System

Temperature, residence time, oxygen concentration and the degree of air-waste mixing are the main factors affecting combustion efficiency. These four factors are not independent variables, but affect each other. The higher temperature, of course, can increase the rate of combustion, but the gas expansion due to heating, the residence time of gas expansion due to heating will be reduced; air input is large, can increase the supply of oxygen and the degree of mixing, but will reduce the residence time, and due to the limitations of the exhaust treatment system, resulting in lower treatment capacity.

1. Input control of waste

The rate of medical waste is the most important factor affecting the operation of the incinerator; therefore, as long as the appropriate hydraulic pressure and the burner or nozzle piping are kept open, the input can be continuous. The combustion conditions and temperature in the combustion chamber are relatively easy to control because of the more uniform mixing and little change in the calorific value. As any object into the incinerator must be heated, volatile, combustion and other processes, volatile and combustion rates directly affect the stability of the combustion chamber. Operation of the rotary kiln incinerator should first change the speed control from automatic to manual, and then adjust the speed down. If the waste input still continues to fall, that is, the calorific value of the waste in the barrel and volatility are very low, not to cause overheating or the risk of rapid combustion, you can gradually increase the amount of high-calorific value waste liquid or auxiliary fuel input of high-calorific value to maintain temperature stability.

2. Monitoring and maintenance of operating conditions

The proper operation of the incineration system is determined by the values displayed by the measuring instruments installed in the main equipment (e.g., indicators or detectors for temperature, pressure, flow rate, oxygen, and carbon monoxide concentration in the flue gas).

The combustion temperature of the incinerator must exceed the minimum temperature sufficient to destroy the waste to achieve the purpose of incineration. The temperature of the furnace wall and combustion gases should be kept stable to avoid damage to the refractory bricks due to overheating or thermal shock, not only because the maintenance of refractory bricks is the largest expense in the operation of the incineration system, but also the main cause of incinerator downtime. Even if the temperature remains stable, the refractory bricks can be reduced in thickness or spall due to friction, failure of adhesives, corrosion from alkaline metal hydrochloric acid in the waste or hydrogen fluoride from fluoride combustion. The easiest way to check is to observe the external equipment of the incinerator at night. If the external equipment is hot, it means that the refractory bricks inside the part have been seriously damaged or spalled and must be stopped for repair. The operator can also use infrared temperature telemetry, every shift regularly tests the temperature of the incinerator peripheral whether it is overheating, and some places even use the infrared scanner with computer connection for long-term detection and recording of the incinerator peripheral surface temperature. The flame should be kept in the incinerator at all times and a flame detector should be installed in the furnace for long-term monitoring.

3. Incinerator operation

(1) Basic requirements: Incineration system operators must be familiar with the disposal plan, operating procedures, incineration system processes, pipelines, equipment functions and locations, as well as emergency situations. Medium and large incinerators have device control systems, and the operation of various equipment should be automatic, so the main job of the operator of the incineration system's is to maintain the stability of operating conditions and to find and deal with abnormalities. The operator should monitor or adjust the operating reference values of the system (temperature, pressure, flow rate, etc.). If an abnormal condition occurs, the operator should determine the cause and promptly solve the problem based on basic mechanical or physical principles. Prior to formal waste incineration, it is advisable to run the system on fuel heat to train the operator to handle hypothetical abnormal conditions under non-automatic control.

(2) Incinerator start-up: Check the main equipment, instrumentation, interlock system, and emergency stop system. The incinerator start should first be blown

into the furnace with air, while opening the flue gas treatment system, after its stable operation, start heating the incinerator.

The incinerator must be heated slowly to extend the use of the refractory material in the furnace. When heating, the use of auxiliary fuel, heating rate and curve should follow the recommendations of the equipment manufacturer. When heating the rotary kiln incinerator, the rotary kiln and the secondary combustion chamber should be heated at the same time, and the exhaust gas temperature of the secondary combustion chamber should be used as a reference. When a new furnace or a new masonry in the furnace has not been calcined at high temperatures, it must be operated according to a strict heating curve. The refractory bricks in the rotary kiln are not actually combined with the shell, and the refractory bricks in the kiln are supported on the surface of the kiln due to heating expansion, and if the heating procedure is not appropriate, the refractory bricks are easy to spall. At the beginning, solid fuel such as wood can be used to bake and after reaching 300–400 °C, liquid auxiliary fuel burner is started for heating. When the temperature of the kiln head reaches 200 °C, the driving device should be turned on to start the kiln body rotation, and the kiln body should be rotated by 30° every 0.5–1 h. When the temperature of the kiln head reaches 500 °C, the kiln body should be rotated continuously, and its rotational speed should be greater than 0.1 r/min.

Regardless of whether it is a new furnace, each cold start must first raise the temperature to 500 °C, and then maintain 6–8 h after the scheduled operating temperature has be raised. Incinerator refractory by high temperature calcination after each start, you can start the liquid auxiliary fuel burner heating. At the beginning of the heating, at the same time, the kiln body continuous rotation, its speed should be greater than 0.1 r/min.

A fixed pyrolysis furnace, mechanical pusher grate type and other furnace types, should be configured with a secondary combustion chamber. The starting chamber should be heated at the first and the second combustion chambers, but the second combustion chamber exhaust gas temperature should be used as a benchmark. A new furnace or new masonry refractory furnace is not calcined at high temperatures, according to the strict heating curve operation. Each cold start furnace must first raise the temperature to 500 °C and then maintain it 6–8 h later, before upgrading to the intended operating temperature.

The auxiliary fuel burners should be able to stop supplying oil combustion, maintain the continued supply of air, prevent high temperatures in the furnace burner front parts.

(3) Incineration operation: The incineration process, temperature, residence time, oxygen concentration, and degree of air and waste mixing are the main factors affecting the combustion efficiency; these four factors are not independent parameters but the influence of each other. High temperatures, although it can increase the rate of combustion, but the gas expansion due to heating, its residence time will be reduced; air input is large, can increase the supply of oxygen and the degree of mixing, but will reduce the residence time, and due to the limitations of the tail gas treatment system, resulting in the reduction of treatment

capacity. Incinerator operation can be based on the type of furnace, the type of waste and its own operating experience, choosing the appropriate combustion conditions.

The combustion temperature of the incinerator must exceed the minimum temperature to ensure waste destruction to achieve the purpose of incineration. The temperature of the furnace wall and combustion gases should be maintained stable to avoid damage to the refractory bricks due to overheating or thermal shock, because the maintenance of refractory materials is an important factor affecting the cost of incineration, and is also the main cause of incinerator downtime. Even if the temperature is kept stable, the refractory material can be reduced in thickness or spalled due to friction, failure of adhesion, corrosion from burning alkaline metals, chlorine or fluoride in the waste, and other factors. The temperature inside the furnace is controlled by adjusting the air distribution and the auxiliary fuel burner. By adjusting the feed rate, rotary kiln to adjust the kiln speed, pusher grate type incinerator to adjust the pusher action time and to control the residence time of waste in the furnace. The general incineration system should be set up with the incineration temperature, waste incineration residence time, and other major process parameters of the automatic control system, set the value of the parameters, and achieve automatic control. The operator should monitor the operating conditions, and the normal operation of the incineration system is based on the installation of the main equipment measuring instruments (such as temperature, pressure, flow, oxygen in the flue gas, carbon monoxide concentration indicators or monitors) to determine the value displayed.

The tail gas outlet temperature of the second stage furnace is an important process parameter, which should be networked with the environmental management department to achieve real-time display and storage records.

The temperature of the kiln wall should be checked regularly during the operation of the rotary kiln to ensure that it is below the limit. Other types of furnaces should also regularly check the temperature of the furnace wall to ensure the safe operation of the equipment. If the temperature of the furnace wall is too high, it means that the refractory bricks inside the part have been spalled or damaged seriously, and must be shut down for repair. Infrared temperature telemetry can also be used to regularly test the temperature of the furnace shell every shift to see if it is too hot, and some systems use infrared scanners connected to a computer to detect and record the temperature of the incinerator shell surface over time.

The incinerator should always maintain the flame inside the furnace, and the furnace should be installed in the flame detector for long-term monitoring.

It is strictly forbidden to burn waste in states where the bypass flue is open and the flue gas is not treated. The emergency vent must be opened in case of an accident, and the feeding must be stopped before opening.

(4) Slag removal and transportation: Before feeding the waste, inert material or incinerated ash is sent into the furnace, check whether the residue removal system is operating properly. The incinerator should maintain the appropriate amount of residue; residue can form a protective film in the furnace bed to buffer

waste into the furnace after combustion generated by the thermal shock, and to prevent material adhesion to the furnace wall, while the solid waste can be used as a medium for heat transfer.

The discharge slag should take mechanical and other continuous automatic transmissions and should not use an artificial method of transmission. Water seal slag or water flushing slag should be equipped with a slag water separation device, the water content of the slag should be lower than the limited value specified in the process. The dry slag method should be through spray water heat absorption and to avoid dust ash slag.

The handling of incinerated fly ash should include fly ash collection, conveying, packaging, temporary storage, etc. The fly ash handling system should be kept airtight. The fly ash handling operation should be mechanical or pneumatic way, and not manual. Should often check the collection of fly ash with ash storage tanks, timely removal, and packaging.

Incineration slag can be sent to a domestic landfill. Slag can be sent to the domestic waste landfill as ordinary waste.

(5) Accident situation and contingency measures. Occasionally, malfunctions can occur during the operation of the incinerator, and Table 8.1 lists a number of malfunctions and contingency measures for reference.

8.3.1.3 Operation and Management of Waste Heat Utilization System

The waste heat utilization boiler in an incineration system must consider issues such as: dust characteristics and content of the incineration exhaust gas, wear and corrosion issues, scale build-up and scale removal, waste calorific value changes, operating temperature of the incineration, and steam utilization methods.

The classification of waste heat utilisation boilers can be classified according to the type of fluid in the tube, furnace water circulation mode, heat transfer mode and structure configuration. According to the type of fluid in the tube, the boiler can be divided into two types namely; flue tube type (or called fire tube type) and the water tube type. The so-called flue tube, which is the boiler heat transfer tube fluid for combustion gases; and the water tube, which is the boiler heat transfer tube fluid for water. According to water circulation, the boiler can be divided into natural circulation, forced circulation and cross-flow circulation. The principle of the natural circulation type for the tube of furnace water heated into a mixture of steam and water, making the fluid density decreases, forming a rising tube, and saturated water, due to the greater density, flowing from the top to the bottom of the tube, forming a descending tube, in the descending tube and rising tube between the two because of the density difference between the natural circulation flow, known as the natural circulation type boiler. The lower the pressure of the boiler, the greater the density difference between saturated water and saturated steam, the better the effect of the water circulation, so the type of natural circulation type is widely used in low and medium-pressure boiler systems. Forced circulation boiler furnace water circulation

Table 8.1 Operational malfunction of the medical waste incineration system and contingency measures

Serial number	Abnormalities	Types of incinerators	Indicator signals for malfunction	Response measures
1	Part (or all) of the liquid waste input is interrupted and the feed is stopped	Fixed bed incinerator	Flow metre indication out of range; pipe blockage; temperature in combustion chamber dropping; feed pump stopped running	Find the cause of the malfunction; add additional auxiliary fuel to maintain temperature; continue to maintain the operation of the exhaust treatment system
2	Waste liquid from a particular burner	Fixed bed incinerator	Flow meter indication out of range; pipe blockage; lowered temperature in combustion chamber feed stop	Stop waste liquid infeed
3	Rotary kiln feed suspension for some or all solid waste	Fixed bed incinerator	Decrease in temperature in the combustion chamber; misconduct of the solid feed system	Find cause of the malfunction; add auxiliary fuel to maintain temperature; continue to maintain operation of the exhaust treatment system
4	Black smoke escapes from the combustion chamber (unstable combustion conditions or poor air tightness)	Fixed bed incinerator	Differential pressure change; black smoke escapes	Stop solid waste feeding for 10–30 min, but continue to maintain the furnace temperature and combustion; staff quickly evacuate the site of the disorder; assess the characteristics of the waste before feeding
5	Forced air supply to the burner is aborted	Rotary kiln incinerator	Flow metre indication out of range; warning signal from automatic flame detector; primary fan malfunction	Stop the waste feed promptly; examine the cause of the malfunction; continue the operation of the exhaust treatment system, but reduce the air extraction volume

(continued)

Table 8.1 (continued)

Serial number	Abnormalities	Types of incinerators	Indicator signals for malfunction	Response measures
6	High combustion temperature	Rotary kiln incinerator	Temperature indication signal; high temperature warning signal	Check that the fuel and waste input is normal; check the temperature indicator sensor; check that the temperature indicators at other locations are also changing; open the emergency vent on top of the combustion chamber
7	Burning temperature is too low	Rotary kiln incinerator	Temperature indication signal; high temperature warning signal	Check for low fuel and waste input; check the accuracy of the temperature sensor
8	Refractory brick spalling	Fixed bed incinerator	High noise is generated; the temperature of the combustion chamber decreases, the amount of dust increases, and the furnace wall overheats	Shutdown
9	The coughing exhaust blackness increase	Rotary kiln incinerator	Visual detector indication exceeds safe operating limit	Check the combustion situation, O_2 and CO detectors; check the exhaust treatment system; check whether the waste feed rate is too high, resulting in poor combustion, whether the waste contains highly volatile substances or sealed containers of gas and liquid suddenly exploded by heat

(continued)

Table 8.1 (continued)

Serial number	Abnormalities	Types of incinerators	Indicator signals for malfunction	Response measures
10	Extractor fan malfunction	Rotary kiln incinerator	The extractor motor overheats; the extractor fan power supply indicates zero or out of range; the fan stops rotating; the differential pressure between the gas inlet and outlet of the extractor fan decreases	Use the backup extractor fan (if there is a backup); if two extractor fans are used at the same time, you can maintain the operation of one of the extractor fans that is not out of order and then repair the one that is out of order; If there is only one extractor fan, the operation of the incineration system must be stopped urgently
11	Filter bag rupture	Rotary kiln incinerator	Increase in flue gas blackness	Progressively isolate the interval inside the filter bag, check whether the filter bag is broken; if there is no interval inside the filter bag, shut down for a complete overhaul

system driven by boiler water circulation pump, mainly used in high-pressure boiler systems.

8.3.1.4 Operation and Management of the Flue Gas Cleaning System

(1) Tail gas cooling

The cooling of exhaust gases can be divided into two types: direct and indirect.

Direct cooling is the use of inert media directly in contact with the exhaust gas to absorb heat, in order to achieve the purpose of cooling and temperature regulation. Water has a high heat of evaporation (about 2500 kJ/kg), which can effectively reduce the exhaust gas temperature, and the water vapor produced will not cause pollution, so water is the most commonly used medium. The cooling effect of air is very poor, must introduce a large amount of air, will cause the tail gas treatment system capacity increased (2–4 times more, depending on the inlet air temperature varies), is rarely used alone.

The indirect cooling method is the use of heat transfer medium (air, water, etc.) through the waste heat boiler, heat exchanger, air preheater, and other heat exchange equipment to reduce the exhaust gas temperature while recovering waste heat to produce water vapour or heating the air required for combustion.

Direct water spray cooling and indirect cooling are the two most common ways to regulate and cool the incineration tail gas, usually applied in the field of domestic waste incineration, for medical waste incineration is mainly applicable to larger-scale incineration and disposal facilities. A comparison of the two methods is shown in Table 8.2.

In general, the use of indirect cooling can improve the efficiency of heat recovery, but investment and maintenance costs are also higher, and the stability of the system is lower; direct water spray cooling can reduce the initial investment and increase the stability of the system, but not only cause water consumption, and waste of energy. Medical waste incineration plants mostly use batch mode or quasi-continuous operation, the heat generated is small, heat recovery is not easy, and the economic

Table 8.2 Comparison of indirect cooling and water spray cooling methods

Serial number	Item	Indirect cooling	Water jet cooling
1	Waste disposal volume	Suitable for waste treatment with a single furnace capacity greater than 150 t/d	Suitable for waste treatment with a single furnace capacity of less than 150 t/d
2	Waste heat generation	Suitable for waste incineration with calorific value of 7500 kJ/kg or more	Suitable for waste incineration with calorific value up to 6300 kJ/kg
3	Exhaust gas cooling effect	Large heat transfer area of boiler tubes and water tube walls, more stable and effective exhaust gas cooling	The cooling effect of the exhaust gas is unstable due to the number of cooling nozzles, water pressure, water volume, and spray direction
4	The amount of exhaust gas and its treatment equipment	Less water vapour content in the exhaust gas, less exhaust gas treatment volume	More water vapor in the exhaust gas, more exhaust gas volume, resulting in the required air pollution control equipment, exhaust fans, flue pipes, chimneys and other required capacity
5	Equipment service life	Less water content in the exhaust gas, less corrosion, longer service life	High water content in the exhaust gas, easier to corrode, shorter service life
6	Waste heat utilization	Co-generation of steam and electricity, high efficiency of waste heat utilization	Low efficiency of waste heat utilization
7	Construction costs	High average construction cost	Low average construction cost
8	Operating management expenses	Higher labor and maintenance costs for operation	Lower labor and maintenance costs for operation
9	Operation management	High requirements, special boiler technicians required	No qualification restrictions for operators

benefits of waste heat recovery are poor, most of the use of water spray cooling to reduce the incinerator exhaust gas temperature.

(2) Exhaust gas cleaning

Incineration flue gas must be treated to meet the standards and then discharged. Emission of flue gas indicators should be in line with the "hazardous waste incineration pollution control standards" in the relevant provisions. Control of dioxin secondary generation is generally used in the rapid cooling process, so that the flue gas temperature is 1.0 s down to 200 °C below, reducing the retention time of the temperature zone of flue gas in the 200–500 °C.

The semi-dry and dry purification process includes a semi-dry scrubber tower, an activated carbon injection device, a bag filter, and other treatment units. Acid pollutants, including hydrogen chloride, hydrogen fluoride, and sulphur oxides, should be neutralised in the reactor by using suitable alkaline substances as neutralising agents. The residence time of the flue gas in the reactor should meet the requirement of a full reaction between the flue gas and the neutraliser. Between the neutralization reactor and the baghouse, activated carbon or porous adsorbent can be sprayed in, and an activated carbon or porous adsorbents absorption tower (bed) can be set up after the baghouse, which can remove most of the residual dioxins in the flue gas and also remove the function of heavy metals. The amount of activated carbon or porous adsorbent sprayed into the process should be controlled according to the requirements, to achieve the removal effect while reducing the load of the bag dust. The temperature of the reacting flue gas should be above 130 °C to ensure no condensation in the subsequent piping and dust removal equipment.

According to the characteristics of the flue gas (temperature, flow rate, and particle size distribution of fly ash, etc.), bag-type dust collectors are generally used. When working, the dust collector should be maintained above the dew point temperature of the flue gas by more than 30 °C. The filter bag used has a temperature limit and the temperature of the flue gas should be controlled in a suitable range without condensation and burning of the filter bag.

The induced draught fan of the flue gas purification system should be a variable frequency speed control device, and the negative pressure of the induced draught fan should be adjusted to a reasonable value to ensure the normal operation of the dragging system. The incineration system should maintain a stable pressure difference between the main equipment, the main combustion chamber, and the secondary combustion chamber pressure. The difference between the pressure inside the furnace and the atmospheric pressure outside the furnace. To prevent the leakage of harmful gases, the pressure of the incinerator should be slightly negative. If the power of the induced draught fan is reduced or the subsequent blockage of treatment equipment, the pressure in the furnace increases, the incomplete combustion gases in the furnace will escape from the furnace, then the input of waste must stop and repair the malfunctioning equipment.

The pressure difference between the input and outlet of the flue gas treatment equipment must be monitored in real time. When the differential pressure increases, it means that the equipment is partially blocked and the gas flow resistance increases. If the differential pressure exceeds the safety limit for normal operation, the equipment must be temporarily shut down for maintenance. The most frequently blocked parts are the gas inlet and outlet pipes of the spray tower, the throat of the text scrubber and the packing and filter bags in the packing tower. If the differential pressure exceeds the upper limit, the filter bags should be cleaned to avoid deterioration of the blockage.

Other major equipment, such as transfer pumps and emergency power supply equipment such as diesel generators, etc., must also be checked regularly to maintain normal operation. Operators should inspect the site at all times to respond to emergencies in a timely manner.

8.3.1.5 Operation and Management of Residue Treatment System

The residue treatment system includes slag treatment system, fly ash treatment system, and fly ash harmless treatment facilities. The slag treatment system should include de-slagging, cooling, conveying, storage, and other facilities. The fly ash treatment system should include fly ash collection, conveying, storage, and other facilities. The operation and management of the residue treatment system should consider the following two aspects: ① waste incineration process generated by fly ash is a hazardous waste, and must be incinerated in the disposal plant for environmentally sound treatment, and then sent to the designated hazardous waste landfill disposal. The harmless treatment consists mainly of thermal curing, curing, chemical stabilisation, and acid or other solvent washing. ② incineration residues are general solid waste which can be sent directly to landfills for domestic waste.

8.3.1.6 Operation and Management of the Auxiliary Fuel Supply System

Medical waste incineration auxiliary fuel is generally applied to light diesel. In order to ensure that in the unexpected external factors affect the normal fuel replenishment still meet the production needs, and the supply system should be able to reserve for the incinerator operation of more than 15d of fuel. Fuel should be replenished when the remaining amount of fuel is not less than 2d. Fuel delivery and unloading should be strictly enforced in safety procedures to prevent accidents.

The storage tank should be set up on the ground according to the requirements of fire fighting, and the oil pump system will be transported to the incinerator upper high level tank and then self-flow to the burner. The oil pump system should generally be linked with the high-level tank level, to ensure that the amount of fuel in the tank does not fall below the specified value. Oil pump system operations must have special operator management, not in the case of unmonitored operation, to prevent accidents. At the same time, warning signs should be set up in the storage tank.

The burner should be set up before the fuel-filtering device. Filtering parts should be inspected and replaced to ensure the smooth oil flow.

8.3.1.7 Operation and Management of Wastewater Treatment System

The wastewater treatment system is mainly centralised. The purpose of wastewater disinfection is to kill various pathogenic bacteria in the wastewater and to improve the water quality to meet national standards. The main method of wastewater disinfection is to add disinfectants to the wastewater. Currently, disinfectants used for wastewater disinfection are chlorinated disinfectants (Cl_2), chlorine dioxide disinfectants (ClO_2), ozone (O_3), ultraviolet light (UV-C), etc. Chlorination disinfection causes chlorine to combine with certain organic substances in water to produce carcinogenic organohalides. Disinfectants such as chlorine dioxide and ozone react chemically with organics by oxidation, not chlorination, and do not produce organohalides when used as disinfectants, and therefore have received attention in recent years. Chlorine dioxide has a stronger oxidising capacity than ozone, and its aqueous solution is nontoxic, odorless, non-volatile, non-combustible, stable, and easy to store and use.

The chlorine dioxide disinfection process compared with sodium hypochlorite (NaClO), ozone, and ultraviolet disinfection process has its spectrum of properties, pathogenic microorganisms spread in the water, including viruses, spores, and anaerobic bacteria in the water system, and sulphate-reducing bacteria and fungi have a good killing effect. The principle of killing microorganisms is that they have good adsorption and permeability to the cell wall, no chlorination, and no carcinogenic substances. Using a complete set of chlorine dioxide disinfectant generator, its sterilization capacity is 3–5 times that of sodium hypochlorite, and less investment, small footprint, stable operation, low operating costs, and easy maintenance. Therefore, the chlorine dioxide disinfectant is the most ideal choice of disinfection chemicals.

From the development of domestic and foreign treatment technology and practical application, the wastewater treatment process of medical waste disposal basically adopts the MBR method whose effluent can reach the secondary standard, and the main processes are: (1) secondary treatment: conditioning tank + MBR + contact disinfection process + discharge; (2) MBR + contact disinfection process uses MBR method to treat wastewater, which consists of grating, conditioning tank, flocculation reaction tank, MBR reaction tank, sludge treatment facility and disinfection system.

8.3.1.8 Operation and Management of Automatic Control and Online Monitoring Systems

1. Automatic control system

The automatic control system of the incineration plant is mainly in the central control room, which can realise the centralized monitoring and decentralised control of the medical waste incineration line and the auxiliary systems through the decentralised control system. The control system obtains and displays important working parameters through instrumentation, and inputs the pre-designed automatic control program to control the safe operation of the incineration system according to the process requirements.

The incineration plant has an automatic control system and an online monitoring system, which provides a real-time display of the real-world of the site in the storage room, the material transfer process, and the important parts of the incineration line, so that the manager can understand the whole site situation on the monitor in the main control room and make the necessary decisions and emergency responses first time.

When the system starts to operate, the online monitoring device starts to monitor, record, and print out as required. Online monitoring is carried out for soot, sulphur oxides, nitrogen oxides, hydrogen chloride, and other pollutants in the incineration flue gas before and after treatment, as well as important process indicators such as oxygen, carbon monoxide, carbon dioxide, and the temperature of the first and second combustion chambers.

The main content of automatic control can generally include feed system control, incineration and flue gas cleaning system control, slag disposal system control. Auxiliary devices that do not affect the overall control system, such as auxiliary fuel supply systems, can be controlled locally while transmitting important information to the central control room.

The control system obtains and displays important working parameters through primary instruments and is operated by pre-designed and prepared automatic control programs to ensure that the incineration system operates according to the process requirements. The operator should always monitor the state of operation, with sound and light signal alarm or find abnormal conditions and should be timely disposed and the results for the record. The system start-up operation and the automatic control device should be important parameters, such as the number of feeds, the temperature of each section of the system, the pressure of each section of the system, gas flow rate, etc., to automatically record and can be printed out as needed.

On-site industrial television monitoring system for storage, material transfer process and important parts of the incineration line of the site live, real-time display on the main control room monitor. Operators should always monitor the operating status of each ministry and keep records when necessary. In the case of an emergency, both the main control room and the local control area should be able to start the emergency stop system.

2. Online monitoring system

The incineration plant online detection system should be able to continuously monitor and record the important parameters of the operation of the main equipment and process systems, the operation status of the main auxiliary machines, the opening and the closing status and opening of automatic valves, the supply status of instruments and power sources for control and the necessary environmental parameters.

The incineration plant online monitoring system should be online monitoring of the incineration flue gas before and after the treatment of soot, sulphur oxides, nitrogen oxides, hydrogen chloride, and other pollution factors, as well as oxygen, carbon monoxide, carbon dioxide, a combustion chamber and the second combustion chamber temperature and other important process indicators, and as required with the local environmental protection department network display. When the system starts running, the online monitoring device should be started at the same time to monitor, record, and print out as required.

The online detection system monitors industrial waste such as exhaust gas, waste water, and residue generated during the incineration process to ensure that the environmental emissions from the production process meet the standards. The specific requirements are as follows:

(1) Record the quantity and weight of each batch of medical waste incinerated.
(2) In accordance with the "Hazardous Waste Incineration Pollution Control Standards", the thermal burn rate of the incineration residue should be monitored at least once every six months.
(3) There should be continuous automatic monitoring of exhaust CO, soot, SO_2, NO_x; for the current is not yet able to use automatic continuous device monitoring "hazardous waste incineration pollution control standards" in the provisions of flue gas blackness, hydrogen fluoride, hydrogen chloride, heavy metals and other compounds, should be in accordance with the "hazardous waste incineration pollution control standards" monitoring management requirements, at least once a quarter sampling and monitoring.
(4) Record the disposal of the final residue of medical waste, including the amount of incineration residue and fly ash, the disposal methods and the receiving units.
(5) Sampling and testing frequency of dioxin not less than 1 time/year.
(6) Medical waste disposal units should be regularly reported on the operating parameters, monitoring data on the effectiveness of disposal. Monitoring data retention period of three years.

Medical waste incineration and disposal is a complex system engineering, to go through many links, such as trial burn, performance testing, safe and stable operation, post-operation supervision and management, etc. In the above links, the safe operation and management of facilities to ensure the long-term safe and stable operation of the core aspects of the equipment. Incineration and disposal facilities must achieve their designed performance indicators, management, and technical means to effectively control the risks of the incineration process to achieve environmentally sound disposal of hazardous waste must be achieved through the adoption of standard operating management means.

Medical waste incineration disposal involves a series of processes such as waste reception, temporary storage, pretreatment, transportation, incineration disposal, residue discharge, waste water and waste gas treatment. To achieve the purpose of standardised management and operation of the facility, we should be considered from both management and technical aspects. To fully integrate the characteristics of the incineration and disposal facilities, from a technical point of view to consider the source of waste classification, waste storage, waste incineration and disposal, the system of equipment maintenance is needed to carry out in-depth work. On the other hand, from the management point of view, consider the basic conditions of the operator, institutional settings and staffing, safety and labour protection, operation and handover records, monitoring and evaluation of disposal operations, accident response, etc.

The implementation of the "National Hazardous Waste and Medical Waste Disposal Facilities Construction Plan" solves the problem of the construction of hardware facilities, but is closely related to the pollution control of facility operation and management and other software issues must also be addressed. For the incineration of medical waste disposal, first of all, according to the characteristics of the life cycle of medical waste, we should first start from the source, to effectively focus on the source of waste reduction and source classification of waste and cut off the source of dioxins and other pollutants; on the other hand, we should effectively focus on the disposal process of disposal facilities operation and management issues, and fully promote the standardisation of facility operation, by improving the disposal unit facilities standardised operation management, and actively use the best environmental practises to promote the solution to the problem of meeting the standards. In addition, we must start from the perspective of whole process management, effectively realize the process control from end-to-end control, and promote the operation of medical waste incineration and disposal facilities to gradually achieve scientific, information, and institutionalization, which is the way to China's medical waste incineration and disposal facilities safe operation and management.

8.3.2 Management of Operation of Non-incineration Medical Waste Treatment Facilities Operation

8.3.2.1 General Requirements for the Operation and Management of Non-incineration Treatment Facilities

The operating units of medical waste non-incineration treatment facilities should be combined with their own process characteristics and the actual scale of operation, etc., to meet the national medical waste pollution control-related policies and to protect environmental safety as the goal, the establishment of an operational management system, sound operational management system, to ensure the continued safe and stable operation of facilities to meet the three aspects.

1. Operation Management System

At least should include the facility operation and management record system, handover record system, medical waste receiving management system, internal oversight management system, facility operation operating procedures, laboratory (laboratory) characteristics of the pollutant detection program and implementation details, treatment facilities in the operation of the emergency plan for accidents, safety and labor protection management system, personnel training system, and environmental monitoring system.

2. Operation management requirements;

   (1) Medical waste non-incineration treatment facilities operating units to receive medical waste should be in accordance with the "Technical Specifications for Centralised Treatment of Medical Waste" requirements for the implementation of the corresponding management system of the transfer coupon.
   (2) Medical waste non-incineration treatment units should promote the interface between source separation of medical waste and non-incineration treatment technology adaptations.
   (3) Medical waste non-incineration treatment facilities operating units should be established in accordance with the management system for the implementation of medical waste treatment, to ensure the safe and stable operation of facilities to meet standards.
   (4) Medical waste non-incineration treatment facilities and management personnel should be trained in the operation of the post, after passing the examination and being licensed to work.
   (5) Medical waste non-incineration treatment facilities operating units should establish operating parameters and pollutant emissions monitoring record system, and should record the final residue of medical waste treatment and secondary pollutant emissions and control, and as required for preservation and reporting.
   (6) Medical waste transport should use special vehicles, containers and identification, and in the processing unit should have a special channel and unloading area.
   (7) Medical waste non-incineration treatment facilities operating units should refer to the "Hazardous Waste Management Unit Emergency Response Plan Preparation Guide" to prepare and approve the medical waste non-incineration treatment facilities accident response plan.

3. Operation management check

   (1) Medical waste non-incineration treatment facilities operation and management inspection can take a combination of on-site inspection or written inspection.
   (2) Medical waste non-incineration treatment facilities. To check the operation and management of medical waste should include medical waste reception, medical waste temporary storage and cleaning, disinfection treatment, and medical waste non-incineration treatment facilities operation.

8.3.2.2 High-Temperature Steam Treatment Facilities Operation and Management

1. Feed

    (1) The use of automated and mechanized feeding should ensure the normal operation of the feeding equipment; the use of manual feeding should be as far as possible to avoid direct contact between the feeding container (or feeding car) and the human body.
    (2) To feed container (or feed cart) should be loosely filled with medical waste, and the maximum loading should not exceed 70% of the sterilization chamber volume.
    (3) When the steam treatment process is adopted after crushing, the air tightness of the feed port should be checked before operation to ensure that the feed unit and crushing unit operate under a certain negative pressure.
    (4) After feeding, check whether the inlet and outlet chain device of the treatment facility is reliable, ensure the sealing performance of the device, and ensure the operation of the facility and safety control requirements.

2. Steam treatment

    (1) The crushing and steam treatment is not carried out at the same time; before the medical waste for steam treatment, a prevacuum or pulsating vacuum should be carried out to sterilize the air out of the room, the specific operation according to the implementation of the "technical specifications for the centralised treatment of high temperature medical waste steam" implementation.
    (2) Crushing and steam treatment process at the same time, treatment of medical waste disinfection before the mandatory prevacuum or pulsating vacuum to discharge the air inside the sterilisation chamber, but there should be appropriate measures to ensure that the air inside the sterilisation chamber does not affect the effect of steam treatment.
    (3) The processing temperature in the sterilisation chamber is not less than 134 °C, the pressure is not less than 220 kPa (gauge pressure), and the processing time should not be less than 45 min.
    (4) To ensure the amount of steam supply, temperature, pressure, steam supply pressure control at 0.3–0.6 MPa, fluctuations of not more than 10%, non-condensable gases (volume fraction) should not exceed 5%, superheat should not exceed 2 °C.
    (5) The residue of high-temperature steam treatment of medical waste after drying the water content should not exceed 20% of the total weight.
    (6) During the operation of the facility, the temperature of the outer surface of the equipment that the operator can contact should be controlled within 40 °C. For the pipeline that transports steam or water above 60 °C and the pipeline that transports cooling water, it should be well insulated.

3. Breakage

   (1) Medical waste crushing equipment must be regularly maintained, so that the particle size of medical waste is stable below 5 cm.
   (2) When crushing unsterilized medical waste, the operation site must be airtight, so that it is carried out under negative pressure.

8.3.2.3 Chemical Disinfection Facilities Operation Management

1. Feed

   (1) The feeding equipment should be opened at the same time as the disinfection crushing equipment and the feeding rate should ensure the stable operation of the disinfection equipment.
   (2) When feeding equipment is opened, the extraction equipment should be opened at the same time to maintain feeding equipment and the disinfection crushing equipment under negative pressure to prevent bacteria-containing dust from escaping from the feeding port during crushing.

2. Chemical disinfection

   (1) Chemical disinfection parameters control. If lime powder is used as disinfectant, the contact reaction time should be greater than 120 min and the dosage should be greater than 0.075 kg/kg.
   (2) Control of chemical disinfection disinfectant concentration, the purity of lime powder should be 88–95%.
   (3) Control of operation of chemical disinfection equipment should ensure adequate contact time and mixing of medical waste and disinfection chemicals, to ensure that the concentration of chemicals, temperature and moisture meet the process requirements to ensure the effectiveness of disinfection.
   (4) Disinfection treatment equipment should be run to prevent human interference, to avoid the disinfection of medical waste before the completion of an artificial stop operation.

3. Breakage

   (1) Medical waste crushing equipment must be regularly maintained, so that the particle size of medical waste is stable below 5 cm.
   (2) Ensure that the primary crushing and secondary crushing of medical waste and chemical disinfectants are in full contact, and the operation site is in a closed negative pressure state.

8.3.2.4 Microwave Disinfection Facilities Operation Management

1. Feed

   (1) After the equipment is opened, the feed should be kept open to prevent the blockage of waste bridges.
   (2) The volume of medical waste packaging should be matched with the size of the input to ensure that medical waste and packaging in the treatment unit are intact.
   (3) When feeding equipment is opened, pumping equipment should be opened at the same time to keep the feeding system running under negative pressure.

2. Microwave sterilisation

   (1) After feeding, open the crushing equipment to crush the waste and the crushing equipment should be operated under negative pressure.
   (2) The crushed waste enters the reaction chamber, injects steam, stirs thoroughly, turns on the microwave generator, and irradiates the waste to complete the disinfection process.
   (3) Control microwave frequency of (915 ± 25) MHz or (2450 ± 50) MHz, processing temperature ≥95 °C, role of time ≥45 min. If the pressure process is performed, the microwave material temperature is <170 °C, to avoid the decomposition of plastic and other chlorine-containing compounds in medical waste caused by secondary pollution. In the joint action of steam and microwave, the temperature is not less than 135 °C, role of time is not less than 5 min.
   (4) Before the operation of the equipment, check the microwave sterilization equipment operating safety measures to prevent electromagnetic wave leakage.

3. Breakage

   (1) Medical waste crushing equipment must be regularly maintained, so that the particle size of medical waste is stable below 5 cm.
   (2) The crushing process should be carried out under a closed negative pressure state.

8.3.2.5 High-Temperature Dry Heat Treatment Facilities Operation and Management

1. Feeding unit
   Feeding equipment should be matched with the subsequent treatment process. The feeding system has good airtightness and safety, the material of the feeding container is chosen reasonably, and the feeding equipment should be automatic feeding equipment. The basic requirements are as follows:

   (1) The feed unit should be matched to the subsequent treatment process unit.

(2) For the equipment with the crushing and then sterilization process, the feeding port of the feeding unit should be kept airtight and equipped with air extraction equipment to maintain the feeding unit and crushing unit under certain negative pressure.
(3) Feeding container (or feeding cart) material should be stainless steel or aluminum alloy and other corrosion-resistant materials, and should have a certain strength. Feed containers (or feed carts) should have measures to prevent the immersion of medical waste. If the feed container (or feed car) used as a dry heat treatment process in the sterilisation room containing medical waste containers, its design should facilitate the process of uniform heat penetration and heat transfer, its material and structure should withstand the high temperature and pressure changes of the high-temperature dry heat treatment process temperature and pressure changes, and the inner wall should be an anti-adhesive treatment.
(4) Medical waste feeding should be mechanized and automated operations as far as possible, to reduce the direct operation of manual. Such as feeding manual operations, measures should be taken to avoid direct contact between the feeding container (or feeding cart) and the human body.

2. Crushing unit
The unit should be strictly broken requirements, and to ensure a certain degree of confinement and negative pressure conditions. The basic requirements are as follows:

(1) The choice of crushing equipment, should follow the principles of reliability, durability, ease of maintenance and safety, no secondary pollution. High-temperature dry heat treatment of medical waste must be crushed, only high-temperature dry heat treatment of medical waste is strictly prohibited, to prevent the phenomenon of high-temperature dry heat treatment of medical waste after recycling.
(2) Crushing equipment should be able to crush hard and soft materials at the same time, the crushed particle size of the material should not be greater than 5 cm, such as crushing cannot meet the requirements, should be set up.
(3) When the crushing unit is located before the high-temperature dry heat treatment unit, crushing should be carried out in a closed and negative pressure state, and the gas inside the crushing unit must be purified before discharge, and at the same time, there should be disinfection measures, and the crushing unit should be disinfected regularly and safely before each maintenance, and the disinfection measures should not produce secondary pollution.

3. High temperature dry heat treatment unit

(1) The unit should be reasonably designed, standardized treatment process, durable and reliable, easy to operate and maintain. The basic requirements are: medical waste dry heat treatment equipment sterilization outside the wall should be close to the external heating layer to ensure uniform temperature inside the sterilization room.

(2) The inner cavity and door of the equipment should be made of corrosion-resistant materials, and stainless steel is generally preferred.
(3) Equipment inlet and outlet doors should be able to meet the requirements of the equipment working pressure on the sealing performance; should set up a chain device, if the door is not locked, high-temperature dry heat treatment equipment cannot raise the temperature, lower the pressure, before the end of the dry heat treatment cycle, the door cannot be opened, in the equipment feeding, discharging and maintenance should be normal in the open state.
(4) The process of treating medical waste dry heat requires that the processing temperature in the sterilization chamber should not be less than 180 °C, the pressure is not greater than 1000 Pa (gauge pressure) conditions, and the corresponding processing time should not be less than 20 min.
(5) The equipment must be installed with a safety valve, the opening pressure of the safety valve should not be greater than the design pressure of the equipment. In the event of a set pressure or in the process of equipment failure, the safety valve should be able to automatically open for pressure relief.
(6) The tightness of each weld and joint of the equipment piping should be able to meet the requirements of equipment pressurization and vacuuming.
(7) High-temperature dry heat treatment equipment should have a drying function, and the water content of the material after drying should not be greater than 20% of the total weight.
(8) Heat insulation measures should be taken on the outer surface of the processing equipment, and the surface temperature of the outer surface of the equipment that may be touched by the operator should not exceed 40 °C. For the pipeline conveying heat conductivity over 60 °C and the pipeline conveying hot disinfection solution, thermal insulation treatment should be implemented.
(9) When sterilization is carried out in the high-temperature dry heat treatment unit, the mechanical stirring device should be stirred at a speed of not less than 30 r/min.

4. Discharge unit
The performance of the discharge unit safety interlocking device and automatic discharge device shall meet the following requirements:

(1) It will not be opened until the set processing conditions are met and the command from the main console is received.
(2) After the disinfection process is completed, the medical waste residue that meets the disinfection requirements is unloaded directly into the waste receiving container through the automatic conveying device, and manual unloading is strictly prohibited.

5. Heating unit

The heating unit should meet the following requirements:

(1) The thermal energy supply should be able to meet the needs of the full load operation of the treatment plant.
(2) The thermal energy supply guarantee rate should not be less than 350d.
(3) The heating source can set the temperature, so that the output heat-conducting oil temperature is maintained at a fixed temperature range.
(4) The pipes of heat-conducting oil should be made of stainless steel.

6. Automatic control unit

Automatic control units should be able to achieve the entire process of automatic control functions, to provide conditions for safe disposal, standardised records, and implementation of environmental supervision and management. The basic requirements are as follows:

(1) High-temperature dry heat treatment system should haveautomatic control of the whole process as far as possible, including vacuum preheating control, temperature rise and fall, self-starting and stopping, high-temperature dry heat treatment, exhaust gas treatment control, crushing, etc.
(2) Automatic control single hardware should include control panels, sensor parts and control regulating valves and other components. There should be a data output interface and communication interface to realize the parameter output and remote monitoring function.
(3) The automatic control software should be programmable to implement automatic control, and its functions should include:
    ① Test air removal effect and equipment sealing performance (only for evacuation type equipment).
    ② Real-time display of the current operating status, including the stage, processing temperature, processing time, sterilization room pressure.
    ③ The display and printing function of the current value of the main parameters during operation. In the process of operation, real-time tracking feedback sterilization room temperature, pressure. After the end of the process, the parameters of the entire process should be stored as a backup record for 5 years, and the automatic control system should have a certain degree of independence and reliability to prevent the stored parameters from being lost, modified and deleted at will.
    ④ The automatic control system, in addition to the steam treatment of each stage of automatic operation, should also have a manual operation mode to achieve the steam treatment of each stage of manual operation. In manual operation mode, do not simplify or avoid any processing links.
    ⑤ The automatic control system should have the function of self-detection of fault, be able to realize the alarm and emergency stop under abnormal

conditions such as over temperature, over pressure, power failure, water failure and misoperation, and be able to realize the chain function of high-temperature dry heat treatment equipment feeding door (discharging door) when the operation is not completed.

⑥ Automatic control software should set the authority to limit the processing time, processing temperature, pressure and other parameters of the modification, and prohibit the processing parameters to reduce the standard parameters below the dry heat treatment of medical waste.

⑦ Automatic control unit in the dry heat treatment process should be able to be based on the temperature and pressure fluctuations in the sterilization room in a timely manner and the processing temperature control in the preset temperature range of ±5 °C.

(4) Processing equipment equipped with instruments and metres, in addition to meeting the requirements of relevant professional standards, should ensure that the accuracy of the temperature control accuracy is not less than ±0.5 °C, pressure control accuracy is not less than ±1.6% and time control accuracy is not less than ±1%. And at least once a year the calibration should be checked, and the corresponding records, the results of the records should be kept as part of the operating records of the treatment plant documents for 5 years.

7. Exhaust gas treatment unit
The exhaust gas treatment unit should meet the following requirements:

(1) The exhaust gas treatment unit must be able to effectively remove microorganisms, volatile organic compounds (VOCs), heavy metals and other pollutants, and be able to eliminate the odor generated during the treatment process.
(2) The exhaust gas treatment unit should ensure that the removal rate of microorganisms, VOCs and other pollutants is more than 99.999%. Exhaust gas treatment units are generally appropriate to set up high-efficiency filtration, adsorption devices, etc., according to specific circumstances, can be considered to add VOC chemical oxidation devices and in high-efficiency filtration devices upstream of the additional medium- or low-efficiency filtration devices. Consider the use of chemicals to remove the odor in the steam treatment process, but also according to the actual set of deodorization devices.
(3) The tail gas efficient filtration device should be hydrophobic mesoporous materials, to meet certain temperature requirements, the filter scale should not be greater than 0.2 μm; filtration device should generally be equipped with an inlet and outlet valve, pressure instrumentation, and drainage valve, the design flow should be compatible with the scale of treatment, the filtration efficiency should be above 99.999%.
(4) There should be technical measures to prevent microbial growth and wet failure in filtering and adsorption devices; if filtering and adsorption devices

do not meet the requirements of processing efficiency, they should be corrected or replaced promptly.

(5) The airtightness of the exhaust gas treatment unit piping and the connection between the piping should be ensured.

8.4 Supervision and Management of the Medical Waste Disposal Process

8.4.1 Medical Waste Disposal Supervision Responsibility System

The process of medical waste disposal involves three departments, namely, medical waste generation units, medical waste disposal units and medical waste supervision and management units. The relationship between the responsibilities of the relevant departments involved in medical waste management is shown in Fig. 8.1.

Source management of medical waste is mainly the internal medical waste management of medical institutions, to reduce the amount of hazardous waste generated through source management to reduce the disposal load, and according to the disposal process to do a good job of waste classification to ensure the safety and stability of subsequent disposal. Medical waste disposal units should be implemented in accordance with the relevant national requirements for the safe and environmentally sound disposal of medical waste. The unit of supervision and management

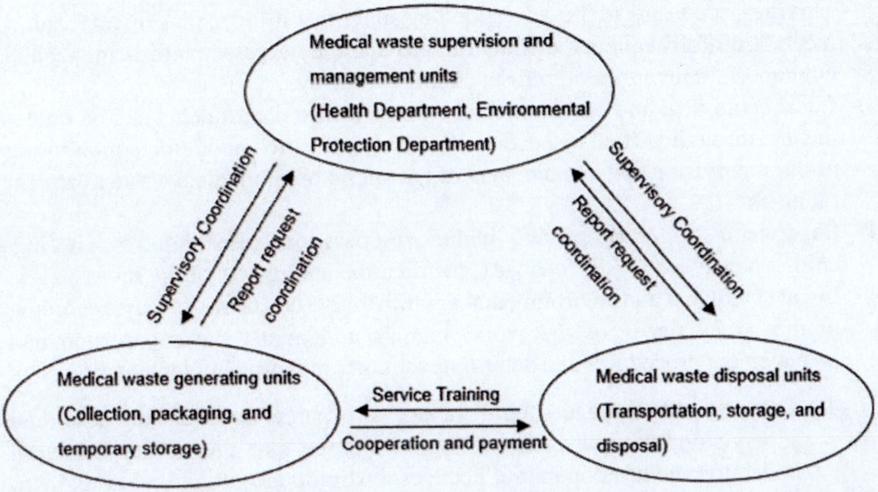

Fig. 8.1 Relationship of responsibilities of relevant departments involved in medical waste management

is mainly the ecological environment administrative department whose responsibility is to take practical supervision and management means to promote the prevention of pollution of medical waste disposal processes. And the responsibility of the health and health administrative departments is to take practical measures to promote medical waste generation units medical waste disease control work. The ecological environment should start with the practical needs of environmental supervision and management of medical waste treatment and disposal facilities, the inspection of the basic operating conditions of medical waste disposal facilities, such as medical waste disposal technology and techniques, medical waste business licence application and use, institutional settings and staffing, regulations and management, pollution prevention facilities configuration and treatment status, safety and labour protection and other implementation, and to ensure the normal operation and management of disposal facilities (National Health and Health Commission, Ministry of Ecology and Environment 2020).

8.4.2 *Medical Waste Disposal Supervision and Management Procedures and Requirements*

(1) Ecological and environmental administrative departments of the people's governments at or above the county level and other medical waste pollution prevention and management departments, and the right to supervise and inspect medical waste treatment facilities under their respective responsibilities.
(2) Medical waste disposal units should actively cooperate with the competent ecological and environmental administrative departments and other supervision and management of medical waste pollution prevention and management activities, according to the corresponding supervision and management requirements, truthfully reflect the situation, provide the necessary information, shall not conceal, misrepresent, refuse, obstruct or delay.
(3) Local ecological and environmental administrative departments can be implemented through written inspections, on-site verification and remote monitoring of the supervision and management of the operation of medical waste treatment facilities.
(4) Supervision and management include preparation, inspection, monitoring, comprehensive analysis, feedback, rectification and review of the seven stages. Local ecological and environmental administrative departments may, according to the actual needs of the work, modify and adjust the supervision and management procedures and determine the corresponding implementation plan.

    ① The preparation stage should include two parts: material collection and supervision, and management implementation plan preparation. Material collection includes operating licences, institutional settings, staffing, institutionalization, facility construction and operation, the overall emission of pollutants, and the commissioning unit that signed the operating contract;

implementation plan should be prepared on the basis of material collection, clear supervision, supervision, procedures, methods, personnel safety and protection measures.

② The inspection phase should be based on the implementation of the plan for medical waste disposal facilities, the operation and management of auxiliary facilities for on-site verification, review of relevant records, accounts, and the problems found should be verified and confirmed.

③ Monitoring phase should be based on the requirements of the implementation plan, monitoring of pollutant emissions (exhaust gas, wastewater, solid waste, noise, etc.) during the operation of the facility, ensuring the quality of monitoring, and keeping monitoring records.

④ The comprehensive analysis stage should be based on inspection, monitoring work, a comprehensive analysis, evaluation of the overall situation of medical waste disposal units, the formation of supervision and inspection conclusions, the existence of the problems to be set out one by one; the need for rectification should be put forward in writing to rectify the content and the rectification deadline; there are violations of the "Solid Waste Pollution Prevention and Control Law" "hazardous waste management licence management" and other laws and regulations behaviour, put forward the corresponding penalties.

⑤ The feedback stage should be the conclusion of the supervision and inspection, rectification notices, penalty notices, etc., according to the prescribed procedures sent to medical waste disposal units.

⑥ Rectification stage should be urged to medical waste disposal units according to the results of supervision, inspection and rectification measures to rectify and submit rectification reports.

⑦ Supervision and management review stage should be the medical waste treatment units to review the rectification of the situation, still does not meet the requirements, should be in accordance with the "Solid Waste Pollution Prevention Law", "hazardous waste management license management", "medical waste management regulations", "medical waste management administrative penalties" on medical waste non-incineration treatment units to impose penalties, such as warnings, rectification, fines, suspension or revocation of business license, etc.

(5) Supervision and management personnel should carefully implement national environmental protection guidelines, policies and relevant laws, regulations and standards when conducting on-site inspections.

① Supervision and management personnel in the on-site inspection, must be more than two people with the appropriate administrative enforcement powers to participate at the same time, carry and show the relevant documents.

② Supervision and management personnel in the on-site inspection. You can take a statement of inquiry, on-site monitoring, sample collection, photography and video, access or copy the information and other means of inspection, and proper storage of relevant information.

③ Supervision and management personnel should strictly enforce the security system when conducting on-site inspections, and conserve technical business secrets for the inspected units.

④ Supervision and management personnel should be inspected for objective, standardized records, and should be requested to check the unit's representatives to confirm. Inspectors and the inspect unit on the inspection records of the content, of the part of the disagreement cannot be resolved in a timely manner, and should be well documented.

Self-inspection and assessment is a kind of supervision and management of medical waste disposal units for the existence of larger problems. According to the actual needs of management, technical assessment of the operation and management of facilities, and combined with the problems in the operation and management of facilities, the corresponding corrective measures to promote the operation of enterprises to meet standards, standardized management, environmental supervision and enforcement to provide management and technical basis.

8.4.3 Medical Waste Disposal Supervision and Management Content

Combined with the "hazardous waste incineration pollution control standards", "medical waste management regulations" and other laws and regulations on the operation of medical waste disposal facilities requirements, in accordance with the "standardized management of hazardous waste indicator system" (Environment Office [2015] No. 99), medical waste treatment facilities to carry out the entire process of operation (that is, from medical waste from the handover to the safe disposal of hazardous waste in the plant) for supervision and management.

Supervision and management of the main content includes four parts: supervision and inspection of the basic operating conditions of the facility operating units, supervision and management of the operation of medical waste disposal facilities, medical waste disposal facilities operating monitoring management and medical waste disposal facilities operating supervision methods.

In the macromanagement status of the facility operations units of supervision and inspection, for medical waste disposal technology, acceptance of technology and engineering acceptance, application and use of medical waste business licence, institutional settings of the medical waste facility operation unit, supervision and management personnel of the medical waste disposal facility, institutionalized construction of the facility operation unit, and the development of emergency plans and other clear requirements.

In the medical waste disposal facility operation supervision and management content, according to the operation of medical waste disposal facilities from the waste into the plant to the waste disposal of the entire process management of the actual needs, for the receipt of medical waste, plant storage and pretreatment, disposal facilities operation, safety and labor protection supervision and management and other links put forward clear requirements. Medical waste disposal facilities operating monitoring management, including incineration, non-incineration treatment of pollutant emissions monitoring and monitoring of the environment around the medical waste disposal facilities operating units.

8.4.4 Medical Waste Disposal Supervision and Management Methods

1. Supervision and inspection of basic operating conditions
   The basic conditions check as the basic basis for the supervision and management of the local ecological and environmental administrative departments, in principle, should be carried out during the initial supervision and inspection to take into account the consistency of the work carried out by the inspection. Through the medical waste disposal technology, process and engineering acceptance, medical waste business license application and use, medical waste facilities operating units of the institutional set-up, staffing, facility operation unit rules and regulations, the development of emergency plans for accidents, system configuration of the review project, review points, inspection indicators and basis, supervision and inspection methods, the basic requirements of the facility operating units Content, such as the basic inspection, to determine the basic operating conditions of supervision and inspection of key elements, inspection methods and inspection programs.
   (1) Medical waste disposal technology, process and engineering acceptance. By the facility operating unit of medical waste pretreatment and disposal technology and process adaptability, the main ancillary facilities, engineering design and acceptance of the situation to provide design documents, environmental impact assessment documents and other supporting materials, supervision and inspection departments for written inspection.
   (2) Medical waste operation license application and use. Supervision and inspection departments through on-site verification of the way to check the disposal facility's medical waste business license, disposal contracts and other medical waste disposal record materials and other information, targeted from the medical waste facilities operating units of the disposal contract business scope of the situation, medical waste business license changes, disposal plans, business license inspection for supervision and inspection.

(3) Medical waste facilities operating units of institutional settings, and staffing. The supervision and inspection department, by way of on-site verification, check the disposal facility disposal contract and other medical waste disposal record materials and other information, especially the need to pay attention to verify that the disposal facility to collect and dispose of the type of medical waste is not beyond the applicable scope of the treatment process. Targeted from the overall staffing of medical waste facilities operating units, professional and technical staffing, personnel training to check the composition of the unit and the division of responsibilities of personnel and personal file materials.
(4) The operation of the facility unit rules and regulations. The supervision and inspection department, through on-site verification, check the disposal facilities of the system text, facility operation records file, facility operation shift records file and implementation.
(5) The development of emergency plans for accidents. The supervision and inspection department, by way of on-site verification, should check the disposal facilities of medical waste storage process when an accident occurs in the emergency plan, medical waste transportation process when an accident occurs in the emergency plan, incineration facilities failure or accident emergency plan, facilities and equipment capacity can not guarantee the normal disposal of medical waste emergency plan.
(6) System configuration. The supervision and inspection department, through on-site verification, check the integrity of the system configuration of the disposal facility, the security of the system configuration, and the requirement to check the hazardous waste treatment design documents.

2. Disposal facility operation process supervision and inspection-receiving, analysis, and storage facilities

(1) Medical waste receiving system. Supervision and inspection departments should check the implementation of the medical waste transfer coupon system, waste into the special channel and marking, waste unloading, and on-site inspection if necessary.
(2) Medical waste storage system. On-site inspection of design documents, mainly to check medical waste temporary storage, medical waste storage facilities, and on-site verification.
(3) Disposal facility operation process inspection. Supervision and inspection department on-site inspection of design documents, mainly to check the configuration of disposal facilities and disposal process operation, and on-site verification.

3. Disposal facility operation process supervision and inspection-supporting disposal facilities

(1) Feeding system inspection. Inspection and supervision department on-site inspection of design documents, mainly to check medical waste transport, feeding devices, and on-site verification.

(2) Incineration system inspection. Thermal energy utilization system inspection, supervision and inspection departments on-site inspection of design documents, mainly to check the configuration and operation of the thermal energy utilization system, and on-site verification. Flue gas purification system inspection, supervision and inspection department on-site inspection of design documents, mainly to check the wet purification process quench scrubber and absorption tower and other unit configuration; check the semi-dry purification process gas washing tower, activated carbon injection, bag filter and other processing unit configuration; check the dry purification process: including dry gas washing tower or dry powder feeding device, bag filter and other processing unit configuration; check the flue gas Purification system configuration, and on-site verification. Slag and fly ash treatment system inspection, supervision and inspection departments on-site inspection of design documents, mainly to check the configuration of the slag treatment system, check the configuration of the fly ash treatment system, and on-site verification.

(3) Non-incineration treatment facilities inspection. High-temperature steam treatment facilities should include feeding units, high-temperature steam treatment units, crushing units, compression units, exhaust gas treatment units, waste liquid treatment units, automatic control units, steam supply units and other auxiliary equipment, supervision and management should include system configuration and operation.

Microwave disinfection processing facilities should include feeding units, crushing units, microwave disinfection processing units, unloading units, automation control units, exhaust gas treatment units, wastewater treatment units and other auxiliary equipment, supervision and management should include system configuration and operation.

Chemical disinfection treatment facilities should include feeding units, crushing units, pharmaceutical supply units, chemical disinfection treatment units, discharge units, automatic control units, exhaust gas treatment units, waste liquid treatment units and other auxiliary equipment, supervision and management should include system configuration and operation.

High-temperature dry heat treatment facilities should include feeding units, vacuum units, grinding units, dry heat sterilization units, exhaust gas treatment units, waste liquid treatment units and other auxiliary equipment, supervision and management should include system configuration and operation.

(4) Automatic control and online monitoring system inspection. The supervision and inspection department on-site inspection of design documents, mainly to check the automatic control system, online monitoring system, operating procedures materials, and on-site inspection.

4. Supervision and inspection of production safety and labor protection
   (1) Safety production requirements. The supervision and inspection departments check the incineration plant safety production, check the relevant safety production materials, and verify on site.
   (2) Labor protection requirements. The supervision and inspection departments check incineration plant labor protection, check labor protection-related materials, and inspect the site.
5. Pollution prevention and control facility configuration and processing requirements
   Supervision and inspection departments should check the monitoring report and on-site verification, which includes:
   (1) Control of atmospheric pollutant emissions and ambient air quality requirements;
   (2) Incineration and non-incineration treatment performance requirements;
   (3) Residue and fly ash disposal requirements;
   (4) Wastewater discharge and surrounding environmental quality requirements (sewage discharge requirements and surface water environmental quality requirements);
   (5) Soil environmental quality and surrounding soil environmental quality requirements;
   (6) Noise emission requirements and surrounding noise environmental quality requirements.
6. Environmental monitoring requirements
   Supervision and inspection departments to check the design documents and monitoring reports, and on-site verification, mainly to check the incineration facility pollutant emissions monitoring, environmental monitoring around the disposal unit, monitoring frequency, monitoring conditions, monitoring sampling and testing methods and other content.
   Medical waste treatment and disposal units should be based on national and local requirements and the establishment of soil and groundwater contamination of potential pollution management systems. Monitoring of heavy metals in the air pollutants, monitoring frequency at least once a month; monitoring of dioxins in the air pollutants not less than two times a year, the arithmetic meaning of three consecutive measurements as the monitoring concentration. Incineration units should be online automatic monitoring facilities, monitoring the main pollutants in the exhaust gas 1 h average value and daily average value, monitoring projects should include at least hydrogen chloride, sulfur dioxide, nitrogen oxides, particulate matter, carbon monoxide and flue gas oxygen content.

8.4.5 *Standardized Management of Medical Waste*

"Twelfth Five-Year" period, China began to implement the "Twelfth Five-Year" national hazardous waste standardized management inspection and assessment work program "and" hazardous waste standardized management index system (Ministry of Ecology and Environment 2013). So far, up till now, through the annual standardised management assessment of medical waste generation and disposal units, the ecological and environmental departments at all levels and related enterprises to improve the level of medical waste management have greatly promoted, the environmental risks of medical waste nonstandard storage, illegal transfer and disposal were reduced, and the environmental safety of medical waste and harmless disposal effect were effectively protected. To implement the "thirteenth five-year plan for national economic and social development" and the "Thirteenth Five-Year" ecological environmental protection plan, strengthen and deepen the prevention and control of medical waste pollution, solidify the responsibility of local governments at all levels and relevant departments on the environmental supervision of medical waste, effectively promote the construction of environmental supervision of medical waste, and promote the legal system and relevant standards and norms in the medical waste system and related standards and norms in the medical waste generation units and business units to be implemented, to enhance the overall standardized management of medical waste. During the "Thirteenth Five-Year Plan" period, the State revised and improved the "standardized management of hazardous waste indicator system" to improve the management system, increase management efforts, promote the refinement of management, strengthen environmental risk prevention and control as the focus of further requirements.

8.4.5.1 Appraisal Method

In accordance with the "standardized management of hazardous waste indicator system" requirements, to carry out the assessment in a graded manner. Municipal organizations to implement the centralized medical waste disposal units within the jurisdiction of the self-examination assessment; national and provincial (district and city) respectively, the national and provincial medical waste disposal units for a comprehensive sampling and sampling assessment. Ecological and environmental departments in the inspection should highlight the problem record, fill out the "basic information of the sampled units record form". Centralized medical waste disposal units in each city belong to hazardous waste management units, where the annual assessment must check the object. Medical health institutions as medical waste generation units, the number of complexes, the daily management of medical waste standardised mainly by the health department, and the disease control department is responsible for, therefore, not included in the scope of the assessment of the ecological environment department.

8.4.5.2 Assessment Content

The assessment of the centralized medical waste disposal unit mainly includes:

(1) Pollution prevention and control responsibility system. Whether to establish and improve the system of responsibility for pollution prevention and control of the environment, whether to take measures to prevent medical waste pollution of the environment, and whether to prominently post information on the responsibility for the prevention of medical waste and hazardous waste.
(2) Signage system. Whether to set up independent medical waste, hazardous waste storage sites, classified storage of medical waste and other hazardous waste; whether based on "environmental protection graphic signs-solid waste storage (disposal) site", "hazardous waste storage pollution control standards". Appendix A shows the corresponding warning signs and signage; whether to set up identification signs on the packaging containers of medical waste and other hazardous waste.
(3) Hazardous waste management plan system. For the treatment and disposal of medical waste generated in the hazardous waste, whether to develop measures to reduce its generation and harm, as well as the corresponding storage, utilisation, and disposal measures. These hazardous waste generation links, types, hazard characteristics, generation, utilization and disposal of the description is clear; whether the annual report to the local people's government at or above the county level ecological and environmental departments for the record and timely declaration of significant changes in the content.
(4) Declaration of registration system. Whether to truthfully declare the receipt and disposal of medical waste types, quantities, sources, storage, disposal methods, etc.; whether the information includes the relevant internal generation of hazardous waste, such as the type, amount generated, treatment and disposal methods and the final flow; secondary hazardous waste can be specifically declared or included in the discharge declaration, environmental statistics together; whether the declaration is complete, true, reasonable; whether the timely reporting of major changes of the matter.
(5) Source classification system. Is the collection of medical waste and internally generated hazardous waste collected in accordance with the characteristics of separate storage, so that there is a clear interval.
(6) The transfer of the joint list system. In the interprovincial transfer of hazardous waste generated internally before, whether the location of the provincial ecological and environmental departments to approve the hazardous waste transfer plan and get approval; intraprovincial transfer of hazardous waste transfer plan without approval. Whether in accordance with the actual transfer of truthful, standardized to fill out the transfer of coupons, and save all.
(7) Operating license system. Whether to hold a medical waste management license qualification, and the scope of the qualification to carry out the relevant activities. Internally generated hazardous waste should be provided or entrusted to the corresponding hazardous waste business license unit for collection,

storage, utilization, disposal, and can provide the relevant disposal contracts signed. This does not match the direct judgment for the assessment of the unqualified level.

(8) Emergency plans for the record system. Whether to develop accident prevention measures and emergency plans (comprehensive emergency plans with relevant chapters or special emergency plans); whether to carry out not less than one drill per year, and save the video archive records; whether the emergency plan in the ecological environment departments above the county level for the record.

(9) Record and report on the operation of the system. Whether in accordance with the "hazardous waste management units to record and report on the operation of the guidelines" requirements, the establishment of a sound medical waste and internally generated hazardous waste management record book, truthfully recording the collection, storage, disposal of various types of waste and the occurrence of accidents and other matters; operation of the record book is kept for more than 10 years; whether in accordance with the medical waste management license and the requirements of the ecological environment department, regular reports medical waste and the internal generation of hazardous waste operations.

(10) Operational training. Whether to develop annual training plans and on schedule to engage in medical waste and internally generated hazardous waste collection, transfer, disposal and other operational and technical personnel to carry out training, and do a good job of training file records.

(11) The management of storage facilities. Whether medical waste and internally generated hazardous waste storage facilities for environmental impact assessment, the completion of the "three simultaneous" acceptance; storage facilities are in line with the "pollution control standards for hazardous waste storage" requirements; whether the establishment of medical waste and hazardous waste storage ledger, and truthful and standardized records of storage.

(12) The management of disposal facilities. Whether the disposal of medical waste disposal facilities is in accordance with the law for environmental impact assessment, the completion of the "three simultaneous" acceptance; the establishment of the use/disposal ledger, and truthful records of the situation; the use/disposal facilities on a regular basis meet the standard discharge of pollutants for environmental monitoring.

(13) Operational safety management requirements. Whether the disposal facilities and supporting monitoring, safety, emergency response equipment for regular inspection; for broken equipment, the cleanup and replacement; whether there is calibration and maintenance of environmental monitoring and analysis instruments; whether the disposal and pollution control facilities can protect the normal operation; whether to do daily inspections or operation and maintenance records.

Various types of medical and health institutions as medical waste generation units should also follow the requirements of the standardised management assessment of hazardous waste management system, the internal generation of medical waste and other hazardous waste such as sludge, waste activated carbon, and other strict management, the establishment of pollution prevention and control responsibility system, marking system, management plan and filing system, declaration and registration system, source classification system, transfer coupon system, business licencing system, emergency plans and filing system; standardise the management of medical waste storage facilities; strengthen medical waste and hazardous waste-related business training, and other work to promote the reduction of medical waste at source, to enhance the standardisation of medical waste management capacity and level within medical and health institutions.

As an important grip on the environmental management of medical waste, centralized medical waste disposal unit standardized management, but also disposal units to enhance the level of internal management, to prevent secondary environmental pollution of the main measures. Adhere to the standardized management of medical waste, will promote the quantification of medical waste and harmless goal, will also play a positive role in the healthy development of the medical waste treatment and disposal industry.

8.5 Performance Evaluation of the Operation of Medical Waste Disposal Facilities

8.5.1 Basic Ideas of Medical Waste Disposal Performance Evaluation

The United States was the first country to introduce performance evaluation theory and methods and apply them to the field of hazardous waste management. Research efforts to conduct trial burns and performance tests were proposed as early as the 1990s, and specific methods were applied to environmental management practices. Standard waste is hazardous waste that is configured according to the performance test requirements of hazardous waste incineration and disposal facilities, and is determined based on the difficulty of disposing organic components by incineration, with the corresponding requirements for their composition, concentration and quantity during the performance test. The standard waste configuration should also be determined on the basis of the type of incineration facility process and the scope of application, etc. The performance test process requires the normal operation of the incineration facility to examine the limits of the operating conditions of the test, on the basis of which to determine the type of waste that can be disposed off by the incineration disposal facility and its corresponding working parameters and the main operating parameters of the facility. The basic idea of medical waste disposal performance evaluation is as follows:

(1) To consider the operating conditions of hazardous waste under both boundary conditions and design conditions. Boundary conditions refer to the operating parameters set to examine the ultimate operating conditions of the incineration facility, such as maximum temperature, minimum temperature, maximum organochlorine feed, maximum heavy metal feed, maximum feed of the main organic and hazardous components (principal organic and hazardous components, POHCs) maximum feed, maximum waste feed, etc. In addition to the boundary conditions, the incineration and disposal facilities are also designed to test the operation of the facility to examine the stability of continuous operation and safety.
(2) To consider the use of standard waste as an evaluation reference for the performance testing and evaluation of incineration facilities. Standard waste should be selected as the background waste with low environmental risk, the waste should have good thermal stability, low toxicity and mature analytical test methods POHCs, according to the actual situation in China, initially consider using at least one of them for carbon tetrachloride (blending the content of chlorine in waste), and another preferably choose polycyclic compounds to determine the DRE, and consider adding copper, lead, mercury, three heavy metal compounds to deploy the content of heavy metals (preferably adding copper, lead, mercury oxides).
(3) To fully consider the waste characteristics indicators, system performance indicators, flue gas emission indicators, equipment operating parameters of the organic combination of four types of indicators, and then determine the specific performance of the incineration facility. The process should be to meet the flue gas emission indicators as the core, to determine the specific values of the other three indicators, for the final approval of the performance indicators of the incineration facility and the issuance of a full license to provide a comprehensive basis.

In the medical waste non-incineration treatment facilities operational performance testing, the United States, Europe, Canada and other Countries also have the relevant basis. In the United States, the U.S. Environmental Protection Agency in its October 2003 proposed medical waste treatment reference technical programme standards, the first edition (UL2334), the use of nonincineration treatment technology for medical waste disinfection effect testing of the basic requirements and procedures, equipment operation, test environment and power conditions, and the killing effect of calculation methods are clearly defined. Performance testing as a set of waste treatment facilities to measure the performance of an important tool covered in the evaluation of the operational performance of hazardous waste and medical waste disposal facilities, and as an important management tool to be applied. In Europe, Italy has prepared and published medical waste disinfection equipment and processes-general requirements for the design, manufacture, operation, maintenance, inspection, testing and supply of medical waste treatment and disposal systems, as well as for medical waste disinfection processes. The requirements also for medical waste treatment equipment instrumentation, such as control instruments, equipment signals, equipment sensors, timing systems, recording and printing and other aspects of the requirements; for equipment control systems, such as the adjustment of process parameters, maintenance of control systems; in testing, the content involves prototype testing, working condition testing, disinfection effect testing (including biological monitoring, physical testing) procedures and requirements, and the different parts of the equipment testing process is specified, such as disinfection room, disinfection load, instrumentation testing, sensor testing, biological indicator requirements; also on the maintenance of equipment, personnel training, operational records and operational file management. In Canada, the Canadian Environmental Protection Agency developed non-incineration technology to deal with medical waste guidelines C-17 (biological testing procedures), combined with high-temperature steam, chemical disinfection and microwave and other non-incineration treatment technology features, the facility operation process of treatment effect testing procedures, testing methods, test results, etc., have made more clear requirements. In addition, the guidelines on the test report and the government approval of the conditions for certification. In terms of medical waste non-incineration treatment technology, because of the variety involved, the process varies greatly, but in the implementation of the technical evaluation process to adopt a similar evaluation model with the evaluation of the performance of incineration disposal facilities. The factors to be considered in the evaluation should include: 1 nonincineration treatment technologies are non spectrum of technologies, any kind of non-incineration treatment technologies has a certain range of application; ② non-incineration treatment technologies are basically integrated technologies, so the characteristics of different technologies to determine the corresponding process parameters and working parameters; ③ non-incineration treatment technologies are key problems. How to ensure the sterilisation effect of medical waste disposal; ④ non-incineration treatment process will also produce secondary pollution, and should ensure that the performance indicators of non-incineration treatment facilities to meet the corresponding pollution control requirements. Therefore, to scientifically and systematically test the performance of a set of non-incineration facilities, it is necessary to combine the characteristics of non-incineration facilities, to comprehensively test the safety performance of a set of facilities, to be evaluated from four aspects: waste characteristics, disinfection effect indicators, pollutant emission indicators and major operating parameters.

Performance testing methods for evaluating the performance of medical waste disposal facilities are applied to the assessment of the issuance of operating permits and the comprehensive evaluation of the performance of disposal facilities. To determine whether a facility meets emission standards and ensures suitability for a particular waste disposal type, a systematic evaluation is required, taking into account the characteristics of the waste and the working conditions of the facility. Therefore, to carry out a scientific and systematic examination of the performance of a set of medical waste disposal facilities, it is necessary to combine the characteristics of medical waste disposal facilities, should be a comprehensive examination of the safety performance of a set of facilities, from the waste characteristics, the performance of the facility working conditions, the performance of pollutant emissions and the combination of facility operating parameters of the whole process of facility performance testing and evaluation (Sun N et al. 2020).

8.5.2 *Performance Evaluation of Medical Waste Incineration and Disposal Facilities*

8.5.2.1 Incineration and Disposal Facilities Performance Evaluation of the Content

The performance testing and evaluation also involves a system of indicators that takes into account four aspects: waste characteristics, performance indicators, flue gas emission indicators and main operating parameters. Among them, waste characteristics indicators include POHCs content (can be proportioned according to the standard medical waste), organic chlorine content, heavy metal content, nitrogen, sulfur and phosphorus content, water content, calorific value; system performance indicators include POHCs DRE, combustion efficiency, flue gas residence time, thermal scorching rate of incineration residue, heavy metal removal rate, hydrogen chloride removal rate, dust removal rate; flue gas emission indicators include "hazardous waste Incineration pollution control standards" in the air pollutant emission indicators; equipment operating parameters indicators, the following three groups of parameters should be considered:

(1) Group I parameters. Process parameters that describe the incineration process and are subject to continuous monitoring, subject to waste feed automatic cut-off system. For continuously operating incineration and disposal facilities, Group I parameters include at least the waste feed rate, heavy metal feed rate, organochlorine feed rate, and POHCs feed rate. For intermittent incineration and disposal facilities, Group I parameters include at least waste input, heavy metal input, organochlorine input, POHCs input.
(2) Group II parameters. The characteristic parameters of the waste feed, which can be obtained from the operating records. Group II parameters should include at least the temperature at the outlet of the second combustion chamber of the incineration system, oxygen concentration before the flue gas emergency cooling, flue gas flow before the flue gas emergency cooling, flue gas flow after the flue gas emergency cooling and the minimum negative pressure at the incinerator inlet.
(3) Group III parameters. They are process parameters describing the operation of the flue gas purification equipment, some of which can be continuously monitored at the facility and interlocked with the automatic cut-off system. Group III parameters include at least the inlet and outlet temperatures of the emergency cooling tower, the inlet gas temperature of the flue gas cleaning facility, the alkaline injection rate, the activated carbon injection rate, and the differential pressure of the baghouse.

8.5.2.2 Incineration and Disposal Facility Performance Testing Scenario Division

Performance tests will be conducted in the following three contexts:

(1) Performance tests of the standard waste in the incinerator at the lower boundary conditions, i.e. at the maximum temperature, maximum heavy metal feed and maximum organochlorine feed (Test 1);
(2) Performance tests of standard waste in the incinerator under upper boundary conditions, i.e. minimum temperature, maximum feed of POHCs and maximum feed of organochlorines (Test 2);
(3) Performance test of conventional waste under normal conditions (Test 3).

8.5.2.3 Incineration and Disposal Facility Performance Testing Requirements

1. Technical requirements for test 1

    (1) The incinerator furnace temperature to maintain the upper temperature (±20 °C) range, the upper temperature according to the design value of the incineration facilities and no load thermal test to determine the situation;
    (2) Into the waste containing heavy metal components for incineration operation. According to the process design or actual operation of the incineration system to set different feed rates, the minimum feed rate should not be less than 75% of the maximum design feed rate, and not less than two different feed rates of the operating phase.

2. Technical requirements for Test 2

    (1) Incinerator furnace temperature to maintain the lower limit temperature (± 20 °C) range, the lower limit temperature according to the design value of the incineration facilities and no load thermal commissioning to determine, but not less than the "hazardous waste incineration pollution control standards" limits.
    (2) The waste containing POHCs was put into incineration operation and tested operationally as required in Test 1.

3. Technical requirements for Test 3

    (1) Incinerator furnace temperature to maintain the normal temperature (±20 °C) within the interval, the normal temperature incineration facilities design values and no load thermal test run to determine the situation.
    (2) Put the conventional waste into incineration operation and perform the operational test as required in Test 1.

Around the above requirements, the performance test index system and test points are proposed, see Table 8.3.

Table 8.3 Performance test index system and test points

Serial number	Category	Code	Test monitoring items	Unit	Test site
1	Waste characteristics (A)	A-a	Calorific value	cal/kg	Waste storage containers
		A-b	POHCs content	g/kg	Waste storage containers, inlets
		A-c	Organochlorine content	g/kg	Waste storage containers, inlets
		A-d	Heavy metal content	g/kg	Waste storage containers, inlets
		A-e	Sulfur content	g/kg	Waste storage containers
		A-f	Water content	%	Waste storage containers
		A-g	Ash	g/kg	Waste storage containers
2	Performance indicators (B)	B-a	Flue gas residence time	s	Before the rapid cooling of the flue gas
		B-b	Heavy metal removal rate	%	Before the flue gas emergency cooling, flue gas cleaning facilities exit
		B-c	Hydrogen chloride removal rate	%	Before the flue gas emergency cooling, flue gas cleaning facilities exit
		B-d	DRE rate	%	Flue gas cleaning facility outlet
		B-e	Combustion efficiency	%	Before rapid cooling of the flue gas
		B-f	Dust removal rate	%	Before the flue gas emergency cooling, flue gas cleaning facilities exit
		B-g	Thermal scorch rate of incineration residue	%	Incineration system ash discharge place

(continued)

Table 8.3 (continued)

Serial number	Category	Code	Test monitoring items	Unit	Test site
3	Flue gas emissions indicators (C)	C-a	Flue gas blackness	Lindgeman	Flue gas cleaning facility outlet
		C-b	Fume and dust	mg/Nm^3	Flue gas cleaning facility outlet
		C-c	Carbon monoxide (CO)	mg/Nm^3	Flue gas cleaning facility outlet
		C-d	Sulfur dioxide $(SO)_2$	mg/Nm^3	Flue gas cleaning facility outlet
		C-e	Hydrogen Fluoride (HF)	mg/Nm^3	Flue gas cleaning facility outlet
		C-f	Hydrogen chloride (HCl)	mg/Nm^3	Flue gas cleaning facility outlet
3	Flue gas emissions indicators (C)	C-g	Nitrogen oxides (in NO_2)	mg/Nm^3	Flue gas cleaning facility outlet
		C-h	Mercury and its compounds (in Hg)	mg/Nm^3	Flue gas cleaning facility outlet
		C-i	Cadmium and its compounds (by Cd)	mg/Nm^3	Flue gas cleaning facility outlet
		C-j	Arsenic, nickel and their compounds (by As + Ni)	mg/Nm^3	Flue gas cleaning facility outlet
		C-k	Lead and its compounds (by Pb)	mg/Nm^3	Flue gas cleaning facility outlet
		C-l	Chromium, tin, antimony, copper, manganese and their compounds (in Cr + Sn + Sb + Cu + Mn)	mg/Nm^3	Flue gas cleaning facility outlet
		C-m	Dioxins	$ngTEQ/Nm^3$	Flue gas cleaning facility outlet
4	Main run parameters (D)	D-a	Temperature at the outlet of the second combustion chamber of the incineration system	°C	Second combustion chamber outlet
		D-b	Feed rate of waste (dosage)	kg/h (kg/time)	Inlet
		D-c	Feed rate of heavy metals (dosing amount)	kg/h (kg/time)	Inlet
		D-d	Feed rate of organochlorine (dosing amount)	kg/h (kg/time)	Inlet
		D-e	Feed rate of POHCs (dosing amount)	kg/h (kg/time)	Inlet

(continued)

Table 8.3 (continued)

Serial number	Category	Code	Test monitoring items	Unit	Test site
		D-f	Oxygen concentration prior to rapid flue gas cooling	%	Before rapid cooling of the flue gas
		D-g	Flue gas flow rate before flue gas emergency cooling	Nm^3/h	Before rapid cooling of the flue gas
		D-h	Flue gas flow rate at the outlet of flue gas cleaning facilities	Nm^3/h	Flue gas cleaning facility outlet
		D-i	Spray rate of activated carbon	g/h	Activated carbon import
		D-j	Gas temperature at the inlet of the flue gas cleaning facility	°C	Flue gas cleaning facility entrance
		D-k	Differential pressure of bag filter	Pa	Bag filter entrance and exit
4	Main run parameters (D)	D-l	Alkaline feed rate	g/h	Acid removal tower inlet
		D-m	Inlet and outlet temperature of the emergency cooling tower	°C	Emergency cooling tower import and export
		D-n	Incineration system negative pressure	Pa	Incinerator feed inlet

Note ① The test items of test 1 include all the items in the table except B-d, B-e, B-f, B-g, D-e, D-g, D-h, D-i, and D-j
② The test items of Test 2 include all the contents of the table except B-b, B-e, B-f, B-g, D-c, D-k, and D-l
③ The test items of Test 3 include all the contents in the table except D-a, D-b, D-c, D-d, and D-e

8.5.2.4 Incineration and Disposal Facility Performance Test Analysis Methods

The analysis methods used in the performance testing process should also be combined with the above four types of tests, as follows:

(1) Waste characteristics indicators testing. Such as POHCs content determination, organic chlorine content determination, nitrogen, sulfur and phosphorus content determination, heavy metal content determination, water content determination and calorific value determination, according to the physical and chemical properties of waste, with reference to the promulgated test methods for environmental protection, chemical and pharmaceutical industries.

(2) System performance index test. Such as the determination of heavy metal removal rate in flue gas, determination of flue gas residence time, determination of hydrogen chloride removal rate, determination of incineration removal rate, determination of thermal scorch rate of incineration residue, determination of combustion efficiency, determination of dust removal rate and flue gas emission index test will be carried out according to the current relevant national standards and calculation methods.
(3) Testing of main operating parameters. The test of temperature parameters, the test of main feed parameters, the determination of oxygen concentration in flue gas, the determination of the amount of activated carbon, the determination of differential pressure of bag filter, and the determination of alkaline feed rate will be carried out with the main instruments and equipment available at present.

8.5.2.5 Incineration and Disposal Facilities Performance Evaluation Method

Based on the monitoring data obtained from the previous performance test index system, the performance data of the facilities in different situations will be calculated and analyzed, and the conclusions on whether the facilities meet the standards will be made by analyzing and studying the situation of pollutant emissions. For the facilities that meet the standards, the waste characteristics, working parameters and main operating parameters should be clarified in the case of meeting the emission standards; for facilities that do not meet the standards, the facilities need to be modified and re-tested and evaluated; for those that do not test, only in accordance with the performance test to determine the type of waste that can be disposed off in the case of meeting the emission standards, the working performance of the facilities and the main operating parameters of the facilities to implement the facilities Operation and management.

The performance evaluation of medical waste incineration and disposal facilities has not been carried out in China, but international experience shows that in order to regulate the operation and management behavior of medical waste incineration and disposal facilities in China, exploring a practical performance evaluation method has gradually become an inevitable choice for medical waste management and disposal in China. On the other hand, the process of medical waste incineration and disposal facilities is complex and diverse. Therefore, different types of incineration and disposal processes should be combined with the implementation of targeted performance testing, in order to provide a basis for mapping the performance of specific facilities, licensing and implementation of the corresponding environmental regulatory actions.

8.5.3 *Performance Evaluation of Non-incineration Treatment Facilities for Medical Waste*

8.5.3.1 Non-incineration Treatment Facilities Performance Evaluation of the Content

The performance evaluation of non-incineration facilities for medical waste should consider the following conditions:

(1) Non-incineration treatment technologies are non- spectrum technologies, and any kind of non-incineration treatment technology has a certain range of application.
(2) Non-incineration treatment technologies are basically integrated technologies, so the corresponding process parameters and working parameters should be determined according to the characteristics of different technologies.
(3) The key problem with non-incineration technologies is how to ensure the sterilization of medical waste disposal.
(4) The non-incineration treatment process will also produce secondary pollution and should ensure that the performance indicators of non-incineration treatment facilities meet the corresponding pollution control requirements.

Therefore, to scientifically, systematically and comprehensively verify the performance of a set of non-incineration facilities, it is necessary to combine the characteristics of non-incineration facilities, taking into account four aspects of waste characteristics, disinfection effect indicators, pollutant emission indicators and main operating parameters for evaluation.

8.5.3.2 Non-incineration Treatment Facility Performance Evaluation Scenarios

Medical waste non-incineration facilities generally usually include waste supply, waste treatment (high-temperature steam, chemical disinfection, microwave, etc.), tail gas purification, wastewater treatment, discharge and other system components. The process parameters of the commonly used medical waste disposal technology, technical practicality, and secondary pollution control measures are shown in Table 8.4.

The performance evaluation should consider the following factors:

(1) Indicators of waste characteristics. Composition, feed volume and number of contaminated bacteria of standard test waste and conventional medical waste.
(2) System performance indicators. Mainly evaluate the completeness of the treatment system and the normality of the unit design. The main body includes units for feeding, crushing, high-temperature steam, microwave, chemical disinfection, and dry heat, waste gas treatment, waste water treatment, discharge, automatic control, etc. Supporting facilities should include the necessary equipment

Table 8.4 Commonly used medical waste disposal technology process parameters, technical applicability and secondary pollution control measures

Technology name	Key process control parameters	Technical applicability and characteristics	Secondary pollution control analysis of measures
High temperature steam processing technology	1. Sterilization chamber processing temperature of not less than 134 °C, pressure of not less than 220 kPa (gauge pressure), processing time of not less than 45 min. if high-temperature steam facilities in alternative processing conditions, such as 115 °C processing 90 min, 121 °C processing 60 min, etc., should be validated testing 2. The filter size of the exhaust gas purification device is not more than 0.2 μm, the heat-resistant temperature is not less than 140 °C, and the filtration efficiency should be more than 99.999%	1. Cannot handle all medical waste 2. The treatment process will produce bad odor, VOCs, etc. 3. Higher operating costs than microwave and chemical disinfection technologies	1. Only for the treatment of infectious waste, part of the pathological waste and sharps in the medical waste classification catalog, for the medical waste can not be treated need to be considered by the local coordination 2. The treated medical waste should meet the corresponding microbial inactivation requirements. Solid residues generated after medical waste treatment can be disposed of as domestic waste (incineration, landfill, other), requiring local supporting the appropriate infrastructure 3. The treatment process will produce bad odor, VOCs and dust, which need to be matched with corresponding exhaust gas purification measures 4. The generated wastewater is discharged or reused after treatment
Chemical treatment technology	1. Chemical disinfection should be preferred to lime powder as disinfectant, purity 88–95%, reaction contact time >120 min, lime powder dosage >0.075 kg/kg, pH control at 11.0–12.5 2. In addition to the evaluation of chemical treatment facilities, the active ingredient content, stability and pH value of chemical disinfectants should be tested	1. To ensure adequate contact time between chemical disinfectants and medical waste, to ensure the disinfection effect 2. Rapid testing methods for treatment effects are not yet available	

(continued)

Table 8.4 (continued)

Technology name	Key process control parameters	Technical applicability and characteristics	Secondary pollution control analysis of measures
Microwave processing technology	1. Microwave generation source frequency using (915 ± 25) MHz or (2450 ± 50) MHz. microwave processing temperature of not less than 95 °C, the action time of not less than 45 min 2. In the joint action of steam and microwave, microwave frequency using (2450 ± 50) MHz, pressure of not less than 0.33 MPa, temperature of not less than 135 °C, the action time of not less than 5 min	1. Not all medical waste can be treated, and it is generally not suitable for metallic medical waste 2. Medical waste disposal process to consider radiation safety protection measures	
High temperature dry heat processing technology	In the sterilization chamber processing temperature of not less than 180 °C, the pressure is not higher than 1000 Pa (gauge pressure) under the conditions of the corresponding processing time should not be less than 20 min	Cannot handle all medical waste	

such as electrical, water supply and drainage, fire protection, heating, ventilation and air conditioning, etc.

(3) Equipment operation parameters. The test includes I III groups of parameters: I group of parameters to describe the nonincineration treatment conditions and must be continuously monitored for the process parameters; II-group of parameters for the disinfection indicators; III-group of parameters for the operation of its purification equipment.

(4) Pollutant emission indicators. Including various air pollutant emission indicators, sewage emission indicators, microbial indicators in residues, etc.

The performance evaluation of non-incineration facilities should also consider the use of standard waste. The configuration of the standard test waste should be noninfectious, and its composition is similar to conventional medical waste items, and can be determined in accordance with the local medical waste in plastic, glass, metal, cotton, waste paper, bamboo, and other material content. Specific reference can be made to the specific types listed in the "medical waste classification catalog" to determine.

8.5.3.3 Non-incineration Treatment Facilities Performance Requirements

(1) High-temperature steam treatment facilities performance testing technical requirements. High-temperature steam processing facilities should be based on the requirements of HJ/T235 working conditions, that is, 134 °C, 45 min and 220 kPa atmospheric pressure conditions for testing. If the high-temperature steam facilities in alternative processing conditions, such as 115 °C for 90 min, 121 °C for 60 min, etc., should be validated testing.
(2) Chemical treatment facilities performance testing technical requirements. For chemical disinfection (dry chemical), priority should be given to dry chemical disinfectants such as lime powder; the purity of the lime powder used should be 88–95%, the contact reaction time should be greater than 120 min, the dosage of the agent should be greater than 0.075 kg of lime powder/kg of medical waste, and the pH of the strongly alkaline environment for reaction control should be within 11.0–12.5. In addition to the evaluation of chemical treatment facilities, the active ingredient content, stability and pH value of chemical disinfectants should be tested.
(3) Microwave processing facilities performance testing technical requirements. Microwave sterilisation of medical waste microwave sterilisation facilities with 24–50 MHz high-frequency electromagnetic waves for sterilization; in the reaction room 2–6 microwave sources should be installed, 2–6 microwave sources and each microwave source has an output power of 1.2 kW; microwave sterilization of medical waste should be at the normal temperature of 95–100 °C; reaction dwell time should not be less than 30 min.
(4) Microwave + high-temperature steam sterilization processing facilities performance test technical requirements. Under the joint action of steam and microwave, sterilization room temperature increases rapidly, when the pressure >0.2 bar, the temperature >135 °C, in the sterilization phase, the timer begins to time. Thereafter, the temperature continues to rise to 140 °C, the pressure rises to 0.33 MPa, and maintained at this temperature and pressure for 3–5 min, after the control system to confirm the validity of the sterilization process is completed.
(5) High-temperature dry heat treatment facilities performance testing technical requirements. The sterilization chamber processing temperature is not less than 180 °C, the pressure is not higher than 1000 Pa (gauge pressure) conditions, the corresponding processing time should not be less than 20 min.

The performance test is the key to the implementation of performance evaluation, and the content and points of the performance test are shown in Table 8.5.

Table 8.5 List of performance test contents and points

Serial number	Category	Code	Test monitoring items	Unit	Test points and methods	Remark
1	Waste features indicators (A)	A-a	Waste disposal volume	kg/h	Instant weighing and calculation	
		A-b	Standard test waste composition	%	On-site testing	
		A-c	Conventional medical waste components	%	Combined with medical waste transfer coupon information for analysis to determine	
		A-d	Waste contaminated with bacteria	g/cfu	Laboratory testing	
2	System performance indicators (B)	B-a	System configuration completeness		Combined with the preliminary design documents on-site verification	
		B-b	Unit design normality		Combined with preliminary design documents on-site verification	
		B-c	Exhaust cleaning efficiency	%	Exhaust gas cleaning facilities import and export	Including malodors, VOCs, particulate matter
2	System performance indicators (B)	B-d	Wastewater treatment efficiency	%	Exhaust gas cleaning facilities import and export	Wastewater pollutants
		B-e	Disinfection efficiency	Killing of pairs of numerical value	Medical waste before and after treatment	

(continued)

Table 8.5 (continued)

Serial number	Category	Code	Test monitoring items	Unit	Test points and methods	Remark
3	Facilities run parameters (C)	C-a1	Processing temperature	°C	Inside the processing equipment	Suitable for high temperature steam treatment facilities
		C-b1	Processing time	min	Inside the processing equipment	
		C-c1	Operating pressure	kPa	Inside the processing equipment	
		C-a2	Effective concentration of disinfectant	g/kg	On-site testing	Applicable to chemical disinfection treatment facilities
		C-b2	Disinfectant action time	min	On-site testing	
		C-c2	Alkalinity	pH	Medical waste after treatment	
		C-a3	Processing temperature	°C	Inside the processing equipment	Suitable for microwave sterilization treatment facilities
		C-b3	Processing time	min	Inside the processing equipment	
		C-c3	Microwave frequency	MHz	Inside the processing equipment	
		C-d3	Microwave power	W	Inside the processing equipment	
		C-a4	Processing temperature	°C	Inside the processing equipment	Suitable for high temperature dry heat treatment facilities
		C-b4	Processing time	min	Inside the processing equipment	
4	Emissions indicators (D)	D-a	Disinfection efficiency	Killing of pairs of numerical value	Medical waste before and after treatment	
		D-b	VOCs	mg/Nm^3	Exhaust gas cleaning facility outlet	

(continued)

Table 8.5 (continued)

Serial number	Category	Code	Test monitoring items	Unit	Test points and methods	Remark
		D-c	Stench concentration	mg/Nm^3	Exhaust gas cleaning facility outlet	
		D-d	Total number of bacteria	Cfu/Nm^3	Exhaust gas purification facility outlet	

Note Serial numbers 1, 2, 3 and 4 represent the four technologies of high-temperature steam treatment, chemical, microwave and high-temperature dry heat treatment, respectively

8.5.3.4 Non-incineration Treatment Facilities Performance Test Analysis Methods

(1) Disinfection effect detection method. High temperature steam treatment technology for thermophilic fatty bacillus bacterium, microwave disinfection chemical disinfection, and high-temperature dry heat to Bacillus subtilis black variant bacillus as the indicator strain to measure the sterilization effect of non-incineration treatment equipment of medical waste, requiring microbial kill log value $\geq$4, the treatment effect detection method in accordance with the national "disinfection technology specifications" implementation.
(2) Process waste gas emission monitoring methods. High-temperature steam, chemical, microwave, and high-temperature dry thermal treatment of four non-incineration technology application process-generated VOCs, odor and dust detection methods can be referred to the "indoor air quality standards" (GB 18883-2002), "air quality determination of odour three-point comparative odour bag method" (GB/T 14675-1993), and the "Comprehensive Emission Standards for Air Pollutants".
(3) Wastewater discharge monitoring method. The wastewater discharge detection method can be carried out in accordance with the "Standard for the discharge of water pollutants in medical institutions" (GB 18466-2005) method. For units with daily wastewater production of less than 1t, the focus should be on microbial indicators.

8.5.3.5 Non-incineration Treatment Facilities Performance Evaluation Method

Based on the monitoring data obtained from the previous performance test index system, the performance data of the facilities in different situations will be calculated and analyzed, and the conclusions on whether the facilities meet the standards will be made by analyzing and studying the situation of pollutant emissions. For the facilities that meet the standards, the waste characteristics, working parameters and main operating parameters should be clarified in the case of meeting the emission standards; for facilities that do not meet the standards, the facilities need to be modified and re-tested and evaluated; for those that do not test, only in accordance with the performance test to determine the type of waste that can be disposed off in the case of meeting the emission standards, the working performance of the facilities and the main operating parameters of the facilities to implement the facilities Operation and management.

8.6 Emergency Management of Medical Waste Treatment and Disposal

8.6.1 *Medical Waste Accident Emergency Management Points*

Medical waste disposal emergency management is to effectively prevent, timely control and eliminate the occurrence of medical waste loss, leakage, proliferation and other emergencies resulting in adverse events, guidance and standardization of medical waste loss, leakage, proliferation after emergency treatment work, improve response capabilities, the establishment of unified command, clear responsibilities, orderly operation, rapid response, powerful disposal, according to the standard, relying on scientific emergency disposal system to minimize the loss of medical waste, leakage, proliferation of public health and national property hazards, to protect the public's physical and mental health and life and property safety, to maintain normal order, to establish the appropriate emergency management mechanism and mode, to promote safe and environmentally sound management and disposal of medical waste to provide security (Zhou SL et al. 2020; Cai HY et al. 2020).

8.6.1.1 Basic Requirements for Emergency Management

(1) In order to implement the "Law of the People's Republic of China on the Prevention and Control of Environmental Pollution by Solid Waste", "Law of the People's Republic of China on the Prevention and Control of Infectious Diseases", "Regulations on the Management of Medical Waste", to strengthen the safety management of medical waste, to prevent the spread of disease, to protect human health, for the production process of possible emergencies must be in accordance with the principles of ensuring personal safety, ensuring the safety of equipment, preventing the leakage of pollutants, preventing derivative accidents, reducing For the emergency situation that may occur in the production process, we must follow the principles of ensuring personal safety, ensuring equipment safety, preventing contaminant leakage, preventing derivative accidents, reducing asset loss, facilitating cause tracing, and facilitating prevention

and education for procedural treatment, and preparing emergency plans as a natural choice.

(2) The preparation unit of the plan is the disposal unit, the approval department is the local environmental protection administrative department, and the execution unit is the medical waste treatment plant. The first responsible person, the second responsible person, and the third responsible person should be implemented. According to the emergency situation of the people present at the time of discovery, according to the preceding sequence, the first is to implement the field command responsibility, do not allow the field command responsibility to fall short, and ensure the effective implementation of the plan measures.

(3) In addition to emergency response measures, this plan should also be used as a mandatory part of the training for production personnel to be proficient. To organize regular drills for production personnel. Training and rehearsals are documented.

(4) The implementation of this plan related to the chart data should be specially stored to ensure access at any time. The relevant signs on site should be regularly checked and maintained to ensure that they are intact. The professional units related to the plan should establish a regular contact system to ensure that the preventive measures are properly set up, the emergency contact is smooth and the rescue treatment is timely and effective.

(5) The plan should be prepared for the production of medical waste disposal site conditions, not applicable to other projects. According to the actual needs, this plan will be improved or adapted and re-approved, when this plan automatically ceases to use.

8.6.1.2 Emergency Rescue Procedures

(1) If there is a medical waste accident, the unit in charge should be developed in accordance with the unit's emergency rescue plan, immediately organize rescue, and immediately report.

(2) Upon receipt of the accident report, immediately in accordance with the medical waste accident emergency rescue plan, good command and leadership work. Relevant departments should immediately take the necessary measures to reduce accident losses, to prevent the spread of accidents expansion.

(3) If the medical waste accident cannot be quickly and effectively controlled, immediately report to higher departments, request the city medical waste emergency rescue command to give support. Command members of the department immediately after receiving notification to the scene of the accident, to carry out rescue work.

The emergency work procedures for medical waste disposal accidents are shown in Fig. 8.2.

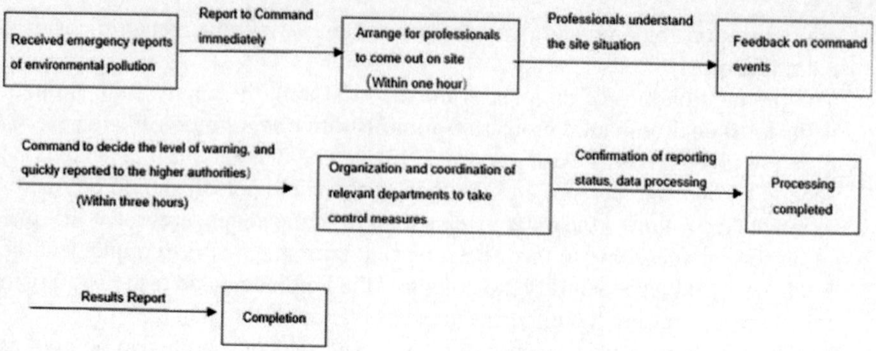

Fig. 8.2 Medical waste disposal accident emergency work procedures

8.6.1.3 Establishment and Responsibilities of the On-Site Rescue Professional Group

Medical waste emergency rescue command according to the actual situation of the accident, the establishment of relevant rescue professional groups, and its responsibilities are as follows:

1. Responsibilities of the members of the command
   (1) They are responsible for the coordination and command of environmental emergency prevention and emergency relief work, responsible for the implementation of this plan in the organization and coordination and unified external relations.
   (2) They are responsible for the construction of the environmental emergency preparedness team and the configuration of equipment and equipment.
   (3) Organization and guidance of emergency drills for environmental emergencies.
   (4) Review the emergency funding budget.
   (5) Participate in the revision of this plan.
2. Responsibilities of each team
   (1) Communication and liaison group. The command department chief is responsible for the communication of the members of the command department and related departments in the emergency process, to ensure smooth communication in the emergency process, and the whole process of the accident to do a good job of processing records and reporting records.
   (2) Emergency repair team. Responsible for the emergency repair of equipment and facilities under various accident conditions.
   (3) Medical care team. Mainly for the emergency process of timely treatment and escort work of the injured.

(4) Mobile alert group. According to the provisions of the command and control of the order of the accident area, evacuation of personnel and guard work in the danger zone, and as mobile personnel on standby.
(5) Logistical support group. Prepare to start the emergency system, responsible for the emergency process of materials and supplies.

8.6.1.4 Accident Emergency Protection System

(1) In order to ensure timely and effective emergency disposal work, emergency equipment and equipment is equipped in advance, and special personnel are responsible for storage, maintenance, and inspection to ensure that all kinds of emergency equipment are in good condition.
(2) Medical waste disposal center should be specified in the emergency state of the alarm communication, notification, traffic security and control, the preparation of environmental emergency contact directory are issued and posted. Maintain contact with the government and related units, in the event of a major emergency, internal cannot be ruled out, and timely request the government to coordinate emergency relief forces.
(3) Implement the responsibility system for environmental emergencies, put the responsibility clearly into practice, and strengthen the sense of responsibility of relevant personnel.
(4) We have established various emergency protection systems, such as the duty system, inspection system, assessment system, training system, environmental management system and emergency drill system, etc.
(5) The appointment of industry experts, the establishment of expert advisory groups, to provide technical support for the emergency response to accidents.

8.6.1.5 Emergency Training Plans and Drills

1. Emergency communication

    (1) Organize staff education on laws and regulations and general knowledge of prevention, risk avoidance, self-help and mutual help. Use bulletin boards and other means to enhance employees' awareness of crisis preparedness and basic emergency knowledge and skills.
    (2) Develop the Environmental Emergency Response Plan and Manual.
    (3) Make a list of environmental emergency plans.

2. Environmental emergency response training
Conduct staff-oriented training on knowledge related to responding to environmental emergency. Environmental emergencies prevention, emergency command, comprehensive coordination and other important training content to improve the ability of staff to respond to environmental emergencies, and actively participate in environmental protection department's related training activities.

3. Environmental emergencies emergency drills

   (1) Organize and carry out emergency plan drills at appropriate times to train emergency teams, implement job responsibilities, familiarize with the command mechanism, decision-making, coordination and disposal procedures of emergency work, test the feasibility of the plan and improve the emergency plan. Thus improve the emergency response and handling ability, and strengthen the sense of cooperation.
   (2) Emergency drills for general environmental emergencies are conducted at least 1–2 times a year.

8.6.1.6 Accident Emergency Disposal Measures

1. Identification of medical waste accident hazards
   Including the medical waste collection, transportation, storage, and disposal processes involved in raw and auxiliary materials, intermediate products, and three waste emissions. The main elements include: collection process that does not use the specified collection containers or is not disinfected, transport accidents caused by medical waste scattered, the spread of pathogenic bacteria, equipment failure and maintenance of medical waste storage excess, more than the capacity of the refrigerated room and cannot be stored, disposal process due to the failure of the main equipment, medical waste residue bacteria indicators failed, because the exhaust gas purification facilities failure caused by excessive emissions.
2. General technical requirements for emergency response to medical waste incidents.

   (1) Plan transportation routes to avoid environmental protection targets as far as possible. The planned transport routes, driving hours and emergency measures are reported to the relevant departments of public security and transportation for approval.
   (2) Medical waste collection, storage, handover, transport of the whole process of collection and transportation should be in strict accordance with the "Medical Waste Management Regulations" "medical waste transfer vehicle technical requirements" and other relevant provisions, in the delivery of medical waste transfer coupon system, by the medical waste disposal center and generate units together to fill out the "hazardous waste transfer coupon" (medical waste special), the transport vehicle to fill out the "medical waste delivery Registration card", and should also fill out a medical waste disposal statement, and the relevant environmental protection departments for the record.
   (3) Medical waste delivery should use special vehicles. The vehicle compartment should be separated from the cab and sealed; compartment should meet the airtight requirements, smooth and flat interior walls, easy to clean and disinfect; compartment materials waterproof, corrosion-resistant; compartment to prevent liquid leakage and a drainage collection device to clean the

sewage. Delivery vehicles should meet the "medical waste transfer vehicle technical requirements" and "centralized medical waste disposal technical specifications" requirements.

(4) Packaged medical waste should be placed in reusable special crates (barrels) or disposable special packaging containers. Special crates (barrels) or disposable special packaging containers should comply with the "medical waste special packaging, container standards and warning labels.

(5) Engaged in road transport drivers, loading and unloading managers, escort personnel by the local municipal people's government competent department of transport examination, to obtain the appropriate qualification certificate. In the process of transport leakage and other accidents, drivers, escorts should immediately report to the local public security departments and disposal centers, explain the accident, hazards and emergency measures, take all possible warning measures at the scene, and actively cooperate with the relevant departments for disposal. Medical waste disposal centres should be transported to professional skills and occupational health protection training.

(6) Medical waste disposal should focus on the content of the underground fuel oil storage tank and oil pump room, the intermediate high fuel oil storage tank, medical waste warehouse, the activated carbon warehouse, waste heat boiler and ancillary facilities, the incineration system equipment, non-incineration system equipment, and the main control system devices.

3. Medical waste emergency disposal measures

(1) Transport process when the occurrence of a rollover, collision resulting in a large amount of medical waste spilled, scattered, transport personnel should immediately contact the unit emergency incident team, request the local public security traffic police, environmental protection or city emergency linkage center support. At the same time, transport personnel should take the following emergency measures:

① Immediately request the public security traffic police to set up a quarantine area in the contaminated area, prohibit other vehicles and pedestrians to cross, to avoid the spread of pollutants and pedestrians from causing harm.

② Collect, clean and disinfect the medical waste spilled and scattered quickly. For liquid spills, use adsorbent materials for absorption treatment and disinfection.

③ Cleanup personnel must wear protective clothing, gloves, masks, boots and other protective equipment when carrying out cleanup work, and after the cleanup work is completed, the equipment and protective equipment must be disinfected.

④ If the body (skin) of the cleaning personnel is inadvertently injured during the operation, treatment measures should be taken in a timely manner, and go to the hospital for medical treatment.

⑤ Cleaning personnel must also disinfect and clean the contaminated site floor.

⑥ Medical waste scattered in the water, according to the specific circumstances of the river, the water conservancy department, environmental protection departments, public security departments, health departments, waterway departments, waterworks downstream of the river, medical waste disposal centers and other units, measures to prevent contaminated water from affecting the health of residents along the line and property damage.

(2) The occurrence of the accident. To take the above emergency measures at the same time, the disposal unit must report the accident to the local environmental protection and health departments. After the accident is completed, the disposal unit to the above two departments to write a written report, and the report includes:

① The time, place, and cause of the accident and its brief history.

② Leakage, scattering the type and amount of medical waste, the cause of contamination and the name of the medical waste generating unit.

③ The hazards and potential effects of medical waste leakage and scattering have been caused.

④ The emergency treatment measures taken and the results of the treatment.

4. Aftercare and recovery measures

(1) An emergency disposal site should be set up as a decontamination station, the emergency disposal process collecting leaks, wastewater, etc. The device used by emergency personnel should be centralised for treatment and the apparatus used by the emergency disposal personnel should be decontaminated. Use disaster relief funds to repair damaged equipment, instruments, vehicles, etc., and actively carry out post-disaster reconstruction work.

(2) Health monitoring or medical examination of rescue personnel.

(3) Actively compensate for environmental damage and damage to the health of the population caused by the accident.

(4) If the spill has been collected, isolated, and decontaminated; casualties are timely rescued; or other conditions should be met, when all are met, the environmental pollution accident emergency rescue team will announce by the end of emergency rescue work.

(5) After the incident, by the emergency command according to the harm and impact of the accident, the formation of the accident investigation team should thoroughly investigate the cause of the accident, clear responsibility for the accident, and summarise the lessons learned, and according to the direct causes of the accident and indirect causes, make recommendations and measures to rectify the accident, and the formation of the accident investigation report.

8.6.1.7 Medical Waste Disposal During Major Infectious Disease Outbreaks

The collection, temporary storage, transport and incineration of highly infectious medical waste must be carried out in accordance with the following requirements:

1. Classified collection and temporary storage

   (1) Medical waste should be collected by a person, double packaging, and the bag should specifically indicate that it is highly infectious waste.
   (2) The temporary storage of medical waste in health care institutions should be dedicated to storage, management, not mixed with general medical waste, household garbage, and mixed packaging.
   (3) Temporary storage sites are disinfected by dedicated personnel using 0.2–0.5% peroxyacetic acid or 1000–2000 mg/L chlorine disinfectant spraying walls or mopping floors once a day in the morning and once in the afternoon.

2. Transportation and disposal

   (1) Disposal units in the transport of medical waste must use a fixed special vehicle, by a person responsible for, and not mixed with other medical waste, mixed transport.
   (2) The delivery time should be staggered to avoid the peak commuting period, and the delivery route should avoid densely populated areas as far as possible; after each unloading of the delivery vehicle, it must be disinfected by spraying with 0.5% peroxyacetic acid.
   (3) Medical waste is disposed off by high temperature incineration, medical waste arrives at the disposal site as far as possible after the disposal, the temporary storage time in the disposal unit should not exceed 12 h.
   (4) Disposal plant must be set up in the medical waste disposal isolation area, the isolation area should be clearly marked, and no unrelated personnel may enter.
   (5) The isolation area must be disinfected by a person using 0.2–0.5% peroxyacetic acid or 1000–2000 mg/L chlorinated disinfectant to spray or mop the walls, floors or object surfaces once a day in the morning and once in the afternoon.

3. Personnel hygiene protection

   (1) Transport and incineration disposal device operator protection requirements should meet the requirements of the health department of the first level of protection, that is, must wear overalls, isolation clothing, protective boots, as well as wearing work caps and protective masks, and close disposal of waste personnel should also wear goggles.
   (2) Hand washing and disinfection immediately after each transport or disposal operation, and bathing. Hand disinfection with 0.3–0.5% iodophor disinfectant solution or rapid hand disinfectant rubbing 1–3 min.

8.6.2 *Development Needs of Medical Waste Emergency Management*

(1) Emergency treatment should be given equal importance with safe disposal

Emergency treatment should be combined with the "National Hazardous Waste and Medical Waste Disposal Facilities Construction Plan" (hereinafter referred to as "the Plan") implementation, medical waste emergency treatment cannot simply meet the needs of a momentary emergency, and give up the requirements of safe disposal goals. It cannot have a "burn, everything is fine" thinking, and ignore the application of other feasible technologies; ignore the secondary pollution that still exists on the threat to the physical and mental health of people and the potential for environmental safety. Emergency treatment measures taken cannot be contrary to the basic principles and requirements of medical waste treatment established by the state, can not reverse the trend of technological development, the use of emergency facilities to fully articulate with the implementation of planning projects, and not to form "two skins". Should fully learn from the "SARS" emergency facilities construction lessons, ahead of time and overall consideration of the current mobile facilities in the completion of emergency treatment tasks after the exit problem. Can be considered for fixed facilities, or continue as mobile facilities, in remote and centralized treatment facilities to continue to play a role in the collection of difficult places, or by the state unified recovery, unified dispatch, to avoid facing the problem of elimination and waste of funds after the completion of the emergency.

(2) Accelerate the construction of an emergency network of hazardous waste in each province

Planning" to determine the prefecture-level cities as a unit to build centralized medical waste disposal facilities is a basic guarantee of urban medical waste disposal, and the "5–12" earthquake within the scope of most municipalities do not timely complete the "planning" to determine the construction tasks, resulting in a passive situation after the disaster. Therefore, we must take this incident as a lesson, increase inspection efforts, and urge local governments at all levels to complete the planning and construction tasks as soon as possible.

The "Plan" to establish a national and 31 provincial hazardous waste registration and exchange and accident response network, the network will be responsible for the coordination and management of hazardous waste transfer, exchange, transport and disposal activities within the jurisdiction, hazardous waste accident response and command and coordination work. This shows that the construction of the network is the main guarantee of accident response, the initial model provinces can be selected to start the network construction as soon as possible, and then gradually expanded to the national scope.

(3) Promote the construction of a national medical waste emergency treatment system

The implementation of the "Plan" for the realization of China's medical waste and hazardous waste safe disposal goals is a historic and important step, but the "Plan" itself does not cover all the problems, in the face of disasters exposed the lack of medical waste emergency disposal system in China and imperfect. The "5–12" earthquake relief work as the starting point for the construction of China's medical waste emergency disposal system, a comprehensive analysis of the problems in the emergency disposal of medical waste in China, the full start of the preparation of emergency preparedness. In particular, to address the emergency work mechanism in the prominent position, through regional demonstration and drive, as soon as possible in the country, to establish a cross-regional, cross-facility coordination of disposal mechanism. Start the construction of decentralized facilities to meet the principle of large centralized premise, to solve the problem of medical waste centralized disposal facilities difficult to cover the safe disposal of medical waste in remote areas.

(4) To carry out medical waste emergency treatment regulation and reserve capacity building

When it comes to the construction of a mobile medical waste integrated treatment system, it is very important to build emergency reserve capacity at the national or regional level to facilitate the dispatch as soon as possible after the occurrence of an emergency to meet emergency needs. It is recommended that a comprehensive analysis of the need for medical waste treatment in various disasters and emergencies, including earthquakes, a proper understanding of the demand for treatment systems in emergency situations, a clear understanding of the specificity of the configuration of functional units, and further handling of the integration of vehicles and treatment systems. The combination of emergency reserve capacity building and normal capacity building. China's remote areas and transportation are not convenient places urgently need to establish a mobile treatment system, can be taken completely mobile small treatment system (processing capacity of a few hundred kilograms per day or so) and semimobile treatment system (processing capacity of 1–2 t/d) combined, usually semi-mobile treatment system can be fixed in the larger medical institutions inside, separate configuration of small collection vehicles for the surrounding urban areas. In case of an emergency, the full set of systems can be quickly installed on the mobile vehicle and quickly transferred to the place of urgent need, forming a mobile treatment system. Uniform design of mobile facilities, processing capacity, and operation mechanism in the country, normal treatment facilities in the event of an emergency, national or regional unified deployment, quickly into emergency treatment capacity, to the most necessary places to carry out the safe disposal of medical waste.

References

Cai Hongying, Zhang Manli, Zhou Lianchuan. Practice of medical waste emergency disposal management for pneumonia outbreak with novel coronavirus infection--a case study in Chongqing city [J]. Environmental Health Engineering, 2020, 28(2): 17–22.

Chen Y, Yang Y, Chen G. Current status and development of hazardous waste pollution control technology in China [J]. China Environmental Protection Industry, 2005(11): 16–17, 45.

Ministry of Ecology and Environment. "The 12th Five-Year Plan for the Prevention and Control of Hazardous Waste Pollution [Z/OL]". [2013–10–08]. http://www.mee.gov.cn/gkml/hbb/bwj/201210/t20121023_240228.htm.

Ministry of Ecology and Environment. Guidance on enhancing the environmental supervision capacity, utilization and disposal capacity and environmental risk prevention capacity of hazardous waste [Z/OL]. [2019–10–16]. http://www.mee.gov.cn/xxgk2018/xxgk/xxgk03/201910/t20191021_738260.html.

Ministry of Ecology and Environment. Epidemic-related medical waste basically realized daily production and daily cleaning [N/OL]. (2020–02–13) [2020–09–26]. Beijing: People's Daily. https://baijiahao.baidu.com/s?id=1658400060841965560&wfr=spider&for=pc.

Ministry of Ecology and Environment. Letter on soliciting comments on the "Guidelines for Best Available Techniques for the Prevention and Control of Pollution from Medical Waste Treatment and Disposal (Draft for Comments)" [Z/OL]. [2011–03–09].

National Health and Health Commission, Ministry of Ecology and Environment, et al. Work plan for comprehensive management of medical institution waste [Z/OL]. [2020–02–24]. http://www.nhc.gov.cn/yzygj/s7659/202002/3643bd749f5945aa97f247a5c695f699.shtml.

Sun N, Liu F P, Zhang Y K, et al. Reflections and suggestions on accelerating the shortcomings of medical waste collection and disposal facilities [J]. Environmental Protection Science, 2020, 46(3): 120–126.

Zhang Shuangchun, Xiao Ai, Hu Chundong, et al. The centralized disposal of medical waste in Hunan Province [J]. China Environmental Management, 2003, 22(3): 31–33.

Zhou S. L., Guo C. X., Shi X. M., et al. Problems and suggestions in emergency medical waste disposal during major epidemics [J]. Journal of Environmental Engineering, 2020, 14(7): 1705–1709.